Porphyrins

Excited States and Dynamics

ACS SYMPOSIUM SERIES **321**

Porphyrins

Excited States and Dynamics

Martin Gouterman, EDITOR
University of Washington

Peter M. Rentzepis, EDITOR
University of California at Irvine

Karl D. Straub, EDITOR
John L. McClellan Memorial Veterans' Hospital and University of Arkansas for Medical Sciences

Developed from
an independently organized symposium
held in Little Rock, Arkansas,
November 16–19, 1985

American Chemical Society, Washington, DC 1986

Library of Congress Cataloging-in-Publication Data

Porphyrins: excited states and dynamics.
(ACS symposium series, ISSN 0097-6156; 321)

"Developed from a conference held in Little Rock, Arkansas, November 16-19, 1985."

Includes bibliographies and indexes.

1. Porphyrin and porphyrin compounds—Congresses. 2. Porphyrin and porphyrin compounds—Spectra—Congresses.

I. Gouterman, Martin, 1931- . II. Rentzepis, Peter M., 1934- . III. Straub, Karl D., 1937- . IV. American Chemical Society. V. Series.

QP671.P6P68 1986 574.19'218 86-14647
ISBN 0-8412-0997-9

PRINTED IN THE UNITED STATES OF AMERICA

FOREWORD

The ACS SYMPOSIUM SERIES was founded in 1974 to provide a medium for publishing symposia quickly in book form. The format of the Series parallels that of the continuing ADVANCES IN CHEMISTRY SERIES except that, in order to save time, the papers are not typeset but are reproduced as they are submitted by the authors in camera-ready form. Papers are reviewed under the supervision of the Editors with the assistance of the Series Advisory Board and are selected to maintain the integrity of the symposia; however, verbatim reproductions of previously published papers are not accepted. Both reviews and reports of research are acceptable, because symposia may embrace both types of presentation.

CONTENTS

PICOSECOND SPECTROSCOPY

RESONANCE RAMAN STUDIES

PHOTOCHEMISTRY

INDEXES

PREFACE

PORPHYRINS HAVE THREE MAJOR ROLES in nature: electron transfer, oxygen transfer, and photosynthesis. In photosynthesis a principal scientific question has concerned the kinetics of chlorophyll excited states after absorbing light. Great progress has been made in understanding this problem, and a symposium volume on this subject has recently appeared[1]. Understanding photosynthesis is based on studies of the excited states of simple porphyrins. In the last decade a major new experimental technique, picosecond laser spectroscopy, has had considerable application to porphyrins[2] and has provided information about excited-state decay processes. Picosecond laser spectroscopy has also helped explain the kinetics of ligand binding to porphyrins. Ligand binding is also involved in the biological activity of porphyrins. A brief review of picosecond pump-probe studies of metalloporphyrins has recently appeared[3], but its scope was very limited.

When molecules absorb a photon and produce an electronic excited state, the energy can be dissipated in several ways: luminescence, radiationless decay to the ground state, and photochemistry. Luminescence dominated the older literature because it was easy to observe. A good review of luminescence is in Volume 3 of David Dolphin's seven-volume series *The Porphyrins*[4]. Picosecond laser spectroscopy allowed for exploration of the radiationless decay pathways, particularly the initial steps that compete with luminescence and lead to photochemistry. Two principal forms of radiationless decay lead to long-term metastables: ligand ejection and electron transfer.

These topics have been studied extensively since the publication of *The Porphyrins* eight years ago. In assembling the International Conference on Excited States and Dynamics of Porphyrins (Little Rock, Arkansas, November 16–19, 1985), we hoped to collect a wide range of scientists engaged in studies on porphyrin excited states to review the developments during the last decade in porphyrin photodynamics. This book, developed from that conference, is concerned with only a limited aspect of porphyrin research, Excited States and Photodynamics. Chapters fall into five categories: theoretical studies on excited states, spectroscopic studies on excited states, picosecond laser kinetic studies, resonance Raman studies, and photochemical studies. In particular, the book examines the nature of metalloporphyrin excited states, solvent perturbations, decay of the S_2 (second excited singlet) state, intersystem crossing rates, ligand ejection, electron transfer, and related areas.

In short, this volume contains a variety of chapters concerned with the fate of porphyrin molecules that have absorbed a photon. Most of the chapters examine changes that occur after this process. In some cases the change is ligand loss, in other cases nuclear motion, and yet other cases electron transfer. These processes also occur in biological systems and have important practical application. Because this book collects works from various groups, we hope that it will find a useful place in many research libraries and laboratories and that it will fill a gap in the current review literature.

We gratefully acknowledge the financial support for the conference provided by the following Arkansas organizations and individuals: Arkansas Electric Power Company; Arkansas Louisiana Gas Company; BEI Electronics, Inc.; First Commercial National Bank; General Dynamics Corporation; John L. McClellan Memorial Veterans' Administration Medical Center; Union National Bank; University of Arkansas for Medical Sciences; Whit Stevens; and Winthrop Rockefeller Foundation. Support for publication of this book was provided in part by the Citadel Research Foundation, Little Rock, Arkansas.

MARTIN GOUTERMAN
Department of Chemistry
University of Washington
Seattle, WA 98195

PETER M. RENTZEPIS
Chemistry Department
University of California at Irvine
Irvine, CA 72717

KARL D. STRAUB
John L. McClellan Memorial
Veterans' Hospital
4300 West 7th Street
Little Rock, AR 72205

March 28, 1986

[1]Michel-Beyerle, M. E. *Antennas and Reaction Centers of Photosynthetic Bacteria;* Springer: Berlin, 1985.

[2]Straub, K. D.; Rentzepis, P. M.; Huppert, D. *J. Photochem.* **1981,** *17,* 419–425.

[3]Holten, D.; Gouterman, M. In *Optical Properties and Structures of Tetrapyrroles;* Blauer, G.; Sund, H., Eds.; Walter de Gryter: Berlin, 1985; p. 63.

[4]Gouterman, M. In *The Porphyrins;* Dolphin D., Ed.; Academic: New York, 1978; Vol. 3, p. 1.

THEORETICAL STUDIES

1

Quantum Mechanical Studies of the Photodissociation of Nitrosoheme Complexes

Lek Chantranupong, Gilda H. Loew, and Ahmad Waleh

Molecular Theory Laboratory, The Rockefeller University, Palo Alto, CA 94304

Initial steps in the photodissociation of nitrosyl ferrous and ferric heme complexes have been investigated by calculation of the ground and excited states and electronic spectra as a function of iron-ligand bond lengthening. The method used is an all valence semi empirical INDO-SCF-CI procedure with transition metal parameterization and the ability to calculate configuration interaction for closed shell systems and, more recently, radical species. The results of these studies allow identification of a common photodissociating state involving $d\pi \rightarrow d^*_{z^2}$ excitations for both systems, the same state found in previous studies of carbonyl and oxy ferrous heme complexes. This result is consistent with the observation of nearly equal initial photodissociation product for all four systems at a picosecond time scale. The low quantum yield at long time (~ 400 μs) observed for the ferrous nitrosyl complex is attributed to the population of charge transfer states with a minimum which allow trapping of the NO radical near the iron. Such states can be involved in both rapid geminate recombination or energy decay to the liganded state. The high quantum yield of the ferric nitrosyl complex is attributed to eventual photodissociation of ferric heme and NO radical on a triplet surface which does not allow population of the low energy charge transfer states also found to be present for these system. By contrast, the high quantum yield of the isoelectronic ferrous carbonyl heme complex is associated with the absence of such charge transfer states.

Photodissociation and recombination studies of diatomic molecules, O_2, CO, and NO from different heme proteins (1-11) and liganded model heme complexes (5, 9, 12-14) have been very valuable in understanding the dynamics of ligand binding and energy relaxation mechanisms. Much of the interest in these studies stems particularly from the need for a better understanding of the physiological function of

0097-6156/86/0321-0002$06.00/0

hemoglobin which involves reversible binding of O_2 at its heme active site. Although photodissociation itself is a nonbiological process, it involves a series of biologically significant events which occur over different time scales, beginning with the metal-ligand bond breakage, leading to changes in the electronic and tertiary structures, and eventually to quaternary conformational changes in the protein.

It is known that some of the early transient effects related to ligand photodissociation in heme proteins evolve in the pico- or subpicosecond range (14-27). Since it is unlikely that any spin, or tertiary-structure changes would take place at such fast rates, these ultrafast events should be closely related to the primary heme-ligand bond breaking process initiated by the activation of the complex to a dissociative state by photon energy. It, therefore, follows that the observed subpicosecond photodissociation properties for various ligands must be strongly dominated by the energy profile of the excited electronic states of their respective complexes during this process. Such a profile will determine the initial dissociative yield of the ligand which, in turn, will influence the subsequent events of photodissociation ranging from the evolution of the intermediate species to electronic and conformational changes accompanying the ligand dissociation and/or recombination beyond the picosecond range (28-39).

The primary requirement for the initiation of the ligand bond breaking process is the existence of photodissociative states in an appropriate energy range such that they can be populated by either direct excitation into such states or through relaxation from states populated by higher energy excitations. The dissociative yield, however, will also depend on the existence or absence of competing mechanisms that may result in the overall deactivation of these dissociative states. In particular, low-lying excited states with paths for energy relaxation from higher-energy states and/or for energy decay to the ligand-bound ground state can effectively depopulate the photodissociative states and thus reduce ligand dissociation yield. It is in the context of these ultrafast processes that theoretical investigation of the ground and excited state of complexes of various ligands with heme proteins can significantly contribute to our understanding of their dissociation and recombination behaviors.

We have previously reported the results of semiempirical INDO-SCF-CI (40-43) studies of the energy profiles of the excited states of closed-shell (S = 0) CO (44) and O_2 (45) ferrous heme complexes as a function of iron-ligand distance. In these studies, the criterion of "decreasing excited state energy as a function of iron-ligand distance and no barrier to dissociation" was used to identify the photoactive states in both complexes. These results were interpreted in terms of a two-channel photodissociation mechanism in oxyheme and a single-channel mechanism in carbonylheme complex. It was shown that although the excited states corresponding to $d_\pi \rightarrow d^*_{z^2}$ were photodissociative in both carbonyl and oxyheme, photodissociation of oxyheme also proceeded through charge-transfer states below the Q-band, which could also provide a competitive decay channel to the ground state. Consideration of the energies and lifetimes of various dissociative and decay channels then provided consistent explanations for the differences in the observed picosecond photodissociation properties of the two ligands.

The INDO program used in the closed-shell oxy- and carbonylheme studies has recently been modified to include a Restricted Hartree-Fock (RHF) and a generalized configuration interaction (CI) treatment for calculation of the excited states of open-shell systems with unpaired electrons (46). This new capability allows us to calculate the excited state profiles and examine the photodissociation properties of heme complexes with a variety of ligands and metals in different spin and oxidation states. Such studies are particularly significant as changes in both ligand and metal can alter the heme electronic structure and thus the photophysics of the primary events of ligand photodissociation. Hoffman and Gibson (5) have shown a good correlation between quantum yield of photodissociation and formal electronic configurations for a number of metals and ligands, which lead them to an overall stereo-electronic classification of these complexes. In particular, they have observed that the iso-electronic systems, [Fe(II)+CO], [Mn(II)+NO], and [Fe(III)+NO] all display ligand dissociation under photolysis with high quantum yield. The flash photolysis of NO and kinetics of its recombination in different ferriheme proteins have also been studied in detail (6-8).

In contrast to ferriheme complexes, ferrous nitrosylhemoglobin (HbNO) shows a negligibly small quantum yield in microsecond flash photolysis studies (4). However, recent picosecond experiments (24, 27) on HbNO show that NO ligand, similar to CO and O_2, is efficiently photolyzed in this time range, although its behavior after photodissociation is strikingly different. In particular, the initial photoproducts of HbNO have a lifetime of 17 ps and the starting material is recovered within 400 ps. This has been interpreted in terms of fast geminate recombination of NO (27).

In this paper, we present the results of the calculations of the excited-state energies and characters for nitrosyl ferrous and ferric heme complexes as a function of iron-nitrosyl distance. The calculated energy profiles are then used to identify the photoactive states for each complex. Similar calculations are also underway to characterize the photoactive states of Mn(II)NO complexes (47).

Method

The calculations were carried out by using a semiempirical INDO (intermediate neglect of differential overlap) program (40-43) which allows for the treatment of transition metal complexes and inclusion of extensive configuration interactions. The geometry of both nitrosyl ferric and ferrous heme complexes were obtained from an X-ray crystal structure of 6-coordinate FeTPP(NMeIm)NO (48) with end-on NO ligand. This geometry was held constant except for variations of the Fe-NO distance along the axis normal to the plane of porphyrin. While an X-ray structure has been reported for a 5-coordinate ferric heme NO complex (49), with a linear NO geometry, the bent geometry of the 6-coordinate ferrous complex should more closely resemble that in the intact protein.

The ground- and excited-state electron distribution and energies, in each case, were calculated at five different iron-nitrosyl distances corresponding to successive displacement of NO in

0.2-Å steps from an initial bonding distance of 1.743 Å. Excited states of the closed-shell ferric complex were calculated by performing single excitation CI calculations using a toal of 207 configurations as described previously (44, 45). For the open-shell ferrous complex, the Self Consistent Field (SCF) calculations were performed by using a generalized open-shell operator (46), and the CI calculations were obtained from a Rumer diagram technique (50-54). At each iron-ligand distance, the energies of doublet excited states were calculated by using a total of 200 configurations corresponding to single excitations spanning an active space of 10 highest doubly occupied, 1 singly occupied, and 9 virtual molecular orbitals of the reference ground state.

Results

The nitrosyl ferric heme complex presented a computational problem at large Fe-NO distances due to the overall closed-shell nature of the liganded complex and the radical nature of the dissociated heme and ligand. The "frozen" (S = 0) spin state imposed on this complex resulted in SCF convergence on a $Fe(II)-NO^+$ ground state at Fe-NO = 2.143 Å and larger. However, at smaller Fe-NO distances, the correct Fe(III) oxidation state of the complex corresponding to a spin paired biradical singlet was obtained. These results indicated that the identification of the photoactive states of the ferric complex should be determined from the energy profile calculations at Fe-NO distances not too far from the equilibrium bonding distance. No such computational problems were encountered in the nitrosyl ferrous heme calculations, since a doublet spin state of the complex is consistent with the radical nature of the NO ligand at all Fe-NO distances.

Table I shows the atomic orbital composition and energies of the ground-state molecular orbitals of the nitrosyl ferric heme complex at three Fe-NO distances of 1.743 (bonding), 1.943, and 2.143 Å. Table II shows the same results for the nitrosyl ferrous heme complex at Fe-NO = 1.743 and 2.143 Å. The singly occupied orbital (79) has primarily NO π and σ character with both d_π and d_{z^2} contributions. Although this orbital is deeply buried, as can be seen from its energy, it is shown between the doubly occupied and virtual orbitals for convenience. For comparison, the ground-state molecular orbitals of carbonyl-(44) and oxyheme (45) complexes are shown in Table III.

Comparison of the molecular orbital energies of the ligand-bound ferric and ferrous complexes in Tables I and II shows that, aside from a general energy shift of about 0.12 au, the relative energies of the porphyrin π and π^* orbitals are little affected by the change in the oxidation state of iron. However, both iron and NO ligand orbitals are significantly affected by this change. The energies of the iron d_π and d_σ orbital are, respectively, 0.14 and 0.18 au lower in the ferric than in ferrous complex. In the ferric complex, the two NO π^* orbitals are the lowest empty orbitals and have nearly equal energies below those of porphyrin e_g^* orbitals. In the ferrous complex, however, these orbitals are split with the singly occupied orbital becoming deeply buried and the virtual orbital energy increasing above the e_g π^* orbitals.

The main effect of lengthening the Fe-NO distance in nitrosyl ferric heme is an increase in the d_σ contribution to virtual orbitals

Table I. Ground State Orbital Description of Nitrosyl Ferric Heme at Fe-NO Distances of 1.743 Å, 1.943 Å, and 2.143 Å.

	Fe(III)-NO = 1.743 Å	
Orb. No.	Energy (au)	Character
92	-0.0017	(54% Imidσ, 18% Porσ, 10% Fe σ)*
91	-0.0097	(100% a_{1u})*
90	-0.0229	(98% e_g)*
89	-0.0235	(96% e_g)*
88	-0.0404	(85% Imidπ, 14% Imidσ)*
87	-0.0522	(99% b_{1u})*
86	-0.0787	(85% Imidπ, 15% Imidσ)*
85	-0.0878	(100% b_{2u})*
84	-0.1026	(53% $d_{x^2-y^2}$, 13% d_{z^2}, 16% Porσ, 1% NOσ)*
83	-0.1029	(50% d_{z^2},15% $d_{x^2-y^2}$, 9% Porσ,8% NOσ,3% NOπ)*
82	-0.1527	(97% e_g)*
81	-0.1533	(97% e_g)*
80	-0.1719	(87% NOπ, 9% d_π)*
79	-0.2111	(32% NOπ, 37% d_π, 14% NOσ, 10% Porπ, 3% d_{z^2})*
78	-0.3444	100% a_{1u}
77	-0.3599	84% a_{2u}, 5% NOπ, 4% d_π
76	-0.4310	57% e_g, 15% Imidπ, 6% NOπ, 8% d_π, 2% d_z^2, 2% NOσ
75	-0.4322	60% Imidπ, 20% Porπ, 9% Imidσ, 2% dπ
74	-0.4347	74% e_g, 13% d_π, 6% Imidπ
73	-0.4470	94% b_{1u}
72	-0.4484	93% a_{2u}
71	-0.4619	96% e_g
70	-0.4628	98% e_g
69	-0.4891	33% d_π, 22% Porπ, 12% NOπ, 8% NOσ, 7% d_{xy}, 1% d_z^2
68	-0.5010	30% Imidπ, 27% d_{xy}, 23% d_π, 3% NOπ
67	-0.5030	57% d_{xy}, 14% d_π, 9% Imidπ, 3% NOπ

Table I. Continued

	Fe(III)-NO = 1.943 Å	
Orb. No.	Energy (au)	Character
92	-0.0002	(75% Porσ, 7% Imidσ, 6% Feσ)*
91	-0.0095	(100% a_{1u})*
90	-0.0218	(97% e_g)*
89	-0.0224	(97% e_g)*
88	-0.0355	(85% Imidπ, 14% Imidσ)*
87	-0.0490	(94% b_{1u}, 2% $d_{x^2-y^2}$)*
86	-0.0640	(69% $d_{x^2-y^2}$, 15% Porσ)*
85	-0.0716	(72% d_{z^2}, 6% NOσ, 6% Imidσ, 2% NOπ)*
84	-0.0732	(82% Imidπ, 15% Imidσ, 2% d_{z^2})*
83	-0.0880	(100% b_{2u})*
82	-0.1471	(55% NOπ, 40% e_g)*
81	-0.1498	(93% e_g)*
80	-0.1518	(54% e_g, 39% NOπ, 3% d_π)*
79	-0.2215	(35% Porπ, 28% NOπ, 15% NOσ, 17% d_π, 2% $d_z{}^2$)*
78	-0.3454	100% a_{1u}
77	-0.3624	61% a_{2u}, 18% NOπ, 7% NOσ, 8% d_π
76	-0.4177	54% e_g, 28 %d_π, 9% NOπ, 4% NOσ, 2% $d_z{}^2$
75	-0.4233	69% e_g, 29% d_π
74	-0.4274	81% Imidπ, 9% Imidσ
73	-0.4431	95% b_{1u}
72	-0.4457	92% a_{2u}
71	-0.4602	93% e_g, 2% d_{xy}, 1% dπ
70	-0.4611	96% e_g, 1% d_{xy}
69	-0.4624	89% d_{xy}
68	-0.4753	38% e_g, 37% d_π, 7% Imidπ, 5% NOπ
67	-0.4810	55% d_π, 21% e_g, 13% Imidπ, 3% NOπ

Continued on next page

Table I. Continued

		Fe(III)-NO = 2.143 Å
Orb. No.	Energy (au)	Character
92	-0.0005	(79% Porσ, 4% Feσ)*
91	-0.0116	(100% a_{1u})*
90	-0.0226	(96% e_g)*
89	-0.0233	(95% e_g)*
88	-0.0312	(84% Imidπ, 14% Imidσ)*
87	-0.0373	(67% $d_{x^2-y^2}$, 11% Porσ)*
86	-0.0489	(55% b_{1u}, 30% d_{z^2}, 2% $d_{x^2-y^2}$)*
85	-0.0494	(48% d_{z^2}, 29% b_{1u}, 2% $d_{x^2-y^2}$)*
84	-0.0684	(85% Imidπ, 15% Imidσ)*
83	-0.0880	(100% b_{2u})*
82	-0.1232	(97% NOπ)*
81	-0.1465	(97% e_g)*
80	-0.1471	(97% e_g)*
79	-0.2215	(53% Porπ, 24% NOπ, 14% NOσ, 6% d_π)*
78	-0.3446	68% a_{1u}, 18% NOπ, 9% NOσ, 2% d_π
77	-0.3614	76% a_{2u}, 13% NOπ, 5% NOσ
76	-0.4078	41% e_g, 49% d_π, 2% NOπ, 1% d_{z^2}
75	-0.4120	48% e_g, 44% d_π,
74	-0.4234	82% Imidπ, 13% Imidσ
73	-0.4349	96% d_{xy}
72	-0.4437	96% b_{1u}
71	-0.4464	92% a_{2u}
70	-0.4591	93% e_g, 3% d_π
69	-0.4601	96% e_g
68	-0.4666	48% e_g, 12% Imidπ
67	-0.4686	56% e_g, 37% d_π, 15 NOπ

Table II. Ground State Orbital Description of Nitrosyl Ferrous Heme at Fe-NO Distances of 1.743 Å and 2.143 Å

	Fe(II)-NO = 1.743 Å	
Orb. No.	Energy (au)	Character
92	0.1013	$(94\%\ \mathrm{Imid}\sigma)^*$
91	0.0999	$(99\%\ a_{1u})^*$
90	0.0913	$(96\%\ e_g)^*$
89	0.0906	$(95\%\ e_g)^*$
88	0.0735	$(43\%\ d_{x^2-y^2},\ 24\%\ b_{1u},\ 9\%\ d_{z^2},\ 7\%\ \mathrm{Por}\sigma)^*$
87	0.0711	$(58\%\ d_{z^2},\ 5\%\ d_{x^2-y^2},\ 5\%\ \mathrm{NO}\sigma,\ 5\%\ \mathrm{Imid}\sigma)^*$
86	0.0672	$(62\%\ b_{1u},\ 23\%\ d_{x^2-y^2})^*$
85	0.0656	$(82\%\ \mathrm{Imid}\pi,\ 14\%\ \mathrm{Imid}\sigma)^*$
84	0.0291	$(84\%\ \mathrm{Imid}\pi,\ 15\%\ \mathrm{Imid}\sigma)^*$
83	0.0226	$(100\%\ b_{2u})^*$
82	-0.0114	$(84\%\ \mathrm{NO}\pi,\ 12\%\ d_\pi)^*$
81	-0.0373	$(97\%\ e_g)^*$
80	-0.0380	$(97\%\ e_g)^*$
79	-0.3285	$62\%\ \mathrm{NO}\pi,\ 26\%\ \mathrm{NO}\sigma,\ 8\%\ d_\pi,\ 3\%\ d_{z^2}$
78	-0.2281	$100\%\ a_{1u}$
77	-0.2390	$94\%\ a_{2u}$
76	-0.2925	$43\%\ e_g,\ 48\%\ d_\pi,\ 6\%\ \mathrm{NO}\pi$
75	-0.3012	$54\%\ e_g,\ 40\%\ d_\pi$
74	-0.3250	$41\%\ d_{xy},\ 21\%\ b_{1u},\ 29\%\ \mathrm{Imid}\pi$
73	-0.3253	$55\%\ d_{xy},\ 21\%\ \mathrm{Imid}\pi,\ 15\%\ b_{1u}$
72	-0.3311	$70\%\ b_{1u},\ 21\%\ \mathrm{Imid}\pi$
71	-0.3320	$83\%\ a_{2u},\ 9\%\ \mathrm{Imid}\pi$
70	-0.3489	$90\%\ e_g,\ 3\%\ d_\pi$
69	-0.3498	$97\%\ e_g$
68	-0.3517	$52\%\ e_g,\ 27\%\ d_\pi,\ 7\%\ \mathrm{Imid}\pi,\ 6\%\ \mathrm{NO}\pi$
67	-0.3599	$40\%\ e_g,\ 47\%\ d_\pi,\ 6\%\ \mathrm{NO}\pi$

Continued on next page

Table II. Continued

		Fe(II)-NO = 2.143 Å
Orb. No.	Energy (au)	Character
92	0.1057	(95% Imidσ)*
91	0.1020	(100% a_{1u})*
90	0.0938	(96% e_g)*
89	0.0931	(93% e_g)*
88	0.0878	(68% $d_{x^2-y^2}$, 12% Porσ)*
87	0.0749	(78% d_{z^2}, 7% Imidσ)*
86	0.0718	(88% b_{1u}, 3% Imidπ, 2% $d_{x^2-y^2}$)*
85	0.0704	(80% Imidπ, 13% Imidσ)*
84	0.0340	(84% Imidπ, 15% Imidσ)*
83	0.0247	(100% b_{2u})*
82	-0.0346	(97% e_g)*
81	-0.0353	(97% e_g)*
80	-0.0451	(89% NOπ, 5% NOσ, 2% d_π)*
79	-0.3725	69% NOπ, 26% NOσ
78	-0.2257	100% a_{1u}
77	-0.2375	94% a_{2u}
76	-0.2883	51% d_π, 38% e_g
75	-0.2912	54% d_π, 41% e_g
74	-0.3087	96% d_{xy}
73	-0.3208	58% Imidπ, 29% b_{1u}, 9% Imidσ
72	-0.3278	72% b_{1u}, 22% Imidπ
71	-0.3309	93% a_{2u}
70	-0.3465	83% e_g, 7% d_π
69	-0.3475	98% e_g
68	-0.3482	63% e_g, 26% d_π, 6% Imidπ
67	-0.3518	55% e_g, 41% d_π

Table III. Ground State Orbital Description of Ligand-bound Carbonyl and Oxyheme Complexes

Carbonylheme (Fe-CO = 1.77 Å)	
Orb. No.	Character
92	(90% Imidσ, 3% d_z2)*
91	(100% a_{1u})*
90	(59% $d_{x}2_{-y}2$, 37% Porσ)*
89	(93% e_g, 3% COπ, 2% d_z2)*
88	(88% e_g, 5% d_z2, 3% COπ)*
87	(43% d_z2, 21% Imidσ, 14% Porπ, 7% COσ)*
86	(65% COπ, 21% Imidπ, 7% d_π)*
85	(55% COπ, 37% b_{1u}, 5% d_π)*
84	(65% b_{1u}, 30% COπ, 3% d_π)*
83	(79% Imidπ, 15% COπ, 1% d_π)*
82	(100% b_{2u})*
81	(99% Imidπ, 1% COπ)*
80	(98% e_g, 2% d_π)*
79	(98% e_g, 2% d_π)*
78	100% a_{1u}
77	97% a_{2u}
76	62% d_π, 31% e_g, 4% COπ
75	62% d_π, 33% e_g, 4% COπ
74	96% d_{xy}
73	70% Imidπ, 30% b_{1u}
72	93% a_{2u}, 3% COπ
71	78% b_{1u}, 22% Imidπ
70	98% e_g
69	99% e_g
68	68% e_g, 17% d_π, 15% Imidπ, 4% COπ
67	70% e_g, 23% d_π, 6% COπ

Continued on next page

Table III. Continued

Orb. No.	Oxyheme (Fe-O_2 = 1.75 Å) Character
92	(99% a_{1u})*
91	(85% Imidσ, 12% Porπ)*
90	(99% e_g)*
89	(88% e_g,10% Imidσ)*
88	(100% b_{1u})*
87	(55% d_{z^2}, 13% Porσ, 13% Imidσ, 7% O_2σ, 4% $d_{x^2-y^2}$)*
86	(58% $d_{x^2-y^2}$, 34% Porσ, 4% d_{z^2}, 1% O_2σ)*
85	(98% Imidπ, 1% d_{z^2})*
84	(100% b_{2u})*
83	(99% Imidπ)*
82	(82% e_g, 12% d_π, 5% O_2π)*
81	(99% e_g, 1% d_π)*
80	(43% O_2π, 36% d_π, 18% e_g)*
79	100% a_{1u}
78	99% a_{2u}
77	47% O_2π, 32% d_π, 21% e_g
76	43% e_g, 34% d_π, 12% O_2π, 7% O_2σ
75	36% Porπ, 32% O_2π, 27% O_2σ, 1% d_π
74	82% b_{1u}, 18% Imidπ
73	87% a_{2u}, 5% O_2π, 3% O_2σ, 1% d_{xy}
72	69% Imidπ, 28% Porπ, 1% O_2π, 1% O_2σ, 1% d_π
71	90% e_g, 3% O_2π, 3% d_π
70	98% e_g
69	79% e_g, 7% d_π, 6% O_2π, 3% $d_{x^2-y^2}$
68	87% d_{xy}, 6% Porσ, 1% O_2π
67	48% d_π, 27% e_g, 15% Imidπ, 7% O_2σ, 3% O_2π

83 and 84 resulting in an appreciable net increase in their energies as well as changes in their relative positions. The NO π^* orbitals are also affected by the iron-ligand bond lengthening, which results in their stronger mixing with the porphyrin e_g π^* and a_{2u} π orbitals.

The effect of bond lengthening in the ferrous complex on the iron d_σ orbitals is primarily an increased localization with smaller changes in their orbital energies and no change in their order. The NO π^* orbitals, on the other hand, are appreciably lowered in energy accompanied by a decrease in the d_σ and d_π contributions to these orbitals.

Simplified, scaled, diagramatic representations of the excited state energies of nitrosyl ferric and ferrous heme complexes at successively increasing Fe-NO distances are shown in Figures 1 and 2, respectively. The excited states of the ferric complex are shown only for Fe-NO distances up to 2.143 Å because of the necessity of relaxing the "frozen" spin approximation beyond this distance. Each excited state calculated at each Fe-NO distance is represented by a horizontal line. However, only the Soret and Q-band, and the states that show large variations in energy with iron-ligand distance or correspon' to important charge-transfer states are identified by character. The percentages shown to the right of some of the excited states specify the contribution of the $d_\pi \rightarrow d_{z^2}^*$ type transitions to these states. The atomic orbital notations used to specify important transitions in Figures 1 and 2 represent only the main character and not the entire composition of the mlecular orbitals involved. The remaining orbital characters can be obtained by examining Tables I and II which contain a complete description of the same orbitals.

Examination of Figures 1 and 2 shows that only the states corresponding to $d_\pi \rightarrow d_{z^2}^*$ and $d_{xy} \rightarrow d_{z^2}^*$ transitions display a dissociative profile according to the criterion of "decreasing energy with increasing iron-ligand distance and no barrier to dissociation." Since the contributions from $d_\pi \rightarrow d_{z^2}^*$ transition are distributed in excited states over the entire energy range between the Q and the Soret band in both ferrous and ferric complexes, the collective behavior of all of these states are indicated by vertical bars and solid arrows. In common with oxy and carbonyl ferrous heme complexes, as shown in previous studies (44, 45), neither the Soret or the Q band show any dissociative behavior.

An important aspect of the results is the significantly different behavior obtained for the charge-transfer states in nitrosyl ferrous and ferric complexes. For the nitrosyl ferric heme, a number of such states primarily a_{1u}, $a_{2u} \rightarrow$ NO π^*, (NO π, $d_\pi)^*$ calculated at energies below the Q-band are similar to those calculated for oxy ferrous heme (45). However, while these states display dissociate profiles in case of O_2, no such pattern can be recognized for the NO ligand. The behavior of the a_{1u}, $a_{2u} \rightarrow$ NO π, NO π^* charge-transfer states in the nitrsyl ferrous heme complex is very different. The energies of these states (Figure 2) are higher, in the range between the Q and Soret band. More interestingly, these state display a minimum corresponding to a 0.4 Å increase in the Fe-NO distance. The minimum persists when the NO ligand is rotated 45 degrees about the Fe-N axis and the CI calculations are repeated. We have determined that this behavior is not due to convergence errors. It is also

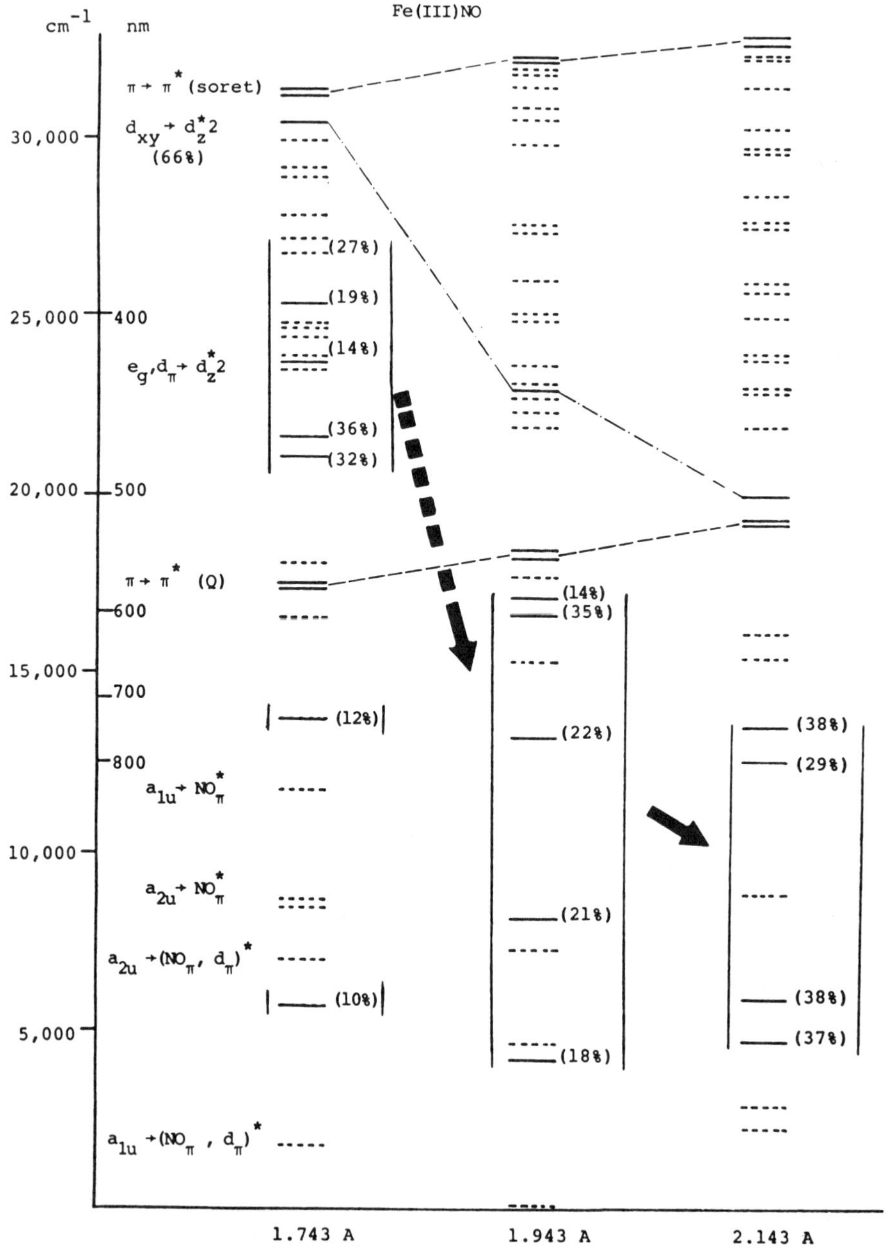

Figure 1. Simplified, scaled, diagramatic representation of the excited-state energies of nitrosyl ferric heme complex at iron-nitrogen distances of 1.743, 1.943 and 2.143 Å showing correlation between similar photodissociating states. The $\pi \rightarrow \pi^*$ transition of the Soret and Q bands are also identified.

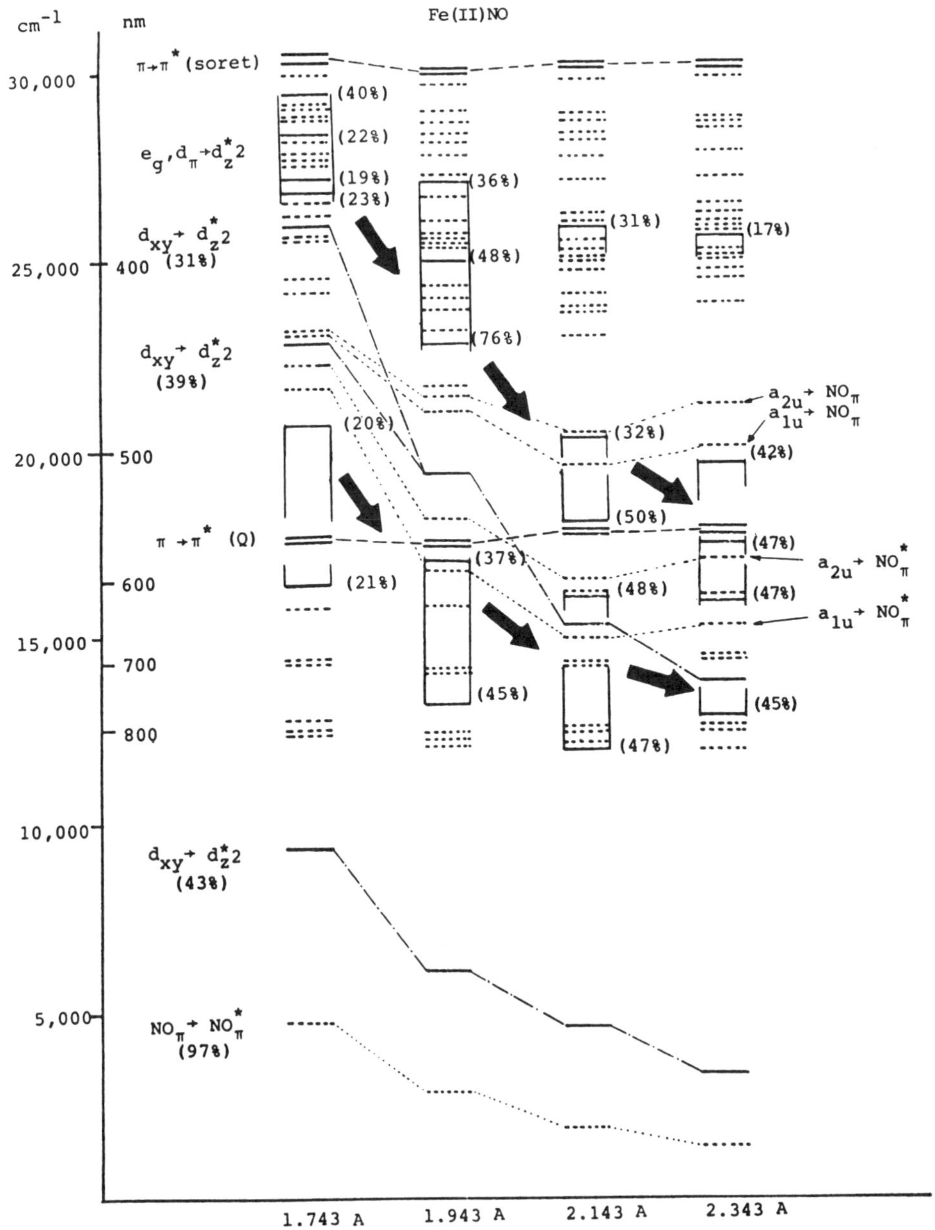

Figure 2. Simplified, scaled, diagramatic representation of the excited-state energies of nitrosyl ferrous heme complex at iron-nitrogen distances of 1.743, 1,943, 2.143 and 2.343 Å showing correlation between similar photodissociating states. The $\pi \rightarrow \pi^*$ transition of the Soret and Q bands are also identified.

unlikely that it arises as an artifact of the parametrization schemes in the INDO methods, since similar charge-transfer states previously obtained for oxyheme (45) do not display such a minimum in the energy profile.

Discussion

The major conclusion of the present study, as can be seen in Figures 1 and 2, is that the primary photodissociating states, in both nitrosyl ferrous and ferric heme complexes correspond to the $d_{\pi} \rightarrow d^{*}_{z^2}$ excitations. The calculated energies also indicate that this dissociative channel can be activated independent of the excitation frequency in the range of Q to Soret band energies. This is the same type of excitation previously identified as the photoactive state involved in CO and O_2 photodissociation from ferrous heme complexes.

It is known that photodissociation of NO from ferric heme proteins occurs in high quantum yield (4-8). Recent picosecond experiments (24, 27) also indicate that photodissociation of ferrous nitrosylhemoglobin is initiated with high yield resulting in transient photoproducts with a 17 ps lifetime. In fact, Cornelius and Hochstrasser (24) have shown that complexes of CO, O_2, and NO with ferrous hemoglobin all are photolized efficiently in the picosecond timescale. They have further shown that, while the initial photodissociations are very similar for the three ligands, their behaviors after photodissociation are not. The similarities in the photodissociation of CO, O_2, and NO are consistent with our finding of a common photodissociating state for all three systems. Our results can also provide insight into their photodissociation differences.

We have previouly suggested (44) that photodissociation of CO occurs via a single $d_{\pi} \rightarrow d^{*}_{z^2}$ dissociative channnel with no mechanism for radiative deactivation of the photoactive state, accounting for the high quantum yield independent of frequency of excitation, leading to long-lived photoproducts. By contrast, we have proposed a two-channel photodissociation mechanism for oxyheme (45), one a $d_{\pi} \rightarrow d^{*}_{z^2}$ type and the other involving charge-transfer a_{1u}, $a_{2u} \rightarrow 0\pi^{*}$ type transitions below the Q-band. The two channel mechanism together with the assumption of a competitive decay path from these charge transfer states to the ground state account for the lowering of the overall quantum yield and its observed excitation frequency dependence in oxyheme.

The present results suggest that photodissociation of NO from ferrous heme also occurs via a single $d_{\pi} \rightarrow d^{*}_{z^2}$ channel similar to that for CO. However, immediately following photodissociation, a cross over to charge-transfer states (Figure 2), which displays a minimum in their energy profile, can result in transient trapping of the radical NO ligand near an equilibrium Fe-NO bond distance. Such a cross over should be efficient because both states are of the same spin multiplicity (S=1/2). Population of such excited states could lead to either geminate recombination or deactivation, via energy decay, both of which could result in small overall quantum yield of NO photolysis. This mechanism together with those suggested for CO and O_2 photodissociation (44, 45) account for the efficient picosecond photolysis of the three ligands from ferrous heme complexes, and also provide an explanation for their different behavior following dissociation.

Our results indicate that NO photodissocation from ferric heme complexes also occurs via a single $d_\pi \rightarrow d_{z^2}^*$ type dissociative channel. Although this initial process seems similar to that suggested for ferrous carbonylheme (44), the excited-state profile of nitrosyl ferric heme is more complicated. Particularly significant is the presence of low-energy, charge-transfer, excited states involving NO π^* orbitals below the Q band which are absent in the carbonylheme complex. While as mentioned above, such charge transfer states in oxyheme complex could provide an energy decay or recombination path to the ligand-bound ground state, the high quantum yield observed for NO photolysis from ferric heme proteins (4-6) suggest that the charge-transfer states in ferric NO complex do not play a role in either geminate recombination or decay to the liganded state. We, therefore, suggest that NO photodissociation from the singlet nitrosyl ferric heme complex, via a $d_\pi \rightarrow d_{z^2}^*$ channel, requires eventual relaxation to a triplet $d_\pi \rightarrow d_{z^2}^*$ channel. The radical nature of both the dissociated NO ligand and the ferric heme system will prevent cross over to a singlet charge transfer state and recombination or decay to the singlet liganded state, since these processes would require a spin change. This argument is consistent with the spin-paired biradical nature of the ground state found for the liganded complex. It is also consistent with the observation of the Fe(III) oxidation state of HRP and myoglobin following NO photolysis (6-8).

In summary, the results obtained in this study and in previous studies indicate a common photodissociative channel for carbonyl, nitrosyl, nitrosyl, and oxy ferrous heme and nitrosyl ferric heme complexes. The low apparent quantum yield of the nitrosyl ferrous system at long times (>400 ps) is attributed to geminate recombination or energy decay from relatively high-energy (near Q) charge-transfer states allowing NO to be trapped in a minimum near the iron. The high quantum yield of the nitrosyl ferric heme is attributed to initial dissociation on a radical pathway which does not allow cross over to low-energy charge-transfer states observed for this system. Such charge-transfer states are absent in the ferrous carbonyl system. Thus the high quantum yield observed for these two "isoelectronic" systems is the result of a common photodissociating state but different properties of the liganded and the dissociated states of each system.

Acknowledgments

We gratefully acknowledge financial support for this work from NSF Grant PCM 8410244. We also thank Professor M. C. Zerner for supplying the INDO program.

Literature Cited

1. Keilin, D.; Hartree, E. F. Biochem. J. 1955, 61, 153.
2. Gibson, Q. H.; Ainsworth, S. Nature (London) 1957, 180, 1416.
3. Brunori, M.; Giacometti, G. M.; Antonini, E.; Wyman, J. Proc. Natl. Acad. Sci. U.S.A. 1973, 70, 3141.

4. Saffran, W. A.; Gibson, Q. H. J. Biol. Chem. 1977, 252, 7955.
5. Hoffman, B. M.; Gibson, Q. H. Proc. Natl. Acad. Sci. U.S.A. 1978, 75, 21.
6. Tamura, M.; Kobayashi, K.; Hayashi, K. FEBS Lett. 1978, 88, 124.
7. Kobayashi, K.; Tamura, M.; Hayashi, K. J. Biol. Chem. 1980, 255, 2239.
8. Kobayashi, K.; Tamura, M.; Hayashi, K. Biochemistry 1982, 21, 729.
9. Bücher, T.; Negelein, E. Biochem. Z. 1941, 311, 163.
10. Noble, R. W.; Brunori, M.; Wyman, J.; Antonini, E. Biochemistry 1967, 6, 1216.
11. Bonaventura, C.; Bonaventura, J.; Antonini, E.; Brunori, M.; Wyman, J. Biochemistry 1973, 12, 3424.
12. Smith, M. H. Biochem. J. 1959, 73, 90.
13. Chang, C. K.; Taylor T. G. Proc. Natl. Acad. Sci., U.S.A. 1975, 72, 1166.
14. Dixon, D. W.; Kirmaier, C.; Holten, D. J. Am. Chem. Soc. 1985, 107, 808.
15. Shank, C. V.; Ippen, E. P.; Bersohn, R. Science (Washington, D.C.) 1976, 193, 50.
16. Noe, L. J.; Eisert, W. G.; Rentzepis, P. M. Proc. Natl. Acad. Sci. U.S.A. 1978, 75, 573.
17. Eisert, W. G.; Degenkolb, E. O.; Noe, L. J.; Rentzepis, P. M. Biophys. J. 1979, 25, 455.
18. Reynolds, A. H.; Rand, S. D.; Rentzepis, P. M. Proc. Natl. Acad. Sci. U.S.A. 1981, 78, 2292.
19. Greene, B. I.; Hochstrasser, R. M.; Weisman, R. B.; Eaton, W. A. Proc. Natl. Acad. Sci. U.S.A. 1978, 75, 5255.
20. Chernoff, D. A.; Hochstrasser, R. M.; Steele, A. W. Proc. Natl. Acad. Sci. U.S.A. 1980, 77, 5606.
21. Terner, J.; Spiro, T. G.; Nagumo, M.; Nicol, M. F.; El-Sayed, M. A. J. Am. Chem. Soc. 1980, 102, 3238.
22. Terner, J.; Stong, J. D.; Spiro, T. G.; Nagumo, M.; Nicol, M.; El-Sayed, M. A. Proc. Natl. Acad. Sci. U.S.A. 1981, 78, 1313.
23. Coppey, M.; Tourbez, H.; Valat, P.; Alpert, B. Nature (London) 1980, 284, 568.
24. Cornelius, P. A.; Hochstrasser, R. M. in "Picosecond Phenomena III", Eisenthal, K. B.; Hochstrasser, R. M.; Kaiser, W.; Laubereau, A. Eds.; Springer-Verlag: Berlin 1982; p. 228.
25. Martin, J. L.; Poyart, C.; Migus, A.; Lecarpentier, Y.; Astier, R.; Chambaret, J. P. in "Picosecond Phenomena III", Eisenthal, K. B.; Hochstrasser, R. M.; Kaiser, W.; Laubereau, A. Eds.; Springer-Verlag: Berlin 1982, p. 294.
26. Martin, J. L.; Migus, A.; Poyart, C.; Lecarpentier, Y.; Antonetti, A.; Orszag, A. Biochem. Biophys. Res. Comm. 1982, 107, 803.
27. Cornelius, P. A.; Hochstrasser, R. M.; Steele, A. W. J. Mol. Biol. 1983, 163, 119.
28. Austin, R. H.; Beeson, K. W.; Eisenstein, L.; Frauenfelder, H.; Gunsalus, I. C. Biochemistry 1975, 14, 5355.
29. Alberding, N.; Austin, R. H.; Chan, S. S.; Eisenstein, L.; Frauenfelder, H.; Gunsalus, I. C.; Nordlund, T. M. J. Chem. Phys. 1976, 65, 4701.

30. Alberding, N.; Chan, S. S.; Eisenstein, L.; Frauenfelder, H.; Good, D.; Gunsalus, I. C.; Nordlund, T. M.; Perutz, M. F.; Reynolds, A. H.; Sorensen, L. B. Biochemistry 1978, 17, 43.
31. Beece, D.; Eisenstein, L.; Frauenfelder, H.; Good, D.; Marden, M. C.; Reinisch, L.; Reynolds, A. H.; Sorensen, L. B.; Yue, K. T. Biochemistry 1979, 18, 3421.
32. Sawicki, C. A.; Gibson, Q. H. J. Biol. Chem. 1976, 251, 1533.
33. Alpert, B.; El Mohsni, S.; Lindqvist, L.; Tfibel, F. Chem. Phys. Lett. 1979, 64, 11.
34. Duddell, D. A.; Morris, R. J.; Richards, J. T. Biochim. Biophys. Acta 1980, 621, 1.
35. Duddell, D. A.; Morris, R. J.; Muttucumaru, N. J.; Richards, J. T. Photochem. Photobiol. 1980, 31, 479.
36. Friedman, J. M.; Lyons, K. B. Nature (London) 1980, 284, 570.
37. Lyons, K. B.; Friedman, J. M.; Fleury, P. A. Nature (London) 1978, 275, 565.
38. Dallinger, R. F.; Nestor, J. R.; Spiro, T. G. J. Am. Chem. Soc. 1978, 100, 6251.
39. Woodruff, W. H.; Farquharson, S. Science (Washington, D.C.) 1978, 201, 831.
40. Ridley, J.; Zerner, M. Theor. Chim. Acta 1973, 32, 111.
41. Ridley, J. E.; Zerner, M. C. Theor. Chim. Acta 1976, 42, 223.
42. Bacon, A. D.; Zerner, M. C. Theor. Chim. Acta 1979, 53, 21.
43. Zerner, M. C.; Loew, G. H.; Kirchner, R. F.; Mueller-Westerhoff, U. T. J. Am. Chem. Soc. 1980, 102, 589.
44. Waleh, A.; Loew, G. H. J. Am. Chem. Soc. 1982, 104, 2346.
45. Waleh, A.; Loew, G. H. J. Am. Chem. Soc. 1982, 104, 2352.
46. Edwards, W. D.; Zerner, M. C., in preparation.
47. Chantranupong, L.; Loew, G. H.; Waleh, A., in preparation
48. Piciulo, P. L.; Rupprecht, G.; Scheidt, W. R. J. Am. Chem. Soc. 1974, 96, 5293.
49. Scheidt, W. R.; Lee, Y. J.; Hatano, K. Inorg. Chim. Acta 1983, 79, 192.
50. Pauncz, R. "Spin Eigenfunctions"; Plenum Press: New York, 1980.
51. Rumer, G.; Teller, E.; Weyl, H. Göttinger Nocher. 1932, 3, 449.
52. Cooper, I. L.; McWeeny, R. J. Chem. Phys. 1966, 45, 226.
53. Sutcliff, B. T. J. Chem. Phys. 1966, 45, 235.
54. Reeves, C. M. Comm. A.C.M. 1966, 9, 276.

RECEIVED July 10, 1986

2

Quantum Mechanical Studies of Charge-Transfer States in Porphyrin Heterodimers

James D. Petke[1] and Gerald M. Maggiora[1,2]

[1]Departments of Chemistry and Biochemistry, University of Kansas, Lawrence, KS 66045

Ab initio quantum mechanical configuration interaction calculations have been carried out on a series of magnesium porphine-porphine (MgP-P) heterodimers to explore the relationship of geometry and environment to the location of charge-transfer (CT) states. In vacuum the calculations show the existence of two singlet MgP → P CT states which lie above the manifold of singlet Q states, the lowest-energy singlet excited states in the visible region of the spectrum. In the presence of polar solvents or chloride ion, the calculations show a dramatic lowering of the CT states, one of which under certain conditions becomes the lowest singlet excited state. The role of these CT states in photo-induced electron transfer processes and their role, in conjunction with the corresponding triplet states, in other photodynamical processes is discussed. Potential design criteria for the development of biomimetic solar energy conversion devices are also discussed.

Porphyrins are probably one of the most well studied classes of organic compounds (1). In addition, they play a number of important roles in biological systems and have been the basis for considerable biomimetic chemistry. Due to their striking similarities to chlorophylls, porphyrins have been used as models for photosynthetic systems. Harriman (2) has reviewed the application of porphyrin model systems in both natural and biomimetic photosynthetic energy conversion, and Gerischer and Katz (3) have edited a now somewhat dated but nonetheless very useful work on the general subject of light-induced charge separation in biology and chemistry. Netzel (4) and Boxer (5) have also considered the application of actual photosynthetic pigments or their closely related analogs to the

[2]Current address: The Upjohn Company, Kalamazoo, MI 49001

0097-6156/86/0321-0020$08.50/0

modeling of these same processes. Recently, Creed and Caldwell (6) have reviewed the general area of photo-induced electron transfer (ET), including material relevant to biologic and biomimetic photosynthetic energy conversion systems.

Two key features of photosynthetic systems is their ability to absorb, transport, and trap electronic excitation energy derived from solar photons, and their ability to use this trapped energy as a means for separating charge irreversibly. Due to a long standing interest in photosynthetic energy conversion, we have carried out a number of theoretical investigations of multimeric porphyrin systems designed to mimic key features of these two important processes. In addition, we have explored the use of these theoretical methods for the development of design criteria for the construction of biomimetic solar energy conversion systems. Specifically, we have investigated the effect of geometry and microenvironment on the formation of low-lying charge-transfer (CT) states in magnesium porphine-porphine (MgP-P) dimers using ab initio quantum mechanical methods (7, 8), and we have investigated the limitations of the point-dipole approximation in exciton theory calculations of multimeric bacteriochlorin systems (9). In the present work we will focus on CT states in porphyrin heterodimer systems, and we will discuss a number of features of these systems not described in our earlier works (7, 8). Theoretical work on exciton states and energy transport will be discussed in forthcoming work (9).

CT States and Photo-induced ET Processes

Electron transfer plays an essential role in many chemical and biological processes (10,11). Most processes involve thermal ET, although a number, such as photosynthetic energy conversion, involve photo-induced ET. Photo-induced ETs are generally characterized by the formation of an initial CT state. Functionally, such highly polar "charge-separated" states act as strong electron donors in an appropriate molecular environment. Thus, photo-induced CT states are likely to play critically important roles in the early stages of the photosynthetic energy conversion process. For example, in purple photosynthetic bacteria ET is thought to proceed from a donor "special-pair" dimer of bacteriochlorophyll molecules to an acceptor bacteriochlorophyll with subsequent transfer of an electron to a nearby bacteriopheophytin molecule all within 10^{-11} sec, and with a quantum yield approaching unity. The resultant radical-pair is sufficiently stable against charge recombination to permit charge separation over larger distances through additional ET steps (12, 13). In addition, Jortner (14) has considered the dynamics of the primary light-induced charge separation in bacterial photosynthesis, and has proposed that ET occurs in competition with vibrational relaxation, from a non-equilibrium nuclear configuration of the donor, initially excited above the crossing point of the potential surfaces of the neutral complex and radical-pair. Warshel (15) has also examined a number of interesting features of this system using a blend of quantum and classical methods to simulate photo-induced ET, and Friesner and Wertheimer (16) have taken a density matrix approach to model the process.

Until a recent x-ray diffraction study (17) provided direct evidence of the arrangement of the pigment species in the reaction center of the photosynthetic bacterium Rhodopseudomonas Viridis, a considerable amount of all evidence pertaining to the internal molecular architecture of plant or bacterial reaction centers was inferred from the results of in vitro spectroscopic experiments and from work on model systems (5, 18, 19). Aside from their use as indirect probes of the structure and function of plant and bacterial reaction centers, model studies have also provided insights into the development of potential biomimetic solar energy conversion systems. In this regard, the work of Netzel and co-workers (20-22) is particularly noteworthy, and in addition, is quite relevant to the material discussed at this conference.

Specifically Netzel et al. (20-22), in studies of "face-to-face", covalently-linked MgP-P dimers, found evidence for the formation within 6 psecs of a low-lying, relatively long-lived intramolecular CT state of the type MgP^{+}-P^{-} in polarizable or highly polar solvents and in solvents where chloride ion coordinates with the magnesium ion of the MgP-macrocycle. These workers also observed the formation of benzoquinone anion radicals as stable photoproducts of the CT formation process when the experiments were carried out in the presence of benzoquinone (21). This approach provides a more direct test for the formation of an intramolecular CT state, and the results are in sharp contrast to those typically observed when porphyrin $^{1}(\pi,\pi^{*})$ states are quenched in the presence of benzoquinone (23).

While Netzel et al. (20-22) have investigated the effect of solvent on the formation of low-lying CT states, they have not extensively investigated the important question of the role of geometry in the formation of these states. In their latest study (22) they did investigate a MgP-P dimer with a face-to-face geometry that was slightly translated from a fully superimposed geometry. The results obtained were quite similar to those they obtained in the usual face-to-face geometry, although the decay kinetics did differ in the two cases.

With the recent x-ray diffraction study of R. Viridis reaction centers (17) it has become clear that non-face-to-face geometries may be important in determining, or at least in not adversely affecting, the photodynamics of MgP-P dimers and related biomimetic systems. However, the difficulty of synthesizing a wide enough variety of structurally well-characterized systems has in the past and will in the future continue to hamper experimental efforts to elucidate this potentially important design feature of biomimetic solar energy conversion systems. It is precisely in this latter area that theoretical calculations are likely to play an important and an expanded role. The results to be discussed in what follows address this point as well as others that relate to the effects that "environmental tuning," taken in its broadest possible sense, can have on the formation of low-lying and relatively long-lived CT states.

Theoretical Methodology

Our approach is based on ab initio quantum mechanics, specifically self-consistent field molecular orbital and configuration interaction (SCF-MO-CI) calculations (24,25). We have employed this technique to elucidate the electronic structural features of MgP-P dimers in a variety of intermolecular geometries and in the presence of "environmental factors" such as chloride ions and polar solvents (7, 8). As in many of our past studies we employed a simple molecular fragment basis set of floating spherical gaussian orbitals (FSGO) (26), which has proven to provide a satisfactory qualitative interpretation of the electronic structure of the ground and low-lying excited states of large, highly conjugated macrocyclic systems (27-29). One effect of the use of an FSGO basis is, however, that calculated transition energies are generally too large in comparison with experimental values. To obtain better estimates of transition energies, ΔE^{est}, we have developed a scaling procedure based on the equation

$$\Delta E^{est} = 0.610\Delta E^{calc} - 441.0\ \text{cm}^{-1}, \qquad (1)$$

which has proved to be quite useful in earlier studies of $\pi \rightarrow \pi^*$ transitions of MgP and P monomers (28) and of a considerable number of related systems (29).

In heterodimer systems, such as those described in the current work, orbitals obtained from supermolecule SCF calculations often bear a close resemblance to monomeric orbitals when the monomeric species are separated by distances greater than or equal to the Van der Waals separation. When the monomers are forced to be closer than this distance, as for example in many cyclophanes, orbital mixing is quite strong, and the monomeric identity of the orbitals is somewhat obscured. This was in fact observed in the case of the face-to-face dimer with an interplanar spacing of 3.35 Å, described in an earlier work (8).

Even in situations where the inter-monomer separation is greater than the Van der Waals separation, orbital mixing can occur which partially obscures the monomeric parentage of particular orbitals. This usually takes place when orbitals from each monomer are approximately degenerate. While CI calculations could, in principle, be carried out with such "scrambled" orbitals, interpretation of the resulting wavefunctions in terms of monomeric contributions becomes virtually impossible. Fortunately, this type of purely accidental degeneracy can be removed by a unitary transformation which "unscrambles" the orbitals. We have developed a method which is analogous to the usual "two-by-two" rotation methods (30) for this purpose. Molecular orbitals resulting from such transformations are quite monomer-like, and have been very useful in producing easily interpretable wavefunctions, since the configurations appearing in the CI wavefunctions have been generated from excitations which clearly maintain their monomeric character. The overall procedure used to obtain the heterodimer CI wavefunctions is summarized in Scheme I.

In the present work the CI procedure (25) employed includes single, double, and higher excited configurations, and can treat in a natural way, using the localized monomer basis orbitals, intra- as well as inter-molecular excitations. States which arise from such "monomer-based excitations" can then be interpreted in terms of their locally-excited, excitonic, CT, or "mixed" character. Details as to the nature of the dimer states have been discussed in our earlier works (7, 8).

Molecular Dimer Geometries Investigated

A considerable variety of molecular complexes have been examined during the course of our investigations, including the following: (1) face-to-face dimers at several interplanar spacings, (2) face-to-face dimers in which the MgP macrocycle has been rotated by 45° about an axis passing through the magnesium atom and perpendicular to the macrocyclic plane, again at several interplanar spacings, (3) a "slipped" face-to-face geometry in which the MgP macrocycle is translated relative to the P macrocycle, (4) a "T-shaped" dimer, (5) an "L-shaped" dimer, (6) a fully planar dimer in which both macrocyclic rings lie in the same plane, and (7) a dimer related

Supermolecule Hartree-Fock
SCF Calculations

↓

Orbital Transformations
(generation of "monomer-like" orbitals)

↓

Generation of Excited Configurations

↓

CI Calculations
(matrix element evaluation
and diagonalization)

↓

Porphyrin Dimer Excited States

Scheme I

to the fully planar one in which one macrocycle is rotated 90 degrees with respect to the other. A number of the complexes are illustrated in Figure 1. In studies involving coordination of the dimers by chloride ion, a Mg-Cl bond length equal to the covalent radius of the magnesium atom and the ionic radius of the chloride ion was used, with the chloride ion positioned along a vertical axis perpendicular to the MgP-macrocyclic plane and passing through the magnesium atom.

Both the face-to-face dimers and the T- and L-shaped dimers have relevance to biomimetic systems, especially in light of the work on the x-ray structure of the R. Viridis reaction center (17). As will be discussed in the following, the results obtained for the face-to-face dimer with an interplanar spacing of 5.35 Å is prototypical of the results obtained for almost all the dimers investigated, and thus it provides a reasonable basis for illustrating a number of important features that arise from the calculations. However, it should be noted that the lack of "covalent links" in our model dimers precludes the possibility of "through-bond" effects.

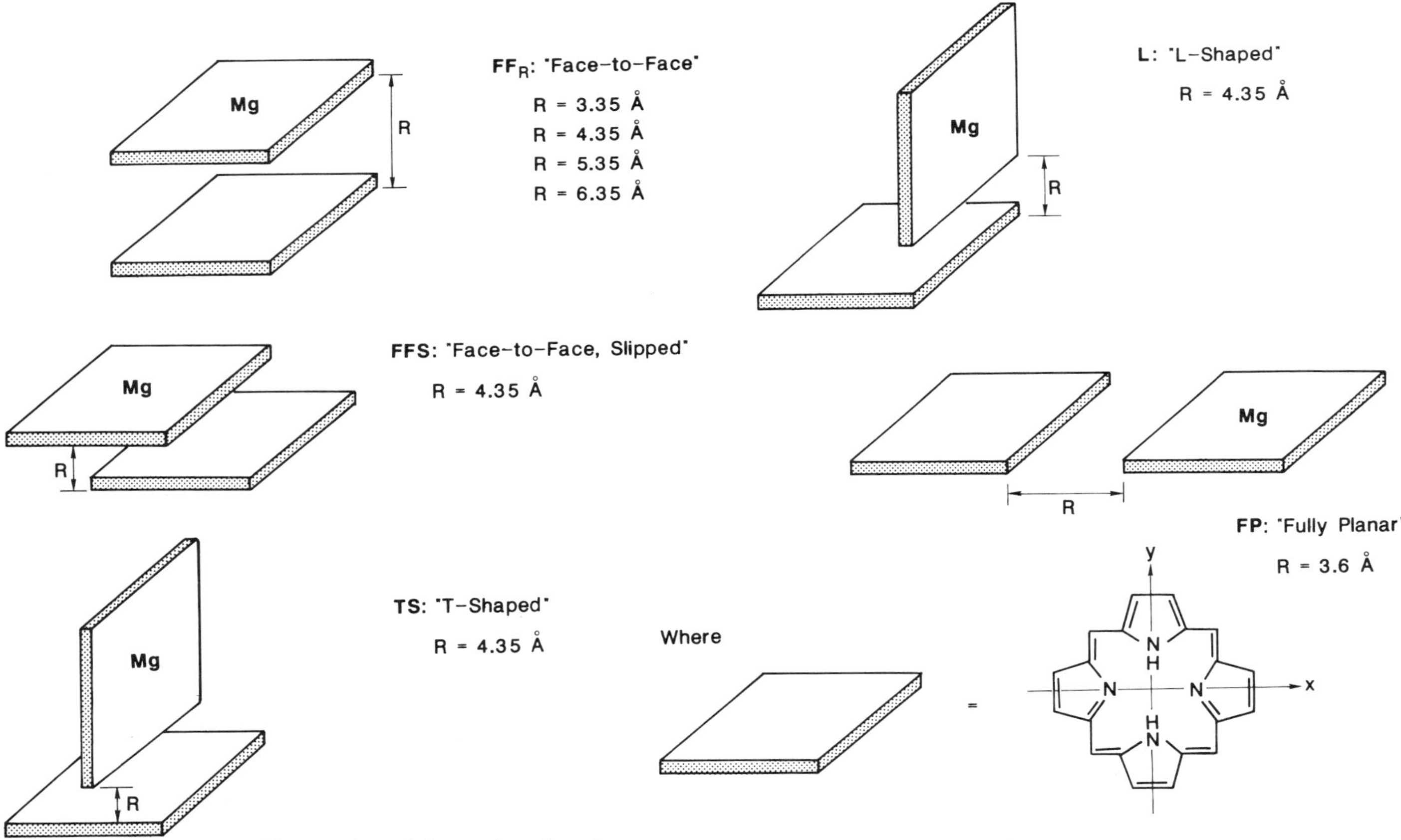

Figure 1. Selected molecular geometries of MgP-P dimers. (Reproduced with permission from Ref. 8. Copyright 1986 American Institute of Physics.)

Whether the omission of such effects would significantly alter the conclusions arrived at as a consequence of our calculations is difficult to say, but we expect that major through-bond effects in photo-induced ET would only become important as the distance between the two porphyrin macrocycles becomes very large. Hoffman and co-workers (31) have discussed several aspects of through-bond effects, and Beratan and Hopfield (32) have approximated them in a recent study of ET in a rigid system.

Identical structures were used in both the ground and excited state calculations, in keeping with the Franck-Condon (FC) approximation (33). Franck-Condon excited states are generally not in equilibrium with respect to their own molecular structure and that of their microenvironments. Thus, structural relaxations could occur that would significantly affect the state ordering and photodynamical properties in these systems (vide infra). Such structural relaxations are not expected to be of major consequence with respect to porphyrin (π,π^*) states, but may be significant for CT states. The latter states can be thought of approximately as being made up of a π-cation and π-anion radical pair. Based on a variety of experimental data Netzel et al. (21, 22) have argued that significant structural relaxations of the individual radical ion structures or of the solvent shell surrounding the radical-pair dimers do not occur. McHale and Simons (34) have, however, expressed an alternative view in which they showed that inclusion of molecular structural changes was necessary to account for the low-lying charge-resonance transition observed in anthracene-dimer anions. While x-ray structures of radical cation and anion porphyrins are scarce, those that are available (see e.g. 35) provide evidence suggesting that significant loss of planarity of the macrocycle does occur. In addition, opposing charges on the two macrocycles, i.e. MgP^+ and P^-, will give rise to an attractive interaction between the macrocycles which will tend to draw them closer together. Finally, even though the dimer system occupies a large volume such that, as argued by Netzel et al. (21, 22), small volume changes should not lead to significant solvent restructuring, it must be borne in mind that very large changes in electron distribution take place upon CT state formation. Thus, it is likely that such large changes in charge distribution would have an effect on the structure of polar/polarizable solvents in regions close to the solute.

Due to the complexity of MgP-P dimer systems, ab initio quantum mechanical studies designed to ascertain the importance of geometric relaxations must be delayed for future investigations. In this regard Warshel (36, cf. 15) has applied a variety of semi-empirical and semi-classical methodologies in order to simulate the effect of solvent reorganization on ET processes.

Results of Theoretical Studies: An Overview

Detailed presentations of the results of our calculations have been given elsewhere (7, 8). In the latter study, special emphasis was given to the relationship of geometry on the location of CT states. Figure 2 summarizes this relationship for the lowest CT state, $^1CT_1$, of MgP-P dimers both in the presence and in the absence of chloride ion. It is seen from the figure that in both cases a substantial variation in CT state energy with dimer geometry occurs. Interestingly, as will be demonstrated shortly, the separation of the macrocycles and not their relative orientations appears to be a dominant factor in determining CT state energy. Thus, in order to provide a basis for a deeper analysis of the electronic characteristics of low-lying singlet and triplet CT states, the face-to-face dimer separated by 5.35 Å will be considered in greater detail. While there are distinct variations due to geometric differences, as noted above, the 5.35 Å dimer is prototypical of most all of the dimers invetigated in our work. Although several features of this system were treated in detail in our earlier works (7, 8), additional discussion is warranted.

Molecular Orbitals. The orbital energies obtained from SCF-MO calculations on the 5.35 Å dimer are displayed in Figure 3 for isolated monomers and MgP-P dimers, both isolated and in the presence of added chloride ion. As discussed earlier, the MgP-P dimer orbitals are essentially "monomer-like" or can be transformed into monomer-like orbitals without loss of generality. From the figure it is clear that only relatively small perturbations to the orbital energies occur upon dimer formation, while in the case of added chloride ion dramatic shifts of the orbital energies are observed due to the presence of the negative charge. In addition, it should be noted that the calculated orbital-energy shifts for a given monomeric component are such that the relative orbital energies remain approximately unchanged in both cases. Thus, it is not surprising that significant shifts in singlet and triplet (π,π^*)-state energies are unlikely to occur, a conclusion that is also supported by the relatively weak intermolecular interaction expected for MgP-P dimers. As will be described in detail in a subsequent section, the effect of the addition of chloride ion is primarily that expected from a purely electrostatic field effect. Moreover, the fact that the energies of the MgP-orbitals are

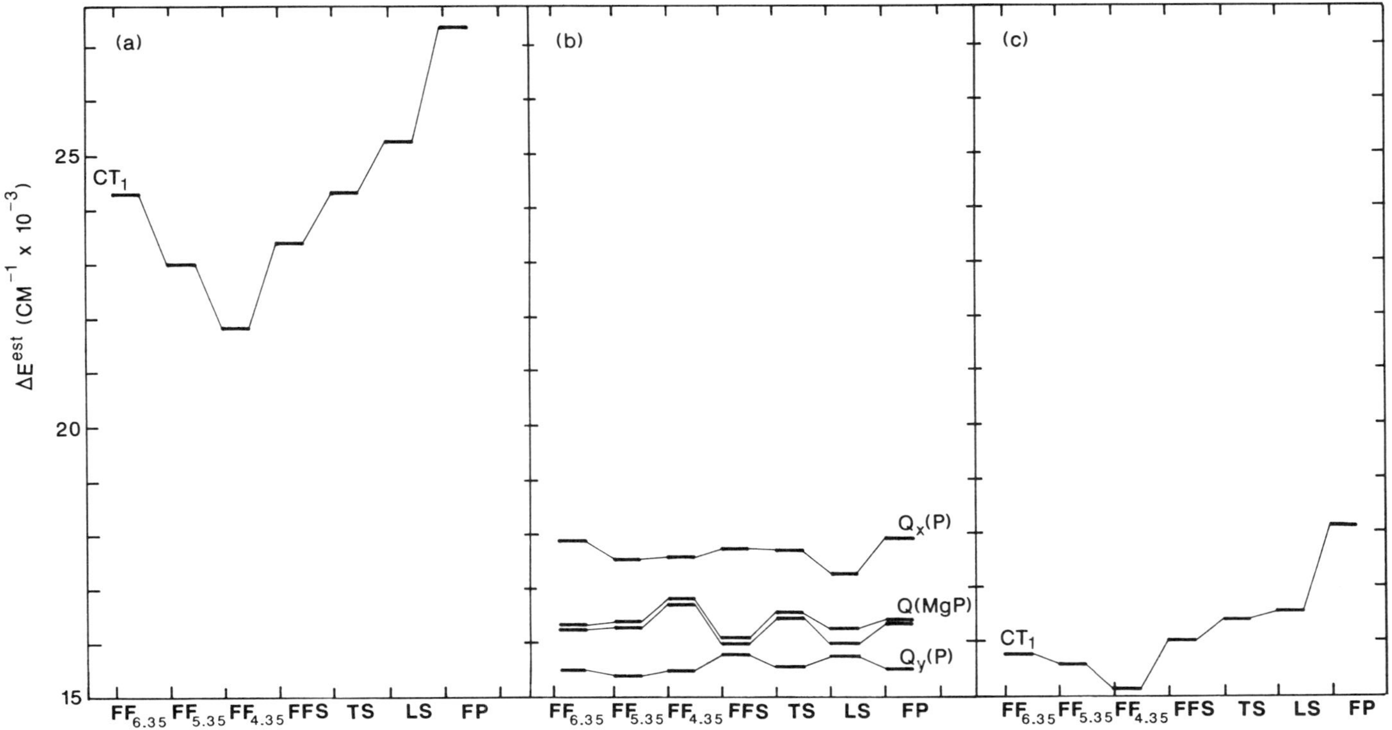

Figure 2. Estimated transition energies of (a) $^1CT_1$ in isolated MgP-P dimers, (b) the lowest four $^1(\pi,\pi^*)$ states in isolated dimers, and (c) $^1CT_1$ in Cl^--MgP-P dimers, for the geometries shown in Figure 1. Reproduced with permission from Ref. 8. Copyright 1986 American Institute of Physics.

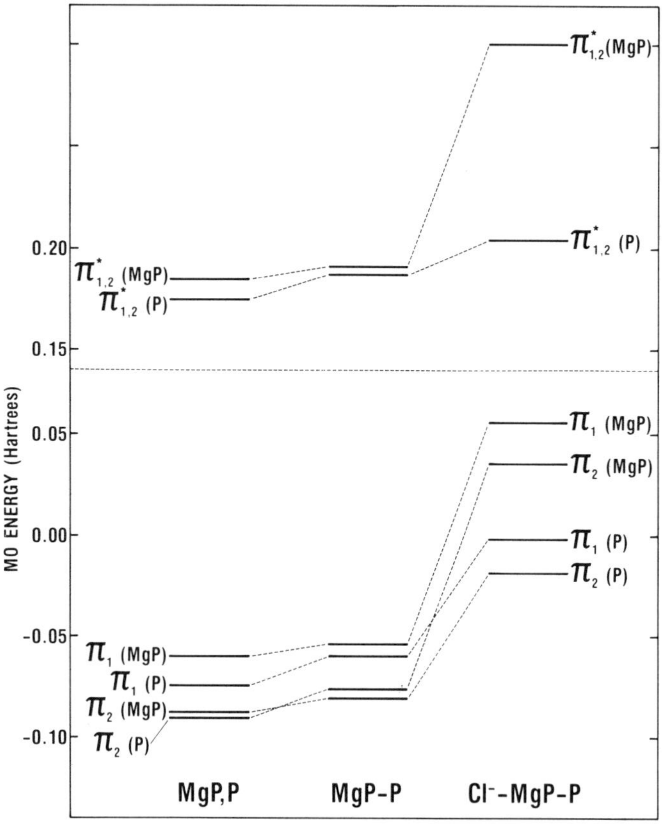

Figure 3. Orbital energies of the four highest occupied and four lowest unoccupied molecular orbitals in isolated monomers (MgP, PO) and in face-to-face MgP-P and Cl^--MgP-P dimers with iterring separations of 5.35Å.

significantly more perturbed than the energies of the P-orbitals is easily explicable in terms of the considerably larger distance of separation of the P-macrocycle from the chloride ion compared to that of the MgP-macrocycle.

Isolated MgP-P Dimers. The spectroscopic consequences of dimer formation are depicted in Figure 4. As shown in the figure, the four lowest-lying states of MgP-P dimers are essentially identical to the corresponding monomeric $^1(\pi,\pi^*)$ states of MgP and P (28, 37), except for small spectral shifts due to electrostatic field effects from the other monomer.

The CT states, the two lowest of which are best described as MgP $\rightarrow$ P charge transfers, all lie considerably above these states, and certainly could not in this case be responsible for any of the photodynamic processes observed by Netzel et al. (20-22). However, the extremely high energy of these states *in vacuo* is not entirely unexpected in light of the dramatic effect that solvent has been shown to exert on the formation of such highly polar states (38, 39). Spectroscopically the CT states have essentially zero oscillator strength with respect to excitation from the ground state, and thus the possibility of radiative emission back to the ground state is practically nil. The lack of oscillator strength observed in these dimers is however, not a general feature shared by CT states in many other systems.

The CT states were characterized by several means including their large dipole moments and Mulliken population analysis (40) of their electron distributions. As reported earlier (7, 8) the dipole moments obtained in our calculations are all in the 25-50 D range, and are well approximated as classical dipoles in which a single positive charge is located at the center of the MgP-macrocycle and a single negative charge at the center of the corresponding P-macrocycle. A summary of quantum mechanically and classically calculated dipole moments for $^1CT_1$ is presented in Table I. As is clear from the table, separation of the two macrocyclic rings plays a dominant role in determining the magnitude of the dipole moments. In addition, Mulliken population analysis of all species examined in the present work showed that the CT states are well represented as the transfer of an electron from one macrocycle to the other, thus providing additional theoretical evidence to support the "classical" view of these states.

Due to the small magnitudes of the relevant exchange integrals (24) ^3CT states are nearly isoenergetic with their corresponding ^1CT states. As the exchange integrals are effectively unchanged by modifications in microenvironment, it follows that singlet-triplet splittings for (π,π^*) and

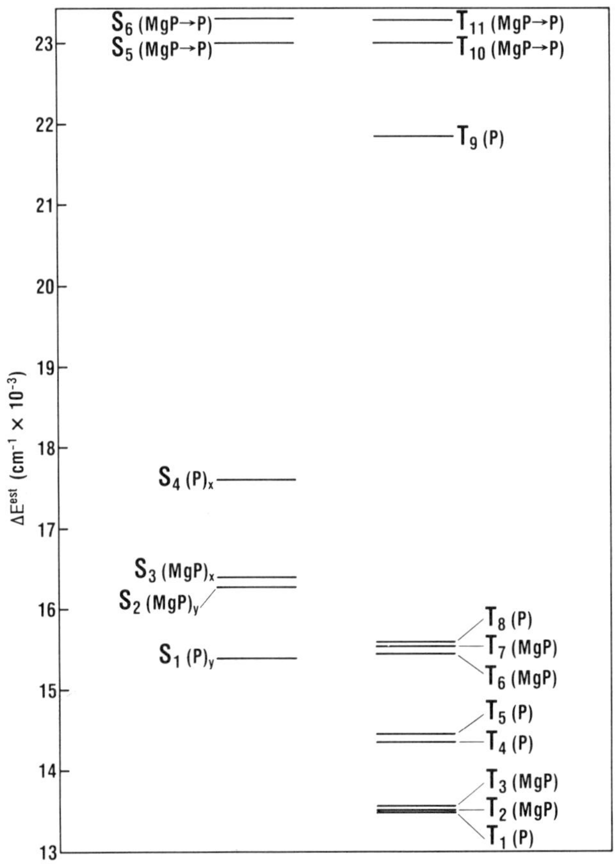

Figure 4. Estimated transition energies of low-lying excited singlet and triplet states in the 5.35Å MgP-P dimer. S_1-S_4 and T_1-T_9 correspond to localized (π,π^*) states on the indicated monomers. S_5 and T_{10} correspond to the lowest-lying singlet and triplet charge transfer states, $^1CT_1$ and $^3CT_1$, respectively.

Table I. Summary of Quantum Mechanically and Classically Calculated Dipole Moments of $^1CT_1$.

Dimer Species[a]	R (Å)[b]	μ^{QM} (D)[c]	μ^{Cl} (D)[d]
Face-to-Face	4.35	24.1	20.9
	5.35	29.2	25.7
	6.35	33.9	30.5
Face-to-Face "slipped"	6.20	31.2	29.8
T-shaped	8.88	41.7	42.6
L-shaped	9.79	49.1	47.9
Fully Planar	12.67	60.4	60.8

[a] See Figure 1 for a description of the geometries.
[b] Measured as "center-to-center" distance.
[c] Quantum mechanically calculated dipole moment.
[d] Classically calculated dipole moment: $\mu^{Cl} = e^2R$, where e is the unit of electronic charge.

CT states will also remain essentially unchanged, even under large variations in microenvironment. Thus, as will be discussed shortly, solvent and other types of microenvironmental stabilizations can exert significant differential stabilization on both singlet and triplet CT states relative to the much less polar singlet and triplet (π,π^*) states.

A qualitative understanding of the energetic features of MgP-P CT states can be gained from the following discussion, which although directed towards MgP → P ^{1}CT states applies to all low-lying MgP-P singlet and triplet CT states. The energy of ^{1}CT relative to the ground state, S_0, is given by

$$\Delta E_{^1CT} = E_{^1CT} - E_{S_0}$$

$$= \langle {}^1CT \mid H \mid {}^1CT \rangle - \langle S_0 \mid H \mid S_0 \rangle \quad (2)$$

where H is the usual non-relativistic Hamiltonian (24). Based on the results of our CI calculations for all of the geometries investigated, it is seen that the ground state and low-lying CT states are dominated by single configurations (8). In particular for S_0 and ^1CT

$$\mid S_0 \rangle \sim \mid SCF \rangle \quad (3)$$

and

$$\mid {}^1CT \rangle \sim \mid {}^1(\pi_{MgP} \rightarrow \pi^*_P) \rangle \quad (4)$$

where π_{MgP} represents an occupied orbital in MgP and π^*_P represents an unoccupied (or virtual) orbital in P (see Figure 3). Upon substitution into Equation 2

$$\Delta E_{^1CT} \approx \mathcal{E}'_{\pi^*_P} - \mathcal{E}'_{\pi_{MgP}} - J_{\pi_{MgP},\pi^*_P} + 2K_{\pi_{MgP},\pi^*_P} \quad (5)$$

where $\mathcal{E}'_{\pi^*_P}$ and $\mathcal{E}'_{\pi_{MgP}}$ are orbital energies of the π^*_P and π_{MgP} orbitals in the dimer, and J_{π_{MgP},π^*_P} and K_{π_{MgP},π^*_P} are the usual coulomb and exchange integrals (24), respectively.

More explicitly, the orbital energies are given by

$$\mathcal{E}'_{\pi^*_P} \approx \mathcal{E}_{\pi^*_P} + \langle \pi^*_P \mid V_{MgP} \mid \pi^*_P \rangle \quad (6)$$

and

$$\mathcal{E}'_{\pi_{MgP}} \approx \mathcal{E}_{\pi_{MgP}} + \langle \pi_{MgP} \mid V_P \mid \pi_{MgP} \rangle \quad (7)$$

where $\mathcal{E}_{\pi^*_P}$ and $\mathcal{E}_{\pi_{MgP}}$ are the corresponding orbital energies of the isolated monomers, and $\langle \pi^*_P \mid V_{MgP} \mid \pi^*_P \rangle$ and $\langle \pi_{MgP} \mid V_P \mid \pi_{MgP} \rangle$ represent the perturbations of a given orbital on one macrocycle due to the effective potential field of the other. Substituting Equations 6 and 7 into Equation 5, and noting that the two intermolecular orbital perturbation terms are nearly equal in the present case (see e.g. Figure 3) and that the exchange term can be safely neglected (7, 8), yields the following approximate expression

$$\Delta E_{^1CT} \approx \mathcal{E}_{\pi^*_P} - \mathcal{E}_{\pi_{MgP}} - J_{\pi_{MgP},\pi^*_P} \quad (8)$$

From the fact that $K_{\pi_{MgP},\pi^*_P} \approx 0$, it follows that $\Delta E_{^3CT}$, the energy of the corresponding ^3CT state (relative to S_0) is effectively equal to $\Delta E_{^1CT}$. Alternatively, the singlet-triplet splitting for CT_1 is, to a very good approximation, equal to zero. As was shown in our earlier work (7, 8), the coulomb integral can be approximated by the simple electrostatic relationship e^2/R, where R is the separation of the two macrocycles measured from their centers and e is the unit of electronic charge. Thus, Equation 8 now can be further simplified to

$$\Delta E_{^1CT} \approx \varepsilon_{\pi^*_P} - \varepsilon_{\pi_{MgP}} - e^2/R \qquad (9)$$

This equation represents a useful relation for the CT-state transition energy, and emphasizes the fact that aside from the obvious and well-characterized importance of donor-acceptor orbital energies, intermolecular separation, not ring orientation, is a major factor in determining $\Delta E_{^1CT}$ and $\Delta E_{^3CT}$ in isolated MgP-P dimers. For $^1CT_1$, $-\varepsilon_{\pi_{MgP}}$ approximates the vertical ionization potential, I, and $-\varepsilon_{\pi^*_P}$ approximates the electron affinity, A, by Koopmans' Theorem (41). Substituting these two terms into Equation 8 yields an expression which is highly reminiscent of the empirical expression $\Delta E_{CT} = I - A - W$ (39), where W is related to the dissociation energy of the CT state. This formula has been quite useful in correlating the energy of CT states with the above independently determined physical quantities (39).

Solvent Effects. As noted earlier, Netzel et al. (20-22) observed striking effects on the photodynamics of electronically excited MgP-P dimers depending on the solvent used. For example, in their earlier work (20) they showed that the photodynamics in CH_2Cl_2 were consistent with the formation of a long-lived singlet CT state which decayed by two pathways, while in THF only a single short-lived (π, π^*) state was observed.

Applying a simple dielectric continuum cavity model (42) as a means of estimating solvent stabilization, it is possible to show that polar solvents, such as CH_2Cl_2 and THF, can sufficiently stabilize MgP $\rightarrow$ P CT states to bring them within the same "energetic region" as the lowest $^1(\pi,\pi^*)$ state (7, 8). For example, for solvents with dielectric constants of ~9 and for CT states with dipole moments ~25 to 30D, the dielectric continuum cavity model gives rise to a differential stabilization of ~6,000-6,500 cm^{-1} for CT relative to (π,π^*) states. Thus, based upon the simple cavity model one would expect similar photodynamical behavior in both solvents, in direct contradiction to experimental observations. As both CH_2Cl_2 and THF possess nearly identical dipole moments [1.60 D (CH_2Cl_2), 1.75 D (THF)] and static dielectric constants [9.08 (CH_2Cl_2), 7.58 (THF)], it would appear that other molecular and/or collective properties of the solvent molecules must be playing a role in determining the differing photodynamic behaviors observed in the two solvents.

Several possible explanations suggest themselves. One possibility is that specific interactions between the ether oxygen of THF and the magnesium atom of the MgP moiety would alter the electronic structure of the dimer such as to increase $\Delta E_{^1CT}$, that is, favorable O····Mg interactions would destabilize $^1CT_1$ relative to the lowest $^1(\pi,\pi^*)$ state. As will be seen in the next section such an interaction is more likely to stabilize $^1CT_1$ rather than to destabilize it. Thus, one would expect that MgP-P dimers in THF should possess similar photodynamical properties to those cases in which $^1CT_1$ lies below $^1(\pi,\pi^*)$ as is suggested by the work of Netzel et al. (20-22). Another possibility may be found in the different electronic polarizabilities of the two molecules: CH_2Cl_2, with its large, highly polarizable chlorine atoms having a larger electronic polarizability than THF. While this may certainly contribute to the observations of Netzel et al. (20), it may not be sufficient to explain the data completely. In fact, recent experimental work by Kosower et al. (43) suggests that the rate of intramolecular excited-state ET, i.e. CT state formation, depends on the detailed motions of solvent molecules, and thus is significantly influenced by solvent response properties. Warshel (36) has shown, using semi-classical computer simulations of an idealized molecular system, that the quantum yield for CT state formation is related to the dielectric relaxation properties of the solvent. A number of other workers have also addressed this question recently using a variety of theoretical approaches (44-46).

Due to the extreme rapidity with which excited CT states may be formed and due to the large changes in charge distribution brought about by their formation, it is certainly not entirely unexpected that high-frequency response properties of the solvent would play a role in CT-state formation as well as in photodynamic processes subsequent to their formation. Moreover, as solvent dynamics is influenced by a number of molecular properties such as charge, volume, shape, and hydrogen bonding propensity, a more detailed evaluation of the nature of the solvents employed in the study of photo-induced CT states, especially solvent response properties, must be carried out if we are to understand the solvent dependent data fully.

Another feature of particular importance in the study of large, spatially-extended molecular systems such as those examined here, is the use of the dipole approximation in assessing solute-solvent interactions.

In such systems the dipole approximation is woefully inadequate. That this is the case can be seen qualitatively if one recalls that the dipole moment of a molecule is generally represented as an infinitesimal point dipole located generally at the center-of-mass of the molecule. Since the dipole moment effectively represents an average over the whole spatially extended molecular charge distribution, namely the second moment of the charge distribution, "local" details of the charge distribution are "washed out". Note that a system with a zero or nearly zero dipole moment by symmetry may still possess regions of significant negative or positive charge, which in a large molecule can exert strong local effects. In small molecules or at great distances from very large molecules the effect of such local charge density variations is insignificant, but in spatially extensive, highly polar systems such as MgP^+-P^- dimers, a more detailed view of local charge distributions is essential for the development of any theory designed to assess the influence of solute-solvent interactions on the photodynamical properties of MgP-P dimer excited states. In this regard, the solvent response and the reorganized, quasi-equilibrium solvent structure which results upon CT-state formation are both likely to play significant roles.

In principle it is possible to gain some assessment of the role of solvent dynamics on the excited-state radiationless processes which lead to CT state formation through the use of solvent isotope experiments. For example, deuteration of non-exchangeable hydrogens appears to be a particularly good way of changing the total mass and mass distribution, and thus of influencing the dynamical response properties of the solvent without significantly affecting other of its properties such as its hydrogen bonding properties, its shape, or its electronic polarizability. Such an approach should thus provide important evidence on the detailed nature of the solvent effects, and will in addition provide information that will be useful in the development of efficient biomimetic solar energy conversion systems.

External Electric Field Effects. As discussed in the previous section, the solvent microenvironment has a profound effect on the photodynamical properties of MgP-P dimers, especially with regard to the formation of low-lying CT states. The focus of the previous discussion was on dynamical rather than on reaction-field factors, but the latter can also bring about significant differential stabilization of the energies and corresponding potential energy surfaces of highly polar CT states relative to (π,π^*) states, which are considerably less polar (see e.g. 47). In addition to solvent reaction-field effects, analogous effects can also be obtained from electric fields generated by other means. For example, molecular Stark effects generated by applied electric fields have been employed in numerous branches of spectroscopy. Electrochromic and related effects generated as a consequence of transmembrane potentials have been attracting the attention of researchers in a number of biological fields, not the least of which has been photosynthesis (3, 48, 49). Electrochromic effects due to the proximity of chromophoric groups to charged groups within the protein have also been attracting attention, particularly with regard to the bathochromic shift observed in the visual pigments (50), but lately also with regard to photosynthetic systems (51; cf. 52).

The shifts observed in both of the latter cases, i.e. those shifts brought about by electric fields produced by "localized charges," are generally considerably larger than shifts due to electric fields generated solely as a consequence of transmembrane potentials. That such is likely to be the case is easily seen if one considers the magnitudes of electric fields generated in various ways. Electric fields generated in the laboratory (e.g. for Stark effect studies) of 60,000 V/cm are quite large, and are in the lower end of the range typically found in membranes which is 50,000 to 500,000 V/cm (53). Based on electrochromic effects it has been suggested by Reich [see pp. 361-387 of (3)] that electric fields in photosynthetic membranes may attain magnitudes on the order of 10^7 V/cm, although such large values exist only transiently. In addition, it should be noted that the electric fields are not totally due to differential charge across the membrane, but undoubtedly also involve electric fields due to charges located on various membrane bound species. Although such fields are large, their ability to produce significant spectral shifts is still quite limited. For example, the energy of an excited state possessing a dipole moment of 25 D, which it should be noted is in the lower part of the range of those calculated for the CT states examined in the present work, will only be stabilized by 420 cm^{-1} per 100,000 V/cm of electric field strength. Thus, the types of spectral shifts observed in, for example, the rhodopsin system, which are on the order of 2,800 cm^{-1} (50), cannot easily be explained as arising from electric fields generated in this way.

The observed shifts become considerably more explicable, however, if one considers the electric field at a distance of about 10 Å from a full negative or positive charge, which is on the order of 1.4 X 10^7 V/cm. Using the above value of an electrochromic shift of 420 cm^{-1} for each 100,000 V/cm of electric field strength, it follows that the single charge could produce a shift of ~ 6,000 cm^{-1}. Thus, if one wants to produce significant alterations in the ordering of electronic states through the use "external" electric fields, it is likely that localized charge perturbations will be necessary as fields developed from transmembrane potentials or from applied external fields will not likely be of sufficient magnitude to produce profound changes to the ordering of electronic states. It must be realized, however, that "bare" charges do not exist in biomolecular systems, and that the presence of counter-charges and the response of the environment to the presence of charges tend to attenuate the strength of the electric field produced (54). Moreover, distributed partial charges rather than full, isolated charges are generally found in biomolecular systems. Finally, it must be borne in mind that electric field effects on the kinetics of thermally mediated ET,and possibly some photo-induced processes as well, may be significant even for much smaller electric field strengths than those necessary for the production of large spectroscopic shifts (3, 48, 49).

Internal Electric Field Effects: Cl^--MgP-P Dimers. Experimentally Netzel and co-workers (20-22) addressed the question of the effect of electric fields by studying MgP-P dimers in the presence of added chloride ion. Assuming that the chloride ion binds to the magnesium atom of the MgP moiety, we have shown through quantum mechanical CI calculations that substantial shifts in CT state energies can be obtained (7, 8). This is illustrated in Figure 5 for the face-to-face dimer with 5.35 Å intermacrocycle separation, which clearly shows that the addition of chloride ion can stabilize the first two CT states, $^1CT_1$ and $^1CT_2$, such that they lie near the lowest excited $^1(\pi,\pi^*)$ state. In this case the large differential stabilization effect of the CT states compared to the (π,π^*) states is entirely explicable in terms of the large dipole moments of the former in comparison to the relatively small dipole moments of the latter states.

It is possible, however, to analyze the situation in greater detail, and thus to obtain additional insights that may be useful in the design of new dimeric and oligomeric porphyrin systems with specified photodynamic properties. Specifically, as discussed in earlier publications (7, 8), the differential effect of chloride ion coordination on $^1CT_1$ and $S_1(\pi,\pi^*)$ can also be analyzed in terms of a suitably modified version of Equation 5. An examination of the four orbitals which form the basis of Gouterman's "four-orbital model" (37), and which are the key orbital contributors to the CT states and the four lowest (π,π^*) states., shows that very little change has occurred to the charge distributions of these orbitals upon dimer formation and upon coordination by chloride ion. The most significant changes are their destabilization by chloride ion coordination as shown in

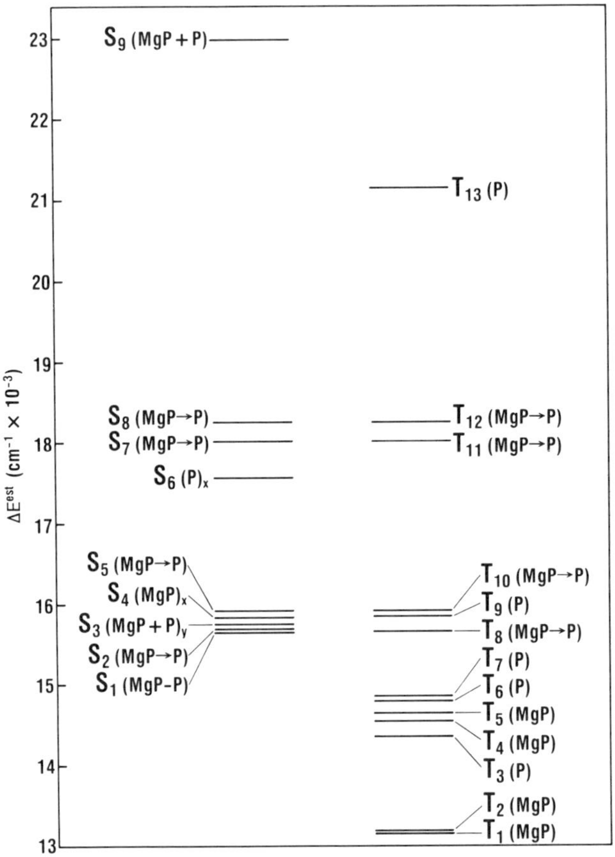

Figure 5. Estimated transition energies of low-lying excited singlet and triplet states of the 5.35 Cl^--MgP-P dimer. S_1 and S_3 correspond to mixtures of localized (π,π^*) states; S_2 and S_5 correspond to $^1CT_1$ and $^1CT_2$, respectively; T_1-T_7 and T_9 correspond to localized $^3(\pi,\pi^*)$ states; and T_8 and T_{10} correspond to $^3CT_1$ and $^3CT_2$, respectively.

Figure 3 [see also the earlier discussion in the "**Molecular Orbital**" subsection].

The orbital energies of π^*_P and π_{MgP}, the key orbitals which contribute to the low-lying MgP → P CT states, can be approximated by (cf. Equations 6 and 7)

$$\varepsilon^{Cl^-}_{\pi^*_P} \approx \varepsilon_{\pi^*_P} + <\pi^*_P | V_{MgP} + V_{Cl^-} | \pi^*_P> \qquad (10)$$

and

$$\varepsilon^{Cl^-}_{\pi_{MgP}} \approx \varepsilon_{\pi_{MgP}} + <\pi_{MgP} | V_P + V_{Cl^-} | \pi_{MgP}> \qquad (11)$$

where V_{Cl^-} is the effective potential due to the chloride ion, and V_{MgP} and V_P are the effective potentials due to the MgP- and P-macrocycles, respectively. Substituting Equations 10 and 11 into Equation 5, where $\varepsilon'_{\pi^*_P}$ and $\varepsilon'_{\pi_{MgP}}$ have been replaced by $\varepsilon^{Cl^-}_{\pi^*_P}$ and $\varepsilon^{Cl^-}_{\pi_{MgP}}$, respectively, and noting that the intermolecular exchange integral K_{π_{MgP},π^*_P} is again nearly zero, and that $J^{Cl^-}_{\pi_{MgP},\pi^*_P}$ remains essentially unchanged, since the "chloride-ion-perturbed" orbitals have nearly identical charge distributions as their unperturbed counterparts, yields

$$\Delta E^{Cl^-}_{1CT} \approx \varepsilon_{\pi^*_P} - \varepsilon_{\pi_{MgP}} - J_{\pi_{MgP},\pi^*_P,}$$
$$+ <\pi^*_P | V_{MgP} + V_{Cl^-} | \pi^*_P> - <\pi_{MgP} | V_P + V_{Cl^-} | \pi_{MgP}> \qquad (12)$$

As was discussed earlier, the two intermolecular orbital perturbation terms,
$<\pi^*_P | V_{MgP} | \pi^*_P>$ and $<\pi_{MgP} | V_P | \pi_{MgP}>$, are nearly equal, so that taking account of Equation 8, Equation 12 becomes

$$\Delta E^{Cl^-}_{1CT} \approx \Delta E_{1CT} + \Delta V_{Cl^-} \qquad (13)$$

where

$$\Delta V_{Cl^-} = <\pi^*_P | V_{Cl^-} | \pi^*_P> - <\pi_{MgP} | V_{Cl^-} | \pi_{MgP}> \qquad (14)$$
$$< 0$$

since the perturbing chloride ion is located closer to the MgP-macrocycle than to the P-macrocycle. Thus, $\Delta E^{Cl^-}_{1CT}$ is less than ΔE_{1CT}.

Employing the same type of analysis, it follows that the positioning of a positive charge closer to the P-macrocycle will differentially stabilize the orbitals on the P-macrocycle relative to those on the MgP-macrocycle. Specifically, π^*_P will be lowered relative to π_{MgP}, and thus the transition energy of the MgP →P CT states will again be lowered. Richardson et al. (55) have synthesized an interesting "capped" porphyrin which is made up of a heterocyclic crown doubly-covalently linked to a free base porphyrin. The heterocyclic crown binds a variety of cationic species. By covalently linking such a capped porphyrin to a MgP moiety, one can obtain a system for investigating cationic perturbations experimentally. If both negative and positive charges are appropriately positioned such that they act primarily to shift the orbital energies without significantly distorting the charge distributions, the energy lowering will be even greater.

In distinction to CT states, in localized $^1(\pi,\pi^*)$ states, the chloride ion uniformly destabilizes the orbitals of a given macrocycle, and thus orbital energy differences are largely unaltered and the transition energies of these states remain essentially unchanged. Note, however, that in an actual experimental situation [see e.g. Netzel et al.(20, 21)] the effect of chloride ion is most likely diminished from that observed in the present work due to the presence of solvent and counter-ions. Nevertheless, using properly positioned charges it is possible to isolate both $^1CT_1$ and $^3CT_1$ effectively from all other singlet and triplet (π,π^*) states, which are now higher lying. Such a system could possess very favorable photodynamic properties with regard to singlet-triplet intersystem crossing, although the rate of radiationless relaxation to the ground state may be increased somewhat over that observed in the Cl^--MgP-P case. Some of the possible consequences of this will be discussed in the following section.

The Photodynamic Role of Triplet States. A detailed picture of the photodynamics of MgP-P dimers requires an understanding of the role of triplet states in the process, as they provide a potential decay channel for the lowest excited singlet state. In fact, the picosecond spectroscopic results of Netzel et al. (20-22) clearly show that in CH_2Cl_2 intersystem crossing (ISC) from $^1CT_1$ to the triplet manifold plays a significant role in depopulating $^1CT_1$. In addition, inhibiting ISC would in most cases enhance the lifetime of $^1CT_1$, and thus would likely enhance the potential of $^1CT_1$ for ET to some suitable acceptor, certainly a desirable design feature of an efficient biomimetic solar energy conversion system.

Spin-orbit coupling (SOC) plays an important role in determining the likelihood of ISC. If SOC is small, then so is the quantum yield for ISC: the lack of SOC thus can inhibit ISC significantly. If, on the other hand

SOC is relatively large, the overall ISC process is no longer inhibited by the need for spin-state changes since spin is no longer a good quantum number; and other factors become rate limiting. Spin-orbit coupling depends essentially on the ratio of the square of the SOC matrix element to difference in energy of the states being coupled (the so-called "energy denominator"). Theoretical work by Lim, et al. (56) has shown that within the "one-center" approximation the appropriate SOC matrix elements satisfy

$$\langle ^1CT \mid H_{SOC} \mid ^3CT \rangle = 0 \tag{15}$$

and

$$\langle ^1CT \mid H_{SOC} \mid ^3(\pi,\pi^*) \rangle \neq 0 \tag{16}$$

where H_{SOC} is the SOC Hamiltonian (57). Thus, when $^1CT_1$ is the lowest excited singlet state, there must exist $^3(\pi,\pi^*)$ states in the neighborhood of $^3CT_1$ for efficient ISC to occur. Examination of Figure 4 shows that this is indeed the case if one includes the differential stabilization of CT compared to (π,π^*) states brought about by polar solvents. In such a case $^1CT_1$ will be nearly isoenergetic to a number of $^3(\pi,\pi^*)$ states, and thus ISC should not be inhibited significantly. Such a situation was observed by Netzel et al. (20-22) in CH_2Cl_2 and several other polar solvents, and in CH_2Cl_2/C_6H_6I, which is known to enhance SOC through the external heavy atom effect (57) of the iodine atom.

That CT states are significantly stabilized relative to (π,π^*) states in the presence of chloride ion is shown in Figure 5. Additional stabilization due to solvent further lowers the CT states, especially CT_1, such that $^3CT_1$ now lies below and is effectively isolated from all $^3(\pi,\pi^*)$ states. From Lim's SOC "selection rule" (56) given in Equations 15 and 16, it follows that ISC should be substantially inhibited, and thus its rate should no longer be competitive with internal conversion (IC) to the ground state. Again, as in the previous case, the theoretical work provides a consistent interpretation of the experimental observations of Netzel, et al. (21), where it was shown that in the presence of chloride ion only a single decay channel from $^1CT_1$ was observed, and no evidence of triplet formation was obtained. Carrying out the experiment in the presence of chloride ion and C_6H_6I may provide an experimental test of this interpretation, as ISC would be expected to remain largely inhibited even in the presence of added C_6H_6I.

It must be borne in mind, however, that vibronic coupling effects (33) among CT and (π,π^*) states may substantially alter the picture portrayed by a given state energy diagram (se e.g. Figures 4 and 5). In addition, geometry relaxation effects also may significantly modify state energy orderings. Both these effects are not accounted for in our present work, and thus conclusions drawn there from must be considered tentative.

Biomimetic Solar Energy Conversion Systems: Design Issues

The development of practical solar energy conversion devices based on principles derived from the study of biological systems has engendered considerable research in the general area of "artificial photosynthesis" (58, 59). And although significant strides are being made in the design and testing of a number of promising biomimetic systems, some of which were discussed above, the problem of how to optimize the "design process" still remains largely unresolved. Molecular quantum mechanical methodologies, both quantitative and qualitative, can certainly play an important role in the design process.

Three key issues must be addressed in the development of effective biomimetic solar energy conversion systems. First, the molecular system should possess a large optical absorption cross-section in the desired spectral region. Second, the system should possess appropriate characteristics to insure formation of a sufficiently long-lived, low-lying state which can initiate the primary ET efficiently. And third, the system should be able to effect the ET process irreversibility, that is electron-hole recombination should be substantially inhibited.

The first issue can be addressed in two ways: a primary ET species which has a large optical absorption cross-section can be chosen or arrays of molecules with large optical absorption cross-sections can be used as "antennas" that will efficiently collect and transport the electronic excitation energy to the primary ET species, in direct analogy to photosynthetic systems. While in the latter case it should be possible to develop systems with more efficient solar photon collection, the number of primary ET species will have to be reduced due to the spatial limitations, which will also reduce the potential electric current that can be produced by the system. Thus, questions related to the detailed molecular architecture of biomimetic solar energy conversion devices will have to address this issue, and it is quite likely that a number of compromises will have to be made before optimal design characteristics are obtained.

The second issue can be addressed both experimentally and theoretically, some aspects of which have already been discussed. For example, we have described, based on the results of quantum mechanical calculations described above, ways of obtaining low-lying CT states of

MgP-P heterodimer systems through the manipulation of microenvironmental factors (7,8), and we have related these results to the picosecond spectroscopic experiments of Netzel et al. (20-22). Thus it can be inferred from the above discussion and from the foregoing studies, that the energies of CT states may be manipulated by the presence of external point charges, that the effect of a single point charge positioned properly may be large, and that the effect may be modulated simply by changing the magnitude and/or position of the charge. The work of Netzel et al. (20) on the effects of chloride ion on the photodynamics of MgP-P dimers described earlier also provides support for this view.

Other possibilities exist for manipulating the location of electronic excited states. One such possibility is through the substitution of appropriate "electron-donating" and "electron-withdrawing" functional groups on the porphyrin macrocycles. Although we have not as yet employed this approach in our work, such an approach has been the basis for the design of numerous CT complexes, which have been extensively studied over the last two decades (38, 39). In the case of porphyrins and related macrocyclic systems, however, caution should be exercised in the application of the "substituent approach." This is due to the fact that ideally substituents designed to lower CT state energies should not lower the energies of (π,π^*) states as well. As discussed above, such differential stabilizations were achieved through changes in the microenvironment, but similar differential stabilizations have not been demonstrated, experimentally or theoretically, for substituent effects. As will be discussed below, substituent groups may also affect properties other than the location of CT states.

State energies depend to a large degree on the energies of the MOs involved in an electronic transition. Thus, by taking proper account of the nodal structure of the relevant MOs it should be possible to determine, at least qualitatively, where substituents should be placed to achieve optimal differential stabilization effects. More detailed CI calculations can then be carried out to determine whether the expected effects are likely in fact to occur. In addition, the results of numerous experimental studies of substituted porphyrins (37, 60) will also provide a useful guide for the design of porphyrin dimers with the desirable properties.

However, it should be noted that the greater flexibility potentially obtainable in ion mediated systems with respect to modulation of the energies of their CT states, is not obtainable with substituted systems. In the former case, changes in ion concentration of the solution containing the ion-modulated system directly affect ion binding, and thus the electrostatic environment of the system. In the latter case, however, substituent effects generally cannot be altered (cf. 52, 61, 62).

The third issue is most difficult to address theoretically since computational methods for treating radiationless and ET processes in large molecular systems are relatively primitive at this time [see the "CT States and Photo-induced ET Processes" section for further discussion]. Several interesting experimental approaches have, however, provided a number of useful insights into possible ways of reducing recombination, and thus enhancing the irreversibility of the photo-induced ET process.

A key factor in this regard is the lifetime of the low-lying CT state, which presumes of course that we have been able to obtain a low-lying state through appropriate molecular and environmental manipulations. Lowering $^3CT_1$ such that it lies sufficiently below the lowest $^3(\pi,\pi^*)$ state to reduce vibronic coupling of $^3CT_1$ with low-lying $^3(\pi,\pi^*)$ states essentially to zero, will substantially inhibit ISC. Thus, if the rate of IC to the ground state remains largely unaffected, a situation that is likely to obtain unless $^1CT_1$ approaches the ground state too closely, the lifetime of $^1CT_1$ should increase due to the inhibition of the ISC decay channel.

Internal conversion from $^1CT_1$ to the ground state can also be inhibited by the proper molecular design. For example, it may be possible to reduce the efficiency of promoting and/or accepting modes in the radiationless relaxation process. In this regard, Lim et al. (63) have studied an interesting trimeric system made up of two identical donors (D) and a single acceptor (A) with the following structure: DDA. Lim found that delocalization of the positive charge over both donors, i.e. $[DD]^+$, tended to inhibit charge recombination and to reduce the role of key accepting modes in the radiationless process. Interestingly, this result may have relevance in photosynthetic systems where chlorophyll and bacteriochlorophyll "special pairs" are thought to play a role in the initial photo-induced ET events (4, 5, 17, 19).

Another, interesting approach to the irreversibility problem was carried out recently by Wasielewski, et al. (58). In their work they synthesized a novel porphyrin species in which donor and acceptor moieties were covalently attached in a well defined geometric relationship to the porphyrin ring. In their system the porhyrin acted as the primary donor, which transferred an electron to the acceptor within a few picoseconds. Charge recombination was inhibited by the subsequent transfer of an electron from the secondary donor. The resulting charge-separated species was stable for up to milliseconds. An analogous approach was carried out somewhat earlier by Moore and his associates

(59). These workers employed a more elaborate system than Wasielewski et al., which included the presence of a membrane to assist in maintaining the initial photo-induced charge separation. Although the complexity of the system did generate a number of technical problems, it probably comes closer to an actual, viable prototype of a workable biomimetic system.

Thus, it appears that several systems have been developed which hold promise as prototypes of biomimetic solar energy conversion devices. And in conjunction with the advances being made in experimental and theoretical methods for investigating molecular excited-state processes, prospects for the development of practical biomimetic devices are now substantially better than they were only a decade ago.

Achnowledgments

This work was supported by the Division of Chemical Sciences, Office of Basic Energy Sciences of The United States Department of Energy. Services and computer time provided by United Information Systems, Kansas City, MO, Argonne National Laboratory, and the National Magnetic Fusion Energy Computer Center have been invaluable in this study. The authors would also like to thank Ms. D. Piper for her assistance in the preparation of this manuscript.

Literature Cited

1. See e.g. "The Porphyrins"; D. Dolphin, Ed.; Academic Press, New York; 1977; Vols. I-VII.
2. Harriman, A. In "Energy Resources Through Photochemistry and Catalysis"; M. Gratzel, Ed.; Academic Press, New York, 1983; p. 163.
3. See e.g. "Light-Induced Charge Separation in Biology and Chemistry"; H. Gerischer and J.J. Katz, Eds.; Verlag Chemie, New York, 1979; Vol. 12 in Life Sciences Research Reports.
4. Netzel, T.L. In "Biological Events Probed by Ultrafast Laser Spectroscopy"; R.R. Alfano, Ed.; Academic Press, New York, 1983; p. 79.
5. Boxer, S.G. Biochim. Biophys. Acta 1983, 726, 265.
6. Creed, D.; Caldwell, R.A. Photochem. Photobiol. 1985, 41, 715.
7. Petke, J.D.; Maggiora, G.M. Chem. Phys. Lett. 1983, 97, 231.
8. Petke, J.D.; Maggiora, G.M. J. Chem. Phys., 1986, 84, 1640.
9. Petke, J.D.; LaLonde, D.E.; Maggiora, G.M. J. Am. Chem. Soc., in preparation.

10. Newton, M.D.; Sutin, N. Ann. Rev. Phys. Chem. 1984, 35, 437.
11. Marcus, R.A.; Sutin, N. Biochim. Biophys. Acta 1985, 811, 265.
12. Parsons, W.W.; Ke, B. In "Photosynthesis"; Govindjee, Ed.; Academic Press, New York, 1982; Vol. I, Chap. 8.
13. Clayton, R.K. In "The Photosynthetic Bacteria"; R.K. Clayton, Ed.; Plenum, New York, 1978; Chap. 20.
14. (a) Jortner, J. J. Am. Chem. Soc. 1980, 102, 6676. (b) Jortner, J. Biochim. Biophys. Acta 1980, 594, 193.
15. Warshel, A. Proc. Natl. Acad. Sci. USA 1980, 77, 3105.
16. Friesner, R.; Wertheimer, R. Proc. Natl. Acad. Sci. USA 1982, 79, 2138
17. Deisenhofer, J.; Epp, O.; Miki, K.; Huber, R.; Michel, J. J. Mol. Biol. 1984, 180, 385.
18. Katz, J.J.; Norris, J.R.; Shipman, L.L.; Thurnauer, M.C.; Wasielewski, M.R. Ann. Rev. Biophys. Bioengineer. 1978, 7, 393.
19. Maggiora, G.M. Int. J. Quantum Chem. 1979, 16, 331.
20. Netzel, T.L.; Kroger, P.; Chang, C.-K.; Fujita, I.; Fajer, J. Chem. Phys. Lett. 1979, 67, 223.
21. (a) Netzel, T.L.; Bergkamp, M.A.; Chang, C.-K.; Dalton, J. J. Photochem. 1981, 17, 451. (b) Netzel, T.L.; Bergkamp, M.A.; Chang, C.-K. J. Am. Chem. Soc. 1982, 104, 1952.
22. Fujita, I.; Fajer, J.; Chang, C.-K.; Wang, C.-B.; Bergkamp, M.A.; Netzel, T.L., J. Phys. Chem. 1982, 86, 3754.
23. See e.g. Bergkamp, M.A.; Dalton, J.; Netzel, T.L., J. Am. Chem. Soc. 1982, 104, 253; and references cited therein.
24. (a) Roothaan, C.C.J. Rev. Mod. Phys. 1951, 23, 69. (b) Hall, G.G. Proc. Roy. Soc. 1951, A205, 541.
25. Whitten, J.L.; Hackmeyer, M. J. Chem. Phys. 1969, 51, 5584.
26. Christoffersen, R.E.; Spangler, D.; Hall, G.G.; Maggiora, G.M. J. Am. Chem. Soc. 1973, 95, 8526.
27. Spangler, D.; Maggiora, G.M.; Shipman, L.L.; Christoffersen, R.E., J. Am. Chem. Soc. 1977, 99, 7470, 7478.
28. (a) Petke, J.D.; Maggiora, G.M.; Shipman, L.L.; Christoffersen, R.E. J. Mol. Spectrosc. 1978, 71, 64. (b) Ibid 1978, 73, 311.
29. (a) Petke, J.D.; Maggiora, G.M.; Shipman, L.L.; Christoffersen, R.E. Photochem. Photobiol. 1979, 30, 203. (b) Ibid 1980, 32, 399, 661. (c) Ibid 1981, 33, 663.
30. Mille, P.; Levy, B.; Berthier, G. In "Localization and Delocalization in Quantum Chemistry"; O. Chalvet, R. Daudel, S. Diner, J.-P. Malrieu. Eds.; D. Reidel Publishing Co., Dordrecht, Holland, 1975.

31. (a) Hoffman, R. Accts. Chem. Res. 1971, 4, 1. (b) Hay, R.J.; Thibeault, J.C.; Hoffman, R. J. Am. Chem. Soc. 1975, 97, 4884.
32. Beratan, D.N.; Hopfield, J.J., J. Am. Chem. Soc. 1984, 106, 1594.
33. Fischer, G. In "Vibronic Coupling", Academic Press, New York, 1984.
34. McHale, J.; Simons, J. J. Chem. Phys. 1980, 72, 425.
35. (a) Barkigia, K.M.; Spaulding, L.D.; Fajer, J., Inorg. Chem., 1983, 22, 349. (b) Fajer, J.; Barkigia, K.M.; Smith, K.M.; Goff, D.A. This Volume.
36. Warshel, A. J. Phys. Chem. 1982, 86, 2218.
37. Gouterman, M. In "The Porphyrins"; D. Dophin, Ed;, Academic Press, New York, 1977; Vol. 3.
38. Davis, K.M.C. In "Molecular Association"; R. Foster, Ed.; Academic Press, New York, 1975; Vol. 1, Chap. 3.
39. Mataga, N.; Kubota, T. In "Molecular Interactions and Electronic Spectra", Marcel Dekker, New York, 1970.
40. Mulliken, R.S. J. Chem. Phys. 1955, 23, 1833, 1841.
41. Koopmanns, T. Physica 1933, 1, 104.
42. Bottcher, C.J.F.; van Belle, O.C.; Bordewijk, P.; Rip, A. "Theory of Electric Polarization"; Elsevier, Amsterdam 2nd Edition; 1973; Vol. I.
43. Kosower, E.M.; Huppert, D. Chem. Phys. Lett. 1983, 96, 433.
44. Mozumder, A. J. Chem. Phys. 1969, 50, 3153.
45. (a) Onsager, L. Can. J. Chem. 1977, 55, 1819. (b) Hubbard, J.; Onsager, L. J. Chem. Phys. 1977, 67, 4850.
46. van der Zwan, G.; Hynes, J.T. J. Chem. Phys. 1982, 76, 2993.
47. Salem, L. "Electrons in Chemical Reactions", Wiley-Interscience, New York, Chpt. 8, 1982.
48. Rackovsky, S.; Scher, H. Biochim. Biophys. Acta 1982, 681, 152.
49. Gopher, A.; Blatt, Y.; Schonfeld, M.; Okamura, M.Y.; Feher, G.; Montal, M. Biophys. J. 1985, 48, 311, and references cited therein.
50. Birge, R. Ann. Rev. Biophys. Bioengineer. 1981, 10, 315.
51. Eccles, J.; Honig, B. Proc. Natl. Acad. Sci. USA, 1983, 80, 4959.
52. Maggiora, L.L.; Petke, J.D.; Gopal, D.; Iwamoto, R.T.; Maggiora, G.M. Photochem. Photobiol. 1985, 42, 69.
53. Davis, T.D.; Christoffersen, R.E.; Maggiora, G.M. J. Am. Chem. Soc. 1975, 97, 1347.
54. Warshel, A.; Russell, S.T. Quart. Rev. Biophys. 1984, 17, 283.
55. Richardson, N.M.; Sutherland, I. Tetrahedron Letts. 1985, 26, 3739.
56. Lim, B.T.; Okajima, S.; Chandra, A.K.; Lim, E.C. Chem. Phys. Letts. 1981, 79, 22.

57. McGlynn, S.P., Azumi, T.; Kinoshita, M. "Molecular Spectroscopy of the Triplet State"; Prentice-Hall, Englewood Cliffs, New Jersey, 1969.

58. Wasielewski, M.R.; Niemczyk, M.P.; Svec, W.A.; Pewitt, E.B. J. Am. Chem. Soc. 1985, 1097, 5562.

59. Seta, P.; Bienvenue, E.; Moore, A.L.; Mathis, P.; Bensasson, R.V.; Liddell, P.; Pessiki, P.J.; Joy, A.; Moore, T.A.; Gust, D. Nature 1985, 316, 653.

60. Falk, J.E. "Porphyrins and Metalloporphyrins"; Elsevier, Amsterdam, 1964.

61 Petke, J.D.; Maggiora, G.M. J. Am. Chem. Soc. 1984, 106, 3129.

62. Hanson, L.K.; Chang, C.K.; Ward, B.; Callahan, P.M.; Babcock, G.T.; Head, J.D. J. Am. Chem. Soc. 1984, 106, 3950.

63. Lim, B.T.; Okajima, S.; Chandra, A.K.; Lim, E.C. J. Chem. Phys. 1982, 77, 3902.

RECEIVED April 23, 1986

3

Consequences of Electron Transfer in Chlorophylls, Chlorins, and Porphyrins

Structural and Theoretical Considerations

J. Fajer[1], K. M. Barkigia[1], K. M. Smith[2], and D. A. Goff[2]

[1]Department of Applied Science, Brookhaven National Laboratory, Upton, NY 11973
[2]Department of Chemistry, University of California, Davis, CA 95616

Crystallographic results are presented for a chlorin, a bacteriopheophorbide _d_, a nickel isobacteriochlorin derived from chlorophyll _a_, and for the cation radical of a magnesium porphyrin. These data form part of an increasing body of evidence that demonstrates the significant conformational flexibility of the porphyrin skeleton in the solid and in solution. Since a wide spectrum of photosynthetic and enzymatic reactions proceed via π radicals of porphyrin derivatives, the consequences of the macrocyles' flexibility are considered for distances and orientations between donors and acceptors, interactions with ligands, nearby charges or polar groups, electronic profiles and the gaps between the a_{1u} and a_{2u} highest occupied orbitals.

The primary events that follow photoexcitation of bacterial reaction centers (R.C.) involve the picosecond transfer of an electron from a bacteriochlorophyll (BChl) special pair to one or more acceptors, also comprised of bacteriochlorophyll derivatives (BChl and a demetallated BChl, bacteriopheophytin, BPheo.)(_1,2,3_) Recent x-ray data for _Rhodopseudomonas Viridis_ R.C.'s (which contain BChl b) show the donor and acceptor chromophores to be held in close proximity (~4A edge-to-edge) by surrounding protein residues.(_4_) Some of these are in position to ligate the central magnesium of the BChls (histidines) or to peripherally hydrogen bond the 9-keto group of one of the BPheos (glutamic acid)(_5_).

We examine here possible structural effects that may result from or accompany the generation of the primary photoproducts, and speculate about the consequences of concomitant changes in distances, conformations, relative orientations and charges on the electronic profiles of and interactions between the BChls, BPheos and their radicals. Because the primary events in green plant photosynthesis also involve a series of chlorophyll donors and acceptors (_6_), similar trends should therefore prevail for chlorophyll radicals as well. Furthermore, radicals of porphyrins and hydroporphyrins (saturated porphyrins such as chlorins and isobacteriochlorins) have been

0097–6156/86/0321–0051$06.00/0

detected or proposed as transients in several enzymatic reactions (7,8); an examination of the structure of porphyrin π radicals is thus also relevant to biocatalytic reactions.

We present first crystallographic results for porphyrins, chlorins and isobacteriochlorins that illustrate the wide range of structural variations that can be assumed by the porphyrin skeleton.

A nickel isobacteriochlorin which combines structural features related to chlorophylls (photosynthesis), isobacteriochlorins (nitrite and sulfite reductases), as well as the nickel-containing F430 corphin (methanogenic bacteria) was recently synthesized (9) by the Raney nickel reduction of nickel (II) methylmesopyropheophorbide a (Figure 1). An x-ray analysis of a single crystal of the compound reveals a considerable buckling of the macrocycle, which is illustrated in Figure 2. These striking distortions of the skeleton are features often observed not only in nickel hydroporphyrins (10,11) but in metal-free derivatives as well (12). Examples of these are provided by the synthetic chlorin, 5, 20-dimethyl, 2, 3- dihydro, 2',3',7,8,12,13,17,18 octaethylporphyrin (dimethyl octaethylchlorin, Me_2H_2OEC), and by the bacteriopheophytin d (methyl 4,5-diethylbacteriopheophorbide d, BPheo d) derived from the antenna chlorophylls of green photosynthetic bacteria (13). Figure 3 illustrates the "propeller" shape of the dimethyl chlorin. Note the large displacements of the β carbons of the saturated ring and the up (+0.47A) and down (-0.57A) tilts of the methyl groups that flank that ring. The flexibility of the chlorophyll skeleton is further demonstrated in Figure 4 which displays the deviations from planarity observed for BPheo d. The molecules crystallize as cofacial dimers (Fig. 5) formed by hydrogen bonding between the oxygens of the 9-keto groups of rings V and the protons of the hydroxy groups on rings I that distinguish these chlorophylls from those found in green plants and algae (13). Homologues of this compound can be crystallized in different aggregation modes and different conformations(14).

The examples cited above represent part of an increasing body of structural information on chlorophylls, chlorins, bacteriochlorins and isobacteriochlorins (10-14 and references therein) that points to the remarkable flexibility of these molecules. This ability of the macrocycle to adjust is not limited to hydroporphyrins but is also observed in porphyrins: 5,10,15,20-tetra-n-propylporphinato lead (II) assumes a "roof" shape by folding along an axis defined by two opposite methine carbons with the two planes of the "roof" inclined at 22° to one another (15). In contrast, triclinic 5,10,15, 20-tetraphenylporphinato cobalt (II) is distinctly saddle shaped with the β carbons of adjacent pyrrole rings lying +0.66 and -0.66A above and below the plane of the four nitrogens (16).

The flexibility of these macrocycles becomes even more significant when the effects of oxidation and reduction following electron transfer are taken into account.

Recent structural studies of single crystals of meso tetraaryl porphyrin cation radicals reveal that their Zn(II), Cu(II), Fe(III)Cl and Mg(II) complexes are all saddle shaped (17-20). Since the unoxidized species are either planar or slightly domed, oxidation to the radicals results in a major conformational change. An example of this effect is shown in Fig. 6 which presents the displacements of the atoms that comprise the skeleton of the cation radical of Mg

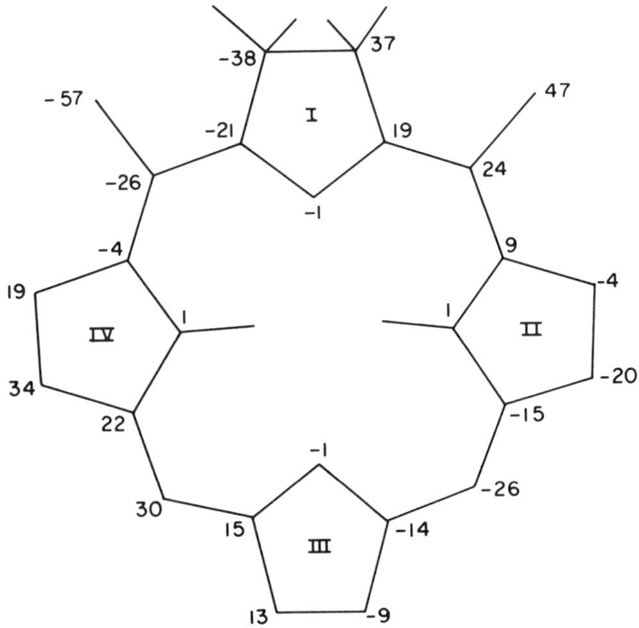

Figure 3. Deviations (Ax10^2) from the plane of the four nitrogens of the dimethyl chlorin 5, 20-dimethyl, 2, 3 trans dihydro, 2', 3', 7, 8,12,13,17,18 octaethyl porphyrin. (Space group C2/c, Z=8.)

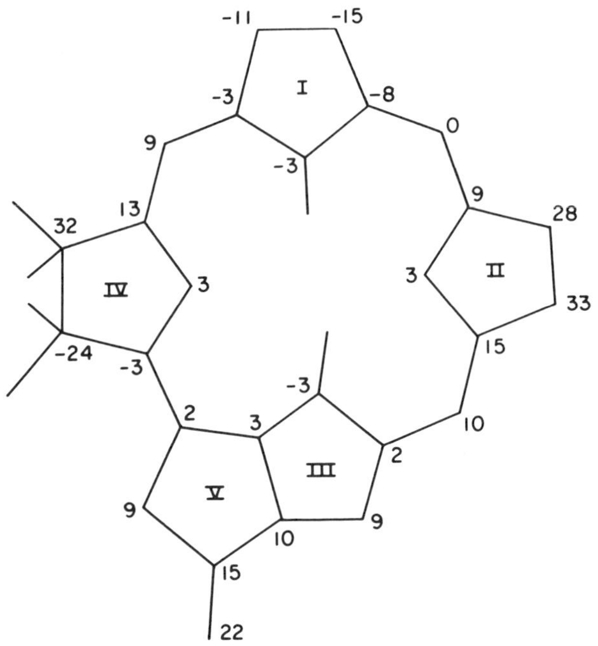

Figure 4. Deviations (Ax10^2) from the plane of the four nitrogens of methyl 4, 5-diethylbacteriopheophorbide d. (Space group P1, Z=2).

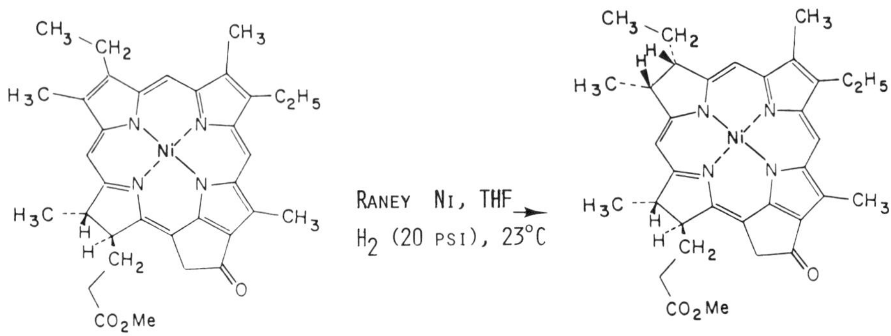

Figure 1. Structure of the isobacteriochlorin obtained by the Raney nickel reduction of nickel (II) methyl mesopyropheophorbide a.

a

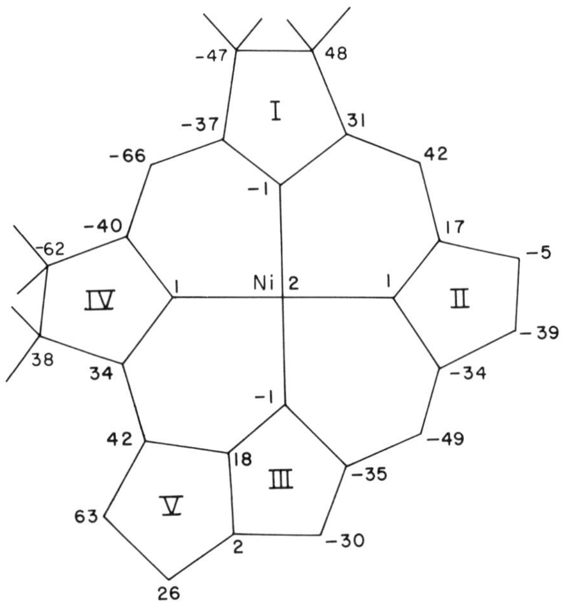

b

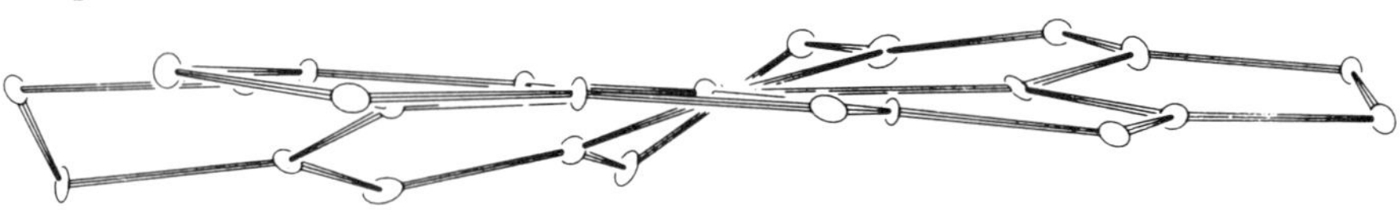

Figure 2. a) Deviations (Ax10^{2}) from the least squares plane defined by the four nitrogens of the nickel isobacteriochlorin. b) An edge-on-view of the molecule that illustrates the puckering of the macrocycle. Substituents have been removed for clarity. (Space group: I4$_1$, Z=16).

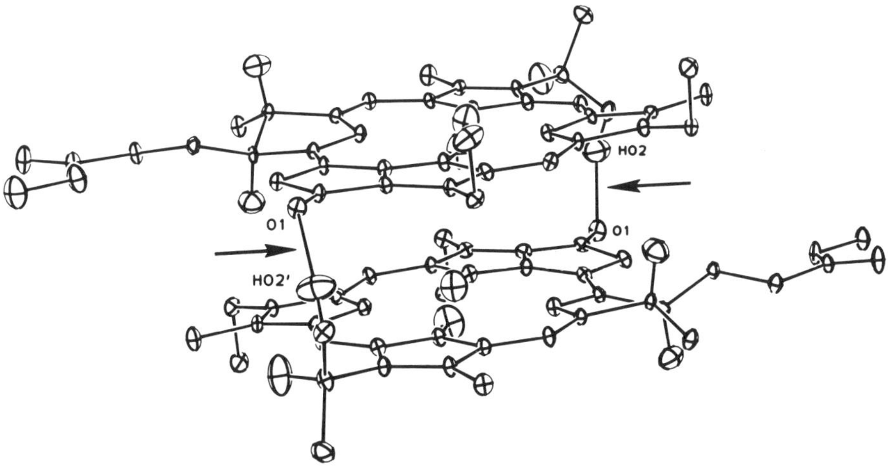

Figure 5. Dimer of the bacteriopheophorbide d. The arrows point to the hydrogen bonds between the protons of the 2-(1-hydroxyethyl) groups and the oxygens of the 9-keto groups.

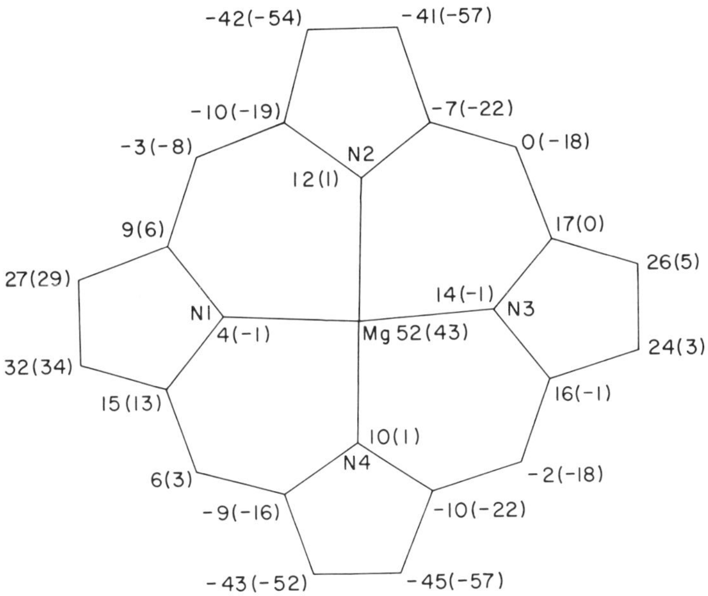

Figure 6. Deviations (Ax10^2) from the least squares plane of the macrocycle and, in parenthesis, from the plane of the four nitrogens of the cation radical of magnesium tetraphenyl porphyrin (Space group $P2_1/c$, Z=4)
(Reproduced from Ref. 20.)

tetraphenylporphin (perchlorato 5,10,15,20-tetraphenylporphinato magnesium (II)) (20).

The question then arises as to whether these deformations of the porphyrin radicals are simply due to packing forces or whether they are found in solution as well. Electron spin resonance (ESR) and double resonance (ENDOR) results for the cation radical of BPheo *a* (21) strongly suggest that the saturated rings of the chromophore can twist, in accord with the x-ray data for methyl-bacteriopheophorbides *a* and *d* (13,21). ESR data for cation radicals of Co(II) nitrosyl chlorins (8) also provide evidence of conformational changes, again, in good agreement with crystallographic steric effects observed in similar compounds (12 and the chlorin results presented above).

If significant variations in conformations can be induced by oxidations, crystals forces, and steric effects in solution, similar effects may presumably also be caused by protein environments. What might be the consequences of conformational changes if they accompany or follow electron transfer reactions in vivo? In general, they might alter interactions with ligands and other protein residues in the immediate vicinity of the porphyrin, i.e. interactions with charged, polar or aromatic groups. In the cases of porphyrin and chlorin cation radicals, in which the highest occupied orbitals (HOMO) are nearly degenerate (a_{1u} and a_{2u} for D_{4h} symmetry)(7), the gap between the two HOMOs could be significantly altered if the radical puckers and the metal-to-ligand distance increases (for an a_{2u} porphyrin cation, iterative extended Huckel calculations (23) suggest that the gap can vary from 0.2 to 0.5 eV). In heme radicals, such as the compound I transients found in catalases and peroxidases, the magnetic coupling between the paramagnetic metal (S=1) and the porphyrin radical (S=1/2) may also depend on the structure of the intermediate. An intriguing in vitro sample of such an effect is observed in high spin Fe(III) tetraarylporphyrin π cations: a planar radical exhibits a magnetic susceptibility of six unpaired electrons whereas in a saddle-shaped complex, S=4, i.e. the metal (S=5/2) and the radical are ferro and antiferromagnetically coupled, respectively (18,19).

Perhaps the most important effect of conformational variations in electron transfer reactions would be to alter the distances and the relative orientations of donors and acceptors. In photosynthetic RC's, where the primary donors and acceptors lie within 4-5A of each other (4), small structural displacements (~0.5A) may significantly affect rates of back reactions, if they occur rapidly (24). (Conformational movements on a picosecond time scale are not inconsistent with resonance Raman data on photo-dissociated heme-CO complexes (25)). On a longer time scale, protein rearrangements triggered by and propagating from the chromophores may also help subsequent reactions such as the transport of protons that is initiated by the primary photochemical event in the R.C. (26).

The close proximity of the photosynthetic donors and acceptors may also account for some of the optical changes observed on short time scales (10^{-12} to 10^{-13} secs) following the primary electron transfer (1-3). In bacterial R.C.'s, this event generates a BChl cation (the special pair) and a BPheo anion (27). According to the x-ray structure of the *R. viridis* R.C. (4), these two radicals are positioned on opposite sides of a BChl b molecule within van der Waals contacts. Theoretical INDO calculations suggest that point charges placed within 3-4A of BChls would induce significant optical shifts (14,28). Interestingly, positive charges placed near rings I and IV or negative charges situated near rings II and III are predicted to cause red shifts whereas reversing the charges should result in blue shifts (29). The primary charge separation in a R.C. should therefore induce a significant electrochromic effect on the optical spectrum of the BChl that "bridges" the primary donor and the BPheo. Shifts of more than a thousand wave numbers are predicted for oppositely placed positive and negative point charges (29). Since the charges in the cation BChl special pair and anion BPheo are delocalized (30), the net effect will be smaller (at a given distance, the shift is proportional to the charge). Experimentally, spectral changes in the bridging BChls have indeed been observed on subpicosecond or picosecond time scales and they have been variously attributed to the formation of a BChl anion, to a charge transfer state or to electrochromic effects, as suggested here (1-3). Again, these effects would be susceptible to additional modulation if changes in distances or orientations between the chromophores follow electron transfer.

In summary, experimental results clearly indicate that porphyrin derivatives are flexible and can undergo rearrangement following electron transfer. If similar conformational changes occur in vivo as the porphyrins fulfill their biological electron transport roles, the changes may help regulate relative orientations and distances between donors and acceptors as well as their interactions with the protein environment with resulting effects on orbital occupancies, and on spectral and magnetic properties.

Acknowledgments

We thank the authors of references 3, 5, 16, 23 and 29 for communicating their results prior to publication.

This work was supported by the Division of Chemical Sciences, U.S. Department of Energy (Contract DE-AC02-76CH00016) at Brookhaven National Laboratory and by the National Science Foundation (Grant No. CHE81-20891) at the University of California.

Literature Cited

1. "Antennas and Reaction Centers of Photosynthetic Bacteria. Structure, Interactions and Dynamics"; Michel-Beyerle, M. E., Ed.; Springer-Verlag, Berlin, 1985.
2. Kirmaier, C.; Holten, D.; Parson, W. W. Biochim. Biophys. Acta, 1985, 819, 49-61.
3. Martin, J. L.; Breton, J.; Hoff, A. J.; Migus, A.; Antonetti, A. Proc. Nat'l. Acad. Sci. U.S.A., 1986, 83, 957-961.
4. Deisenhofer, J.; Epp, O.; Miki, K.; Huber, R.; Michel, H. J. Mol. Biol., 1984, 180, 385-398.
5. Deisenhofer, J.; Epp, O.; Miki, K.; Huber, R.; Michel, H. Nature, 1985, 318, 618-624.
6. Fajer, J.; Fujita, I.; Davis, M. S.; Forman, A.; Hanson, L. K.; Smith, K. M. Adv. Chem. Ser., 1982, 201, 489-513.
7. Hanson, L. K.; Chang, C. K.; Davis, M. S.; Fajer, J. J. Am. Chem. Soc., 1981, 103, 663-670.

8. Fujita, E.; Chang, C. K.; Fajer, J. J. Am. Chem. Soc., 1985, 107, 7665-7669.
9. Smith, K. M.; Goff, D. A. J. Am. Chem. Soc., 1985, 107, 4954-4964.
10. Kratky, C.; Waditschatka, R.; Angst, C.; Johansen, J. E.; Plaquevent, J. C.; Schreiber, J.; Eschenmoser, A. Helv. Chim. Acta, 1985, 68, 1312-1337.
11. Suh, M. P.; Swepston, P. N.; Ibers, J. A. J. Am. Chem. Soc., 1984, 106, 5164-5171.
12. Barkigia, K. M.; Fajer, J.; Chang, C. K.; Williams, G. J. B. J. Am. Chem.Soc., 1982, 104, 315-317.
13. Smith, K. M.; Goff, D. A.; Fajer, J.; Barkigia, K. M. J. Am. Chem. Soc., 1982, 104, 3747-3749.
14. Fajer, J.; Barkigia, K. M.; Fujita, E.; Goff, D. A.; Hanson, L. K.; Head, J. D.; Horning, T.; Smith, K. M.; Zerner, M. C. In reference 1, 324-338.
15. Barkigia, K. M.; Fajer, J.; Adler, A. D.; Williams, G. J. B. Inorg. Chem., 1980, 19, 2057-2061.
16. Hanson, L. K.; Kabuto, C.; Silverton, J. V.; Kon, H. Private communication.
17. Spaulding, L. D.; Eller, P. G.; Bertrand, J. A.; Felton, R. H. J. Am. Chem. Soc., 1974, 96, 982-987.
18. Scholz, W. F.; Reed, C. A.: Lee, Y. J.; Scheidt, W. R.; Lang, J. J. Am. Chem. Soc., 1982. 104, 6791-6793.
19. Buisson, G.; Deronzier, A.; Duee, E.; Gans, P.; Marchon, J. C.; Regnard, J. R. J. Am. Chem. Soc., 1982, 104, 6793-6796.
20. Barkigia, K. M.; Spaulding, L. D.; Fajer, J. Inorg. Chem., 1983, 22, 349-351.
21. Horning, T. L.; Fujita, E.; Fajer, J. J. Am. Chem. Soc., 1986, 108, 323-325.
22. Barkigia, K. M.; Fajer, J.; Smith, K. M.; Williams, G. J. B. J. Am. Chem. Soc., 1981, 103, 5890-5893.
23. Hanson, L. K. Private communication.
24. Marcus, R. A.; Sutin, N. Biochim. Biophys. Acta., 1985, 811, 265-322.
25. Findsen, E. W.; Scott, T. W.; Chance, M. R.; Friedman, J. M.: Ondrias, M. R. J. Am. Chem. Chem., 1985, 107, 3355-3357.
26. Dutton, P. L.; Prince, R. C. In "The Photosynthetic Bacteria"; Clayton, R. C.; Sistrom, W. R., Eds.; Plenum Press, NY 1978; p. 525.
27. Fajer, J.; Brune, D. C.; Davis, M. S.; Forman, A.;. Spaulding, L. D. Proc. Nat'l. Acad. Sci. U.S.A., 1975, 72, 4956-4960.
28. Eccles, J.; Honig, B. Proc. Nat'l Acad. Sci. U.S.A., 1983, 80, 4959-4963.
29. Hanson, L. K.; Thompson, M.; Zerner, M. C.: Fajer, J. Unpublished results.
30. Davis, M. S.; Forman, A.; Hanson, L. K.; Thornber, J. P.; Fajer, J. J. Phys. Chem., 1979, 83, 3325-3332.

RECEIVED April 3, 1986

4

Spin-Orbit and Spin-Polarization Effects in Metalloporphyrins

David A. Case[1]

Department of Chemistry, University of California, Davis, CA 95616

Some general features of spin-orbit and spin-polarization effects in paramagnetic metalloporphyrins are outlined. Spin-orbit effects are modelled through a four-component relativistic molecular orbital formalism. Results for copper porphine show significant effects of unquenched orbital angular momentum on the Zeeman and hyperfine tensors at both the copper and nitrogen sites. Spin-unrestricted calculations on high-spin iron complexes that are models for the active site in deoxyhemoglobin illustrate the effects of exchange-correlation (polarization) forces. These split the spin-up and spin-down iron d-orbitals by 3-4 eV. I discuss how this splitting affects orbital populations and the identification of charge-transfer transitions in the near-infrared region.

The use of molecular orbital theory to interpret the electronic structure and spectra of metalloporphyrins has a long and generally successful history.(1) The early work was based on Huckel and extended Huckel models, while more recently calculations employing other semiempirical or *ab initio* techniques have become feasible.(2) For open-shell systems, two limitations of the extended Huckel model become important even for qualitative discussions. First, spin distributions are altered in important ways by exchange polarization forces, which favor configurations in which electrons of the same spin are in the same region of space. Second, spin-orbit coupling mixes spin-up and spin-down character and allows for non-vanishing expectation values of orbital angular momentum. This has important effects on magnetic resonance parameters, phosphorescence intensities, and so on. In this paper, I discuss one way in which these effects can be included in molecular orbital calculations, using copper porphine and the high spin form of iron porphine as examples.

[1]Current address: Research Institute of Scripps Clinic, Department of Molecular Biology, University of California at San Diego, La Jolla, CA 92037

0097-6156/86/0321-0059$06.00/0

Details of the Calculations

The $X\alpha$ multiple scattering method generates approximate single-determinant wavefunctions, in which the non-local exchange interaction of the Hartree-Fock method has been replaced by a local term, as in the Thomas-Fermi-Dirac model. The orbitals are solutions of the one-electron differential equation (in atomic units)

$$[-(1/2)\nabla^2 + V_c + V_{X\alpha}(\uparrow)]\psi_i(\uparrow) = \varepsilon_i(\uparrow)\psi_i(\uparrow) \quad (1)$$

Here ψ is an orbital of an electron with $M_s = 1/2(\uparrow)$, ε is its one-electron energy, V_c is the classical Coulomb potential (including electron self-interaction terms), and $V_{X\alpha}$ represents the effects of electron exchange. In Slater's model, this is related to $\rho\uparrow$, the local density of electrons of the same spin

$$V_{X\alpha}(\uparrow) = -3\alpha(3\rho\uparrow/4\pi)^{1/3} \quad (2)$$

Here α is an adjustable parameter, usually determined by comparing Hartree-Fock and $X\alpha$ atomic calculations. In the spin-unrestricted version, the spin-up and spin-down orbitals are distinct, so that in general the resulting wavefunction is not a spin eigenfunction. These approximations, and additional ones involving the muffin tin potential, have been described in recent reviews.(3-4) In particular, the work reported here is entirely analogous to our earlier calculations on porphyrins,(5-10) where we discuss the method of calculation in some detail.

The relativistic (DSW) version incorporates the same approximations but starts from the Dirac rather than the Schroedinger wave equation,(11)

$$[c\underset{\sim}{\alpha}\cdot\underset{\sim}{p} + \beta mc^2 + (V_c + V_{X\alpha})I_4]\psi = W\psi \quad (3)$$

where $\underset{\sim}{\alpha}$ and β are Dirac matrices, I_4 is the 4x4 unit matrix, ψ is now a four-component spinor, and the average of the spin-up and spin-down charge densities is used in calculating the exchange potential. The top two components of ψ correspond to spin-up and spin-down character, so that the molecular orbitals in general will be spin mixtures. Each orbital, however, as well as the overall wavefunction, transforms according to the molecular double point group.

Two new results are reported in this paper. In the next section, relativistic calculations on copper porphine are presented, that incorporate parameters identical to those we have used earlier(5) for nonrelativistic calculations. (This calculation assumes a planar porphyrin and uses a Cu-N bond distance of 1.98 Å). These are the first fully relativistic molecular orbital calculations to be reported for a metalloporphyrin, or indeed for any molecule of this size. The following section gives details of calculations on a high-spin form of iron porphine, using a geometry and parameters reported earlier.(7) This is not the ground state for this molecule, but has some features relevant to the high-spin, five coordinate structures found in the deoxy forms of hemoglobin and myoglobin.

Spin-Orbit Effects in Copper Porphine

In many respects the results of a relativistic molecular orbital calculation on copper porphine are similar to the corresponding non-relativistic results. (5,8) The porphyrin levels are virtually unaffected by spin-orbit mixing (e.g. the ratio of spin-down to spin-up character is always less than 10^{-5}), and the energy level spacings are unchanged, although all of the levels are higher in energy by 0.5 eV than in the corresponding non-relativistic calculation. The "crystal field" orbitals having significant copper 3d character are more affected, and the relativistic results are given in Table I. The orbitals are expressed in a coordinate system in which the methine carbons are on the x and y axes. As required by relativistic symmetry, the orbitals are complex (i.e. are expanded in complex shperical harmonics $Y_{\ell m}$) and are mixtures of spin-up and spin-down character.

In both relativistic and non-relativistic calculations there are four relatively low-lying filled orbitals, formed from the dx^2-y^2, dxz, dyz, and dz^2 atomic orbitals. Their energies range from -11.80 to -11.64 eV in the non-relativistic calculation, and from -11.09 to -10.69 eV in the present results. The fifth crystal field orbital is partially occupied and has about 60% dxy character (this orbital points toward the nitrogen atoms in our coordinate system); its energy rises from -8.95 to -8.09 eV upon incorporation of relativistic effects. As a result of these changes, the copper d-levels are about 0.5 eV higher in energy relative to the prophyrin π levels than was the case in the non-relativistic calculation reported earlier.

Table I also contains an analysis of the orbital character of these five energy levels. These were determined from the four-component spinors by neglecting the two lower, "small," components, and by assuming that the radial functions depend only upon ℓ, i.e. that the radial functions for $p_{1/2}$ and $p_{3/2}$, or for $d_{3/2}$ and $d_{5/2}$, are the same. The orbitals may then be written in "Pauli" form as products of (complex) spherical harmonics and spin functions. Populations are equal to the squares of the absolute magnitudes of the coefficients listed in Table I. [For all but $17e_{3g}$, an additional orbital (not shown) is occupied which has the same energy but the opposite spin pattern (i.e. α and β are interchanged).] It may be seen that there is significant spin mixing in the lower four crystal field orbitals, e.g. that the $13e_{3g}$ orbital has 41% spin-up (α) character on copper (mostly in $dx^2-y^2 = (Y_{22} + Y_{2-2})/\sqrt{2}$), 45% $d\pi$ character which is spin-down (β), and smaller contributions of both spins from the porphyrin. Unlike $[Pt(CN)_4]^{2-}$ that we looked at earlier, (12) there is no nearly pure dz^2 (Y_{20}) orbital, although most of this character is in the $13e_{2g}$ orbital.

The partially occupied $17e_{3g}$ orbital is of greatest interest in the interpretation of magnetic resonance parameters. As Table I shows, this orbital is split off from the other copper 3d levels, and is almost pure α spin, with only 0.14% population in the spin-down $d\pi$ atomic orbital. As written, the orbital is nearly pure imaginary: for example, the coefficients of Y_{22} and Y_{2-2} are nearly equal and opposite, and $(Y_{22} - Y_{2-2})/i\sqrt{2} = dxy$. This could be made into a real orbital by multiplication by i, and it should be

Table I. Crystal field orbitals for Cu Por

state	energy, eV	Pauli decomposition[a]	
13 e_{3g}	-11.09	Cu:	$-.460\ Y_{2-2}\alpha + .672\ Y_{2-1}\beta - .444\ Y_{22}\alpha$
		N:	$-.010\ (1+i)\ Y_{1-1}\alpha + .018(1+i)\ Y_{10}\beta$
			$+.028(1-i)\ Y_{11}\alpha$
		Cα:	$(.033 + .071i)\ Y_{10}\beta$
		Cβ:	$(.045 + .055i)\ Y_{10}\beta$
		Cm:	$.070i\ Y_{10}\beta$
12 e_{2g}	-11.03	Cu:	$.117\ Y_{00}\alpha - .347\ Y_{20}\alpha + .816\ Y_{21}\beta$
		N:	$-.016(1-i)\ Y_{1-1}\alpha + .016(1+i)\ Y_{11}\alpha$
			$-.029(1-i)\ Y_{10}\beta$
		Cα:	$(-.045 + .093i)\ Y_{10}\beta$
		Cβ:	$(-.058 + .071i)\ Y_{10}\beta$
		Cm:	$.090i\ Y_{10}\beta$
14 e_{3g}	-10.80	Cu:	$-.526\ Y_{2-2}\alpha - .590\ Y_{2-1}\beta - .519\ Y_{22}\alpha$
		N:	$-.018(1+i)Y_{1-1}\alpha + .023(1-i)\ Y_{11}\alpha$
		Cα:	$(-.025 - .057i)\ Y_{10}\beta$
		Cβ:	$(-.038 - .047i)\ Y_{10}\beta$
		Cm:	$-.058i\ Y_{10}\beta$
13 e_{2g}	-10.69	Cu:	$.281\ Y_{00}\alpha - .865\ Y_{20}\alpha - .343\ Y_{21}\beta$
		N:	$-.034(1-i)\ Y_{1-1}\alpha + .034(1+i)\ Y_{11}\alpha$
		Cα:	$(.013 - .032i)\ Y_{10}\beta$
		Cβ:	$(.022 - .028i)\ Y_{10}\beta$
		Cm:	$-.034i\ Y_{10}\beta$
17 e_{3g}	-8.12	Cu:	$-.543\ Y_{2-2}\alpha + .587\ Y_{22}\alpha - .037\ Y_{2-1}\beta$
		N:	$-.130i\ Y_{00}\alpha + .116(1+i)\ Y_{1-1}\alpha$
			$+\ .1151(1-i)\ Y_{11}\alpha$
		Cα:	$.041i\ Y_{00}\alpha - (.047 + .012i)\ Y_{1-1}\alpha$
			$-(.047 - .011i)\ Y_{11}\alpha$
		Cβ:	$-.021\ Y_{00}\alpha + (.016 + .030i)\ Y_{1-1}\alpha$
			$+(.016 - .030i)\ Y_{11}\alpha$

[a]Coefficients multiply a normalized radial functions (not shown), complex spherical harmonics $Y_{\ell m}$, and spin functions as indicated. Values for the ligand are for a single atom. Coefficients smaller than 0.01 are not shown.

realized that our choice of phase factor is an arbitrary one within each irreducible representation. Hence the $17e_{3g}$ orbital can be considered to be "almost" at the non-relativistic limit in which all of the orbitals may be taken to be real.

The small deviations from the non-relativistic symmetry, however, are quite important for understanding the magnetic behavior of CuPor. It is of interest to note, for example, that the coefficient of the $d\pi$ orbital (0.037) is almost exactly equal to the value of 0.035 that one would obtain from first-order perturbation theory,(13) λ/Δ, where λ is the spin-orbit coupling constant for Cu(*ca.* -830 cm^{-1}) and Δ is crystal field splitting between the dxy and $d\pi$ orbitals (from Table 1, 2.9 eV or 23,500 cm^{-1}). As a result there are few surprises in properties depending on the electron distribution around copper--the results presented below are in good accord with values predicted from conventional crystal field theory (modified to include effects of covalency). On the other hand, the nature of the spin-orbit mixing in the vicinity of the ligand atoms is much harder to understand since a large number of excited states can mix into the ground state *via* spin-orbit coupling, and it is a most difficult task to account for all of them. In fact, it is often assumed that spin-orbit effects at first-row atoms may be ignored. Our recent DSW results for fluorines bound to Xe and Np cast doubt upon this assumption,(14-16) and the nitrogen results presented here also suggest that spin-orbit effects may be significant for first-row atoms bound to heavier elements.

Copper porphyrin is one of the best-characterized of the metalloporphyrins, and its electron spin resonance (ESR) spectrum has been known for a quarter of a century.(17) More recently, electron nuclear double resonance (ENDOR) investigations have provided the complete hyperfine tensors for the metal, the nitrogens and the pyrrole protons.(18) We have used this detailed knowledge earlier(5,8) to assess the quality of scattered-wave calculations. In those calculations, the contributions from electronic orbital motion (induced by spin-orbit mixing) were estimated from crystal field theory (for the copper atom) or were neglected (for the nitrogen and hydrogen atoms). Here I discuss for the first time direct calculations of these contributions to the copper and nitrogen hyperfine tensors, as well as to the molecular g-tensor.

The theory of including magnetic perturbations has been discussed earlier.(11,14-16) In Dirac theory, external fields appear through the operator

$$H' = e\,\underset{\sim}{\alpha}\cdot\underset{\sim}{A} \tag{4}$$

where $\underset{\sim}{\alpha}$ is again a vector of Dirac matrices and $\underset{\sim}{A}$ is the vector potential of the external field. In the case of a constant external field B (the Zeeman effect), $\underset{\sim}{A} = (\underset{\sim}{B} \times \underset{\sim}{r})/2$. For the hyperfine term, $\underset{\sim}{A} = (\underset{\sim}{\mu} \times \underset{\sim}{r})/r^3$ where μ is the nuclear magnetic moment. Matrix elements of these operators are evaluated in the basis spanning the two rows of the e_{3g} representation of the highest occupied orbital. The angular integrals required may be performed analytically, and the radial integrals are done numerically. The resulting energies are then fit to the spin Hamiltonian

$$H_{spin} = \underset{\sim}{S}\cdot\underset{\sim}{g}\cdot\underset{\sim}{B} + \underset{\sim}{S}\cdot\underset{\sim}{A}_n\cdot\underset{\sim}{I}_n \tag{5}$$

where S = 1/2, $\underset{\sim}{I}_n$ is the nuclear spin operator, and n = Cu or N.

Table II gives calculated results for copper porphine and experimental values for copper tetraphenylporphine. Results at the non-relativistic limit (c = ∞) were obtained from the same program by setting the speed of light to a very large number (10^{15} a.u.). As we have discussed elsewhere,(14-16) numerical errors in computing the matrix elements of the Zeeman operator are significant, so that in the non-relativistic limit one does not obtain a g-tensor equal to 2. In other work, we have found it to be a useful approximation to assume that these numerical errors cancel in the non-relativistic and relativistic calculations, so that the computed difference may yield a good estimate of $\Delta g \equiv g - 2.0023$. Results using this approximation are given in part A of Table II, and show approximate, although not quantitative, agreement with experiment. For square-planar copper complexes, equally good agreement is generally possible using ligand field theory, so that the present results provide little new insight. We expect, however, that if this level of agreement with experiment carries over to more complicated cases (e.g. with lower symmetry or more than one metal atom,) the DSW model may provide useful qualitative models for molecular Zeeman effects, particularly for cases with heavy atoms and large spin-orbit mixings.

Hyperfine tensors are given in parts B and C of Table II. Although only the total hyperfine interaction is determined directly from the procedure outlined above, we have found it useful to decompose the total into parts in the following approximate fashion: a Fermi term is defined as the contribution from s-orbitals (which is equivalent to the usual Fermi operator as c → ∞); a spin-dipolar contribution is estimated as in non-relativistic theory from the computed expectation value of $3(\underset{\sim}{S}\cdot\underset{\sim}{r})(\underset{\sim}{I}\cdot\underset{\sim}{r})/r^5$; and the remainder is ascribed to the "spin-orbit" contribution, i.e. to that arising from unquenched orbital angular momentum.

On the basis of their ENDOR results and crystal field theory, Brown and Hoffman(18) have made empirical estimates of these three factors for the copper hyperfine interaction; these are shown in Table II. The spin dipolar contributions may be estimated from α^2, the population of the unpaired electron in the Cu dxy orbital:

$$A^{dip}_{\parallel} = -(4/7)P\alpha^2 \ ; \quad A^{dip}_{\perp} = (2/7)P\alpha^2 \tag{6}$$

where $P = g_e\beta_e g_N\beta_N\langle r^{-3}\rangle$, and the expectation value is over the 3d copper orbital. Both our earlier results(15) and the empirical analysis of Brown and Hoffman suggest that for the copper porphyrin, this value is about 10% larger than the free ion value.(19) The proper value to use for α^2 depends upon how overlap is treated, but for the present qualitative purposes it is sufficient to set it equal to 0.62, the fraction of unpaired spin on the metal estimated from both Xα and crystal field calculations.(5,18)

The spin-orbit effect contributes both to the deviation of the molecular g-tensor from the spin-only value, and to the metal hyperfine interaction. We can use experimental values of the former to estimate the latter:

Table II. Magnetic Resonance Parameters for CuPor

	$c = \infty$	$c = \alpha^{-1}$	diff.	exp.
A. Zeeman interaction				
$g_{\parallel}$	1.761	1.915	0.155	0.188[a]
$g_{\perp}$	1.799	1.858	0.059	0.043
B. ^{63}Cu hyperfine interaction (MHz)[b]				
$A^{dip}_{\parallel}$	-499	-492		-453[c]
$A^{dip}_{\perp}$	249	246		227
$A^{so}_{\parallel}$	0	290		263[d]
$A^{so}_{\perp}$	0	86		47
C. ^{14}N hyperfine interactions (MHz)[b]				
A(Fermi)	52.2	52.8		
A^{dip} xx	-3.7	-3.9		
yy	7.4	7.6		
zz	-3.7	-3.8		
A^{so} xx	-0.2	-3.7		
yy	-0.1	1.1		
zz	-0.2	-1.6		
A^{tot} xx	48.2	45.2		42.8[e]
yy	59.5	61.5		54.2
zz	48.2	47.5		44.1

[a]Value of g (observed) -2.0023, from Reference 9.
[b]See text for method of calculation.
[c]Empirical estimate from Equation (6) using P = 1280 MHz and α^2 = 0.62.
[d]Empirical estimate using Equation (7), P = 1280 MHz and the observed g-tensor.
[e]Source: Reproduced from Ref. 10. Copyright 1985 American Chemical Society.

$$A_{\parallel}^{SO} = -P[(3/7)(g_e-g_{\perp})+(g_e-g_{\parallel})];\ A_{\perp}^{SO} = -P[(6/7)(g_e-g_{\perp})] \quad (7)$$

Table II compares these empirical estimates with those obtained from the DSW calculation. Relativistic contributions have little effect on the spin-dipolar interactions, and both calculations are in reasonably good agreement with the empirical estimates. The spin-orbit contributions are also in moderately good agreement with the empirical estimates, showing that electron currents about the *z*-axis are considerably more important than those about axes in the plane of the ligand. Indeed, in view of the approximations that enter into Equation 7, (5,8), one might have as much confidence in the DSW result as in the empirical estimate given in the final column.

Since core polarization effects are not included in the present DSW calculations, no Fermi contribution to the metal hyperfine interaction arises from the present wavefunction, although its contribution to the experimental tensor is significant. We have discussed such core polarization effects elsewhere.(5,8)

For the nitrogen hyperfine tensors, there is no satisfactory empirical scheme for estimating the various contributions, so that Table II compares the total observed tensor to the DSW result. The tensors are given in their principal axis system, with *z* perpendicular to the plane of the heme and *y* along the Cu-N bond. The small values (0.1 - 0.2 MHz) found for A^{SO} in the nonrelativistic limit are not a consequence of orbital motion (which must vanish in this limit) but are the result of inaccuracies in the decomposition of the total tensor into its components, as described above.

Although all of the calculated values are in moderately good agreement with experiment, of particular interest here is the deviation of this interaction from being axial about the Cu-N bond, i.e. that $A_{XX} \neq A_{ZZ}$. In a spin-restricted, nonrelativistic theory, this tensor will be nearly axial--see, for example the $c = \infty$ results of Table II. Each of the two mechanisms considered in this paper--spin polarization and spin-orbit coupling--might contribute to deviations from axiality. Indeed, non-relativistic calculations including spin-polarization effects(8) illustrated one possible mechanism: exchange effects can alter the spin populations on the nitrogens such that the population in the out-of-plane pπ orbital is greater than that for the in-plane pπ orbital by about 0.01. The spin-dipolar interaction will then yield values of $A_{ZZ} - A_{XX}$ of about 1.3 MHz, in agreement with experiment.(18) Furthermore, examination of the nature of the molecular orbitals suggests that the spin-orbit interaction should be unable to induce pπ populations that are this large,(8,18) leading to the notion that spin-orbit effects may be unimportant for a qualitative understanding of the nitrogen hyperfine tensors in metalloporphyrins.

The results in Table II suggest that this conclusion may be incorrect: orbital motion of the electrons can contribute to the hyperfine tensor amounts comparable to the spin-dipolar interaction. This does not arise from significant changes in spin populations--the in-plane and out-of-plane pπ populations in the partially occupied orbital are 1×10^{-6} and 4×10^{-4}, respectively. As a result, the values for A^{dip} are nearly the same as in the non-relativistic calculation. Rather, a new mechanism is at work, in which the magnetic field associated with orbital motion interacts with the magnetic dipole moment of the nitrogen nucleus. Since this interaction is neither isotropic nor traceless, it can greatly complicate the analysis of the observed results. In particular, the association of the isotropic and anisotropic portions of the observed tensor with Fermi and spin-dipolar mechanisms is no longer appropriate, and it is clearly risky to attempt to derive bonding information from such an analysis.

Unfortunately, the present results are also limited. In addition to errors arising from the exchange potential and from the muffin tin approximations, we have no means at present to include both exchange polarization and spin-orbit effects into these calculations, although each is clearly of potential importance if accuracy greater than 5 Mhz or so is to be achieved. All of the calculations and empirical analyses agree that 7 - 9% of the unpaired spin in CuPor resides on each nitrogen atom; more detailed conclusions will probably have to await the development of reliable computational schemes that can incorporate all of the small effects that contribute to ligand hyperfine interactions.

Exchange Polarization in Iron Porphine

Figure 1 shows the one-electron energies of iron(II) porphine in both spin-restricted and spin-unrestricted (different orbitals for different spins) form. The geometry used has a planar porphine with an Fe-N distance of 2.01 Å and no axial ligands. This calculation assumes a high-spin quintet state with the configuration $(dxy)^2 (dxz)^1 (dyz)^1 (dz^2)^1 (dx^2-y^2)^1$, giving $^5B_{2g}$ symmetry. The differences between the left- and right-hand-sides of this diagram are striking and important in interpreting the properties of this and other high-spin iron complexes. Since the spin density is predominantly located on the iron atom, exchange forces split the iron *d* levels by about 3.5 eV, reflecting the favorable exchange interactions among the majority spin (α) electrons compared to those of spin β. (There is little spin density on the ligand, so that the α and β energies are nearly the same; in the figure, only one line is used to represent both energies.) The α orbitals (except for dx^2-y^2) are low in energy and are always occupied in the ground and low-lying excited states. The other β spin *d* orbitals are near the Fermi level, and are important in determining spectra and reactivities. It is these orbitals, rather than the average of α and β orbitals seen in spin-restricted theories, whose characters and energies are important for qualitative considerations.

As an example, consider the excited state created by a promotion from the iron *dxy* orbital into one of the unoccupied orbitals on the porphyrin ligand. In a spin-restricted theory (such as extended Huckel theory or the left-hand side of Figure 1) the excitation energy is not approximated by the corresponding difference orbital energies; rather one must correct this with an exchange term. If a β spin electron is promoted, the exchange correction is about -1.2 eV,(*20-22*)i.e., the favorable exchange interactions in the resulting high-spin Fe(III) species make the excitation energy smaller than the one-electron energy difference. On the other hand, if an α electron is promoted, the correction is +1.8 eV, a reflection of the loss of exchange correlation in the resulting ferric state. The right hand side of Figure 1, however, clearly shows that the promotion of an α dxy electron is more costly than promotion of a β one by about the 3 eV difference in the two "correction" factors cited above--to a large extent the exchange corrections are already included in the one-electron energies.

Table III shows the populations for the "crystal field" orbitals, i.e. for those occupied orbitals with greater than 10% population on the iron atom. The comparison in this table is between the populations for the $^5B_{2g}$ state (upper line in the table) and a $^3A_{2g}$ state (lower line) with occupation $(dxy)^2(dz^2)^2(d\pi)^2$, which is likely to be a major component of the ground state.(9) The triplet→quintet change thus corresponds to the orbital promotion $dz^2\beta \rightarrow dx^2-y^2\alpha$. The results illustrate the ways in which spin-exchange affects the electronic structure of this complex. The energies of the β-spin orbitals are nearly unchanged but the strongly localized α orbitals ($6b_{2g}\alpha$, $8a_{1g}\alpha$ and $7b_{1g}\alpha$) are lower in energy in the quintet state by about 1.5 eV due to more favorable exchange interactions in the high-spin state. The situation is more complex for the e_g orbitals, which can mix strongly with prophyrin π orbitals of the same symmetry. In the quintet state, the $d\pi\alpha$ orbitals are stabilized by this interaction, and are the lowest of all the d-like levels, appearing mainly in the $2e_g\alpha$ level, with about 9% character in the $4e_g\alpha$ orbital. In the triplet, both $2e_g$ and $4e_g$ have significant $d\pi$ population, a feature that makes it difficult to apply crystal field theory to this complex.(9)

We have discussed the optical absorption spectrum of triplet FePor in our earlier paper.(9) Similar results for the quintet state, particularly for charge transfer transitions, are of interest as possible explanations for absorption bands seen the near-infrared region of deoxy heme proteins.(20-22) Both metal-to-ligand and ligand-to-metal transitions need to be considered. A typical ligand-to-metal promotion would be from the highest occupied porphyrin $a_{2u}(\pi)$ orbital into $4e_g\beta$, which is 76% $d\pi$; a characteristic metal-to-ligand transition is from $dxy\beta$ to the lowest unoccupied porphyrin $e_g(\pi)$ orbital. Since the energies of both the porphyrin orbitals and the metal spin-β orbitals are nearly the same in the triplet and quintet states, the excitation energies should also be about the same. For the triplet state, we have shown that $X\alpha$ calculations predict both metal-to-ligand and ligand-to-metal transitions to lie in the 1.5 to 2.0 eV range,(9) and the same general conclusions should hold for high-spin states. We shall report elsewhere(23) results for five-coordinate high spin complexes that support this conclusion. The calculations of charge transfer excitation energies, however, are sufficiently uncertain to make it hazardous to attempt detailed assignments of observed optical spectra.

Unlike the situation for charge transfer energies, *d-d* excitations are expected to be quite sensitive to spin state (triplet *vs*. quintet) and the presence of axial ligands. Calculations of *d-d* transitions for high-spin five-coordinate iron porphyrins will be presented elsewhere.

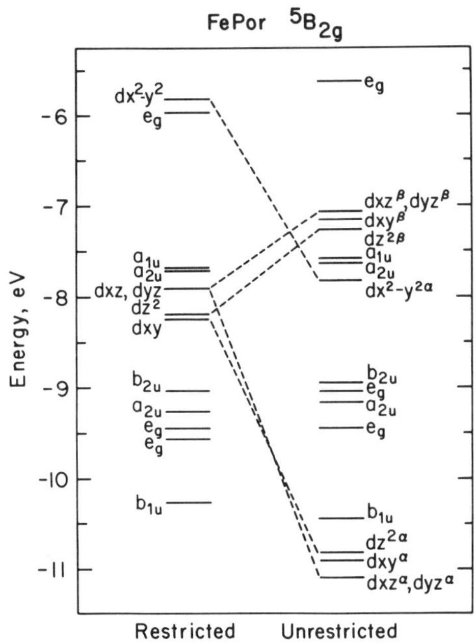

Figure 1. Orbital energies for iron porphine.

Table III. Crystal Field Orbitals for FePor[a]

Orbital	Energy, eV	Populations
$6b_{2g}$ β	− 7.18	.973 Fe dxy
	− 7.31	.970 Fe dxy
$7b_{1g}$ α	− 7.82	.665 Fe dx^2-y^2 + .049 N ρσ + .018 Ns
	− 6.31	.744 Fe dx^2-y^2 + .036 N ρσ + 0.16 Ns
4eg α	− 8.86	.088 Fe dπ + .090 N + .061 Cβ
	− 8.45	.382 Fe dπ + .062 N + .037 Cβ
3eg β	− 9.24	.112 Fe dπ + .066 N + .078 Cβ
	− 9.28	.110 Fe dπ + .049 N + .083 Cβ
$8a_{1g}$ α	−10.67	.105 Fe s + .845 Fe dz^2
	− 9.15	.107 Fe s + .871 Fe dz^2
$6b_{2g}$ α	−10.89	.950 Fe dxy
	− 9.26	.965 Fe dxy
2eg α	−11.07	.804 Fe dπ + .009 Cα + .011 Cβ + .009 Cm
	− 9.93	.557 Fe dπ + .033 N + .021 Cβ + .020 Cm

[a]Populations per atom. All orbitals shown are occupied. For each orbital, the upper line gives results for the $^5B_{2g}$ state, and the lower line for the $^3A_{2g}$ state.

Conclusions

In this paper I have tried to outline some of the qualitative features of spin-orbit mixing and exchange polarization effects in metalloporphyrins. Although the computational models involved are crude ones, the general features found here are expected to persist in more accurate calculations. One improvement that would clearly be desireable is to be able to incorporate both effects in a single model. It is not easy to see how spin-exchange effects can be incorporated into the Dirac model, since the orbitals and states are spin mixtures. In principle, both effects can be included through perturbation theory or by configuration interaction,(24,25) but it remains to be seen if simple physical pictures can be derived from these calculations. Other approximate methods of including both effects have been tried,(26,27) and it is to be hoped that useful models along these lines can be developed in the next few years.

There is also a great need for calculations of greater intrinsic reliability than those reported here. In both of the specific instances discussed above (hyperfine tensors in CuPor and optical excitation energies in FePor) the general features of the Xα results should be reliable but the errors are sufficiently large that detailed interpretations of the experiments are of dubious validity. *Ab initio* calculations have great potential, but suffer from the large number of basis functions required, and from the fact that the Hartree-Fock model is not a very good one for open-shell transition metal complexes--for example, it overestimates by a large amount the stability of high-spin complexes relative to intermediate or low-spin configurations. Until methods of including and understanding correlation effects in these systems are developed, the connection between theory and experiment will continue to be exciting and illuminating, but not definitive.

This work was supported in part by a grant from the National Science Foundation.

Literature Cited

1. Zerner, M.; Gouterman, M. *Theor. Chim. Acta* 1966, *4*, 44. See also Gouterman, M. in *The Porphyrins*, *Vol. III*. *Physical Chemistry*, D. Dolphin (ed.) Academic Press, 1977.
2. For reviews see: Loew, G.H., in *Iron Porphyrins*, *Part I*, A.B.P. Lever and H.B. Gray, eds. Addison-Wesley, 1983, p. 1. Dedieu, A.; Rohmer, M.-M.; Veillard, A. *Adv. Quantum Chem*. 1982, *16*, 43.
3. Johnson, K.H. *Adv. Quantum Chem*., 1973, *7*, 143.
4. Case, D.A. *Annu. Rev. Phys. Chem*. 1982, *33*, 151.
5. Case, D.A.; Karplus, M. *J. Am. Chem. Soc*. 1977, *99*, 6182.
6. Huynh, B.H.; Case, D.A.; Karplus, M. *J. Am. Chem. Soc*., 1977, *99*, 6103.
7. Case, D.A.; Huynh, B.H.; Karplus, M. *J. Am. Chem. Soc*., 1979, *101*, 4433.
8. Sontum, S.F.; Case, D.A. *J. Phys. Chem*. 1982, *86*, 1596.
9. Sontum, S.F.; Case, D.A.; Karplus, M. *J. Chem. Phys*. 1983, *79*, 2881.
10. Sontum, S.F.; Case, D.A. *J. Am. Chem. Soc*., 1985, *107*, 4013.

11. For a review, see Yang, C.Y.; Case, D.A. in Local Density Approximations in Quantum Chemistry and Solid State Physics, J.P. Dahl and J. Avery, eds. Plenum, 1984, p. 643.
12. Lopez, J.P.; Yang, C.Y.; Case, D.A. Chem. Phys. Lett. 1982, 91, 353.
13. Atherton, N.M. Electron Spin Resonance, Theory and Applications. (John Wiley, 1973), Section 6.4.
14. Arratia-Perez, R.; Case, D.A. J. Chem. Phys. 1983, 79, 4939.
15. Case, D.A.; Lopez, J.P. J. Chem. Phys. 1984, 80, 3270.
16. Case, D.A. J. Chem. Phys. 1985, 83, 5792.
17. Manoharan, P.T.; Rogers, M.T. Electron Spin Resonance of Metal Complexes, Yeh, T.F., Ed. Adam Hilger: New York, 1969, p. 143, and references therein.
18. Brown, T.G.; Hoffman, B.M. Mol. Phys. 1980, 39, 1073.
19. McGarvey, B.R. J. Phys. Chem. 1967, 71, 51.
20. Eaton, W.A.; Hanson, L.K.; Stephens, P.J.; Sutherland, J.C.; Dunn, J.B.R. J. Am. Chem. Soc. 1978, 100, 4991.
21. Zerner, M. Ph.D. thesis, Harvard University, 1966.
22. Makinen, M.W.; Churg, A.K. Iron Porphyrins, Part I, Lever, A.B.P. and Gray, H.B., eds. (Addison-Wesley, 1983), p. 141.
23. Sontum, S.F.; Case, D.A.; Karplus, M. manuscript in preparation.
24. See, e.g. Mishra, K.C.; Mishra, S.K.; Das, T.P. J. Am. Chem. Soc. 1983, 105, 7729.
25. Vajed-Samii, M.; Andriessen, J.; Das, B.P.; Ray, S.N.; Lee, T.; Das, T.P- J. Phys. B 1982, 15, 1379.
26. Weinert, M.; Freeman, A.J. Phys. Rev. B 1983, 28, 6262.
27. Ellis, D.E.; Goodman, G.L. Int. J. Quantum Chem. 1984, 25, 185.

RECEIVED March 28, 1986

STEADY-STATE ABSORPTION AND EMISSION SPECTROSCOPY

5

The Role of NH-Tautomerism in the Formation of Excited States

Porphyrins with Nonsymmetrical Substitution

Georgii P. Gurinovich, Edward I. Zenkevich, and Alexander M. Shulga

The Institute of Physics, BSSR Academy of Sciences, Minsk, 220602, USSR

It has been shown that for porphyrins with isocycles (synthetic and natural compounds), bisporphyrins linked covalently via isocycles and for 2,3,12,13-tetraalkylporphyrins additional bands due to the existence of the second tautomer are discovered in absorption and fluorescence spectra of isotropic solutions over a wide temperature range (77 to 500 K). The spacing between the S_0 - S_1 transitions of tautomers is as great as 380 cm^{-1}, while for S_0-S_2 transitions it is 800 cm^{-1}. The difference has been discovered in the types of visible absorption spectra caused by the inversion of $Q_x(0,0)$ and $Q_y(0,0)$ band intensities when the NH-proton migration takes place: while the situation $I_x(0,0) < I_y(0,0)$ is observed for one tautomer, $I_x(0,0) > I_y(0,0)$ is characteristic for the second one and vice versa. NMR^{1}H spectra show that in the basic (short-wave) tautomer of cyclopentanporphyrins NH-protons are located at the opposite pyrrole rings which are not bonded with the isocycle whereas for 2,3,12,13-tetraalkylporphyrins they are fixed at rings having alkyl groups. The investigation of dichroism of photo-induced changes in the intensities of tautomer transitions S_0-S_1 under solution excitation by linearly polarized light permits us to conclude that the system of molecular oscillators X and Y mainly remains fixed at the same pyrrole rings as the H-H axis rotates through 90°. The experimental results and MO (CNDO/2) calculations explain the inversion of the $Q_x(0,0)$ and $Q_y(0,0)$ intensities of electronic transitions with NH-protons jumping as a consequence of the inversion of LUMO's c_1 and c_2 and, therefore, of the change in the configuration interaction of the Q and B states.

0097-6156/86/0321-0074$06.00/0

Introduction

It is well known that in free-base porphyrins two inner hydrogens, which, in their stable conformation, are bound to opposite nitrogens, may jump from one pair of nitrogens to another. According to NMR^{1}H spectroscopy data (1-4), these hydrogens migrate relatively quickly at room temperature, but at 77 K the above tautomerism ceases when the porphyrin molecule is in the ground electronic state (5). The migration of the center protons may still be induced even at liquid helium temperature by photoexcitation of the porphyrin molecule. Such phototransformation between two tautomers at low temperatures has been observed in two types of experiments: i) photo-induced reversible conversion of admixture centers into one another for porphyrin-free bases incorporated in a n-octane Shpol'skii matrix (6,7); ii) variations in the emission intensity and fluorescence depolarization of glassy solutions of porphyrins at 77 K during excitation by constant polarized monochromatic light (5).

In symmetrical porphyrins NH-tautomers do not distinguish in structure and have identical electronic spectra. Correspondingly, in most cases the spectral separation between the individual tautomers (Δ E) does not exceed $\leq$ 100 cm^{-1} and is due to crystalline matrix effect (6,7). Therefore, NH-tautomerism can be effectively detected at 77 K and 4.2 K in the Shpol'skii matrix only at selective monochromatic excitation (5-7). In chlorins, the spectral separation between the two tautomers is large: $\Delta E \simeq$ 800-1500 cm^{-1} (8-11). But the second tautomer which can be produced by photoexcitation can be stable only at 4.2 K due to the fast reverse photochemical relaxation (10).

It has been shown for the first time by us recently that for a whole class of isocycle-containing porphyrins spectral characteristics of individual NH-tautomers greatly differ because of non-symmetrical substitution caused by the isocyclic ring (12-14). Comparison of the fluorescence and absorption spectra of two tautomers for one compound has revealed the spectral separation between their S_0-S_1 transitions to be as great as 380 cm^{-1}, while that for S_0-S_2 transitions can amount to ~ 800 cm^{-1}. Therefore, individual tautomers of these compounds may be easily investigated using normal electronic spectra in a wide temperature range (77 - 500 K). Moreover, when four methyl or ethyl groups are attached to the opposed pyrrole rings of porphyrin molecules similar pecularities of electronic spectra of two tautomeric forms are observed in isotropic solutions at different temperatures (77 - 400 K) (15-17). It should be emphasized that the visible absorption spectra of tautomers show noticeable inversion of $Q_x(0,0)$ and $Q_y(0,0)$ band intensities in every compound in going from one tautomer to another.

Up to now the four-orbital model-based theory of porphyrin absorption electronic spectra (18) has not taken into account the influence of NH-tautomerism in non-symmetrical porphyrins on the positions and intensities of electronic transitions in the visible region. The theoretical consideration of this problem involves solving the fundamental question of the absolute orientation of electronic transition oscillators for each tautomer. First of all, the correspondence between the direction of the long-wave Q(0,0) transition oscillator and that of the H-H axis connecting NH-protons must be found in individual tautomeric forms.

In this paper, we present the detailed spectral information obtained for individual NH-tautomers and straightforward experimental arguments which permit us to relate the real position of the H-H axis to the molecular oscillator axes determined by substituents of different types (isocycle or alkyl groups) in two NH-tautomers. Then we used this structural information as a basis for analyzing the inversion of electronic $Q_x(0,0)$ and $Q_y(0,0)$ band intensities using the four-orbital model.

Experimental Section

Synthesis, purification and identification of cyclopentanporphyrins and their chemical dimers covalently linked via isocycles have been described in previous papers (17,19-23). The porphyrins with the cyclopentanone ring have been prepared by familiar methods (24-26). Our experimental results were obtained when investigating the following compounds: 3^1, 5^1-cyclo-3^1-methyl-2,7,8,12,13,17,-18-heptaethyl-22H,24H-porphine (1); 3^1,5^1-cyclo-3^1-exomethyl-en-2,7,18,12,13,17,18-heptaethyl-22H,24H-porphine (2); 3^1,5^1-cyclo-3^1-methyl-5^1-oxy-2,7,8,12,13,17,18-heptaethyl-22H,24H-porphine (3); phylloerythrin methyl ester (4); deoxophylloerythin methyl ester (5); 10-ethoxy-phylloerythrin methyl ester (6); pheoporphyrin-a_5 dimethyl ester (7); 2-vinylpheoporphyrin-a_5 dimethyl ester (8); 4-vinylprotopheophytin a (9); 3^1,5^1-cyclo-5^1-ethyl-10,15,20-tripropyl-22H,24H-porphine (10); $3^1_,5^1$-cyclodimer (11) which consists of two molecules 1 and 2; 2,3,12,13-tetraethyl-21H,23H-porphine (12); 3^15^1-cyclo-3^1-methyl-2,12,13-triethyl-22H,24H-porphine (13); 2,3,12,13-tetraethyl-7-bromo-21H,23H-porphine (14) and 2,3,12,13-tetramethyl-21H,23H-porphine (15). Structural formulas are shown in Figure 1. Compounds 1,2,3,5,10,11,12,13,14 were dissolved in EPIP (diethyl ether, petroleum ether, isopropanol 5:5:2)whereas compounds 4,6,7,8,9,15 were dissolved in THF-DE (tetrahydrofurane, diethyl ether 1:1). These solvent mixtures can be frozen as glassy samples at 77 K. The absorption spectra were recorded on a standard spectrophotometer SF-10 or Beckman-5270. The measurements of fluorescence excitation and emission spectra were made with the aid of a spectrofluorometer SLM-4800 with automatic correction of spectral response. Fluorescence lifetimes were measured with the aid of a pulse fluorometer PRA-3000. Magnetic circular dichroism (MCD) measurements were carried out in a 8 kG magnetic field using a JASCO J-20 circular dichrometer. Triplet state formation was observed for investigated compounds at the experimental set up, whose detailed description can be found in our paper (27). The optical experiments were carried out with a porphyrin concentration of 4.10^{-6} - 4.10^{-5} $mol.l^{-1}$. In NMR investigations (Bruker WM-360) we used higher concentrations (5.10^{-3} $mol.l^{-1}$) and dried solvents ($CDCl_3$, CS_2 and toluene-d_8).

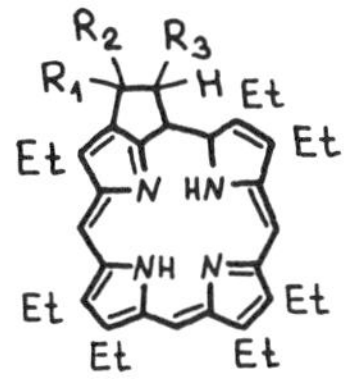

(1) $R_1, R_3 = H, R_2 = CH_3$
(2) $R_1, R_2 = {=CH_2}, R_3 = H$
(3) $R_1 = H, R_2 = CH_3,$
$R_3 = OH$

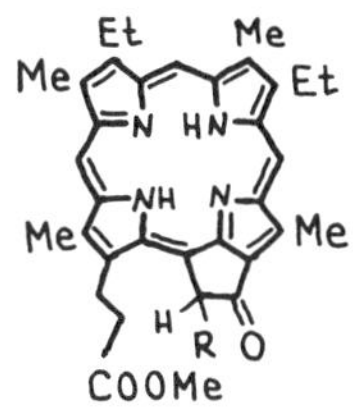

(4) $R = H$
(6) $R = OC_2H_5$
(7) $R = COOCH_3$

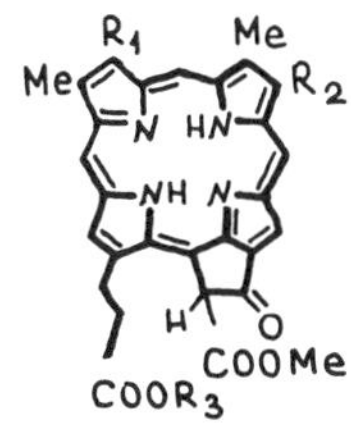

(8) $R_1 = CH{=}CH_2, R_2 = C_2H_5,$
$R_3 = CH_3$
(9) $R_1, R_2 = CH{=}CH_2, R_3 = H$

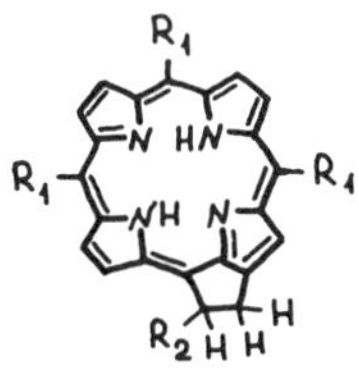

(10) $R_1 = C_3H_7, R_2 = C_2H_5$

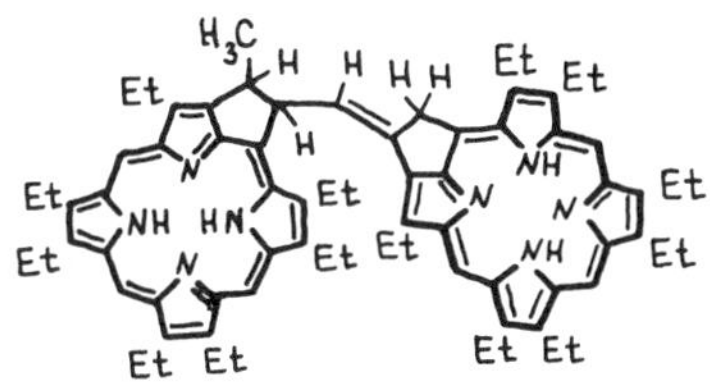

(11)

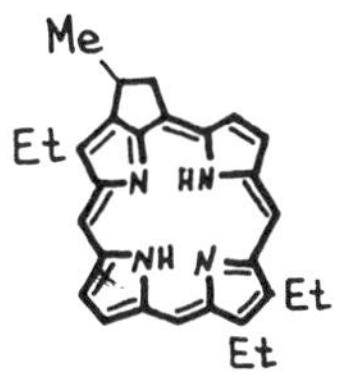

(13)

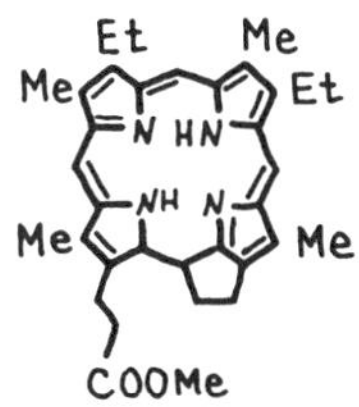

(5)

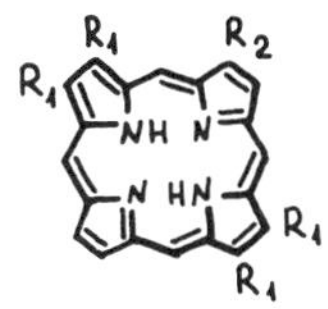

(12) $R_1 = C_2H_5, R_2 = H$
(14) $R_1 = C_2H_5, R_2 = Br$
(15) $R_1 = CH_3, R_2 = H$

Figure 1. Structures of investigated porphyrins with non-symmetrical substitution.

Results and Discussion

When analyzing in detail the spectroscopic observations of the investigated compounds and symmetrical porphyrins some spectral pecularities should be emphasized, which are characteristic only of porphyrins with non-symmetrical substitution. Figures 2-9 give some experimental results obtained for porphyrins with different side and isocycle substituents.

Electronic Spectra. In most cases, along with the four bands in the visible region of the absorption spectra, which are normal for porphyrin free bases (28,29), an additional real band or shoulder can be observed on the long-wave side of the main 0-0 absorption band at 293 K. The relative intensity of this new band becomes greater on heating the solutions up to 500 K. In contrast, the intensity of the additional band decreases as the solutions are cooled down to 77 K but this band becomes sharper and narrower (Fig. 2,5).

The fluorescence spectra exhibit features common to all compounds; besides the two bands of the main product there is a new system of bands which corresponds to the additional absorbing center (Fig. 3,5). The relative intensity of these new bands is strongly dependent on the exciting light wavelength. It should be noted that the fluorescence lifetimes of the 0-0 bands for the two luminescent centers, measured at 77 K, are rather close (for example, in solution of compound 2 τ_1 = 21.9 ns at λ_1 = 622 nm and τ_2 = 20.3 ns at λ_2 = 637 nm). At room temperature the spectral pattern is not so clear due to the temperature broadening and shift of the fluorescence bands. Nevertheless, in some cases additional bands can be observed even at 293 K (Fig. 3b). The analysis of fluorescence excitation spectra recorded at different emission wavelengths shows that the system of fluorescence bands may be assigned only to two fluorescent centers (Fig. 4,5), and permits us to conclude that in all the cases the two absorption centers are noninteracting. Clearly, at 77 K the existence of two centers which have their own fluorescence and absorption spectra can be revealed for all investigated porphyrins (Table 1). The spectral separation between the S_0-S_1 transitions of these centers is different as can be seen from Table 1. Therefore, it is difficult to find the second center at 293 K for some compounds, but for compounds 2 and 10, for example, this center can easily be detected even at higher temperatures (413 K and 500 K) (Fig. 2b and 3b).

It should be noted that the existence of different centers is also found in covalently-linked dimer 11. Moreover, the analysis of all data obtained for dimers points clearly towards the efficient transfer of the excited singlet state energy from two centers of compound 1 to two acceptor centers of compound 2 in dimers (14,30). Increase in the porphyrin concentration by 300-700 times (say, for compounds 1 and 2) does not cause additional changes in electronic spectra as against diluted solutions. If the results obtained from temperature experiments (Fig. 2b) and measurements of fluorescence lifetimes in different bands are taken into account, one may conclude that the additional centers observed in

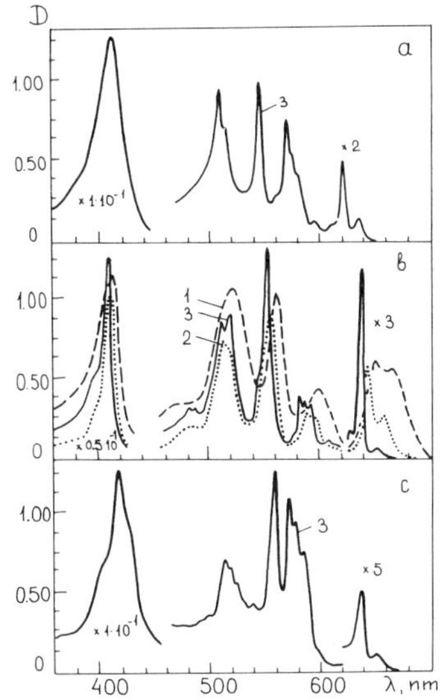

Figure 2. Absorption spectra of compounds 2(a) and 10(b) in EPIP and 6(c) in THF-DE. 1 - 500 K (in methylnaphthalene), 2 - 293 K, 3 - 77 K. (Reproduced with permission from Ref. 12. Copyright 1984 North-Holland Physics Publishing Company.)

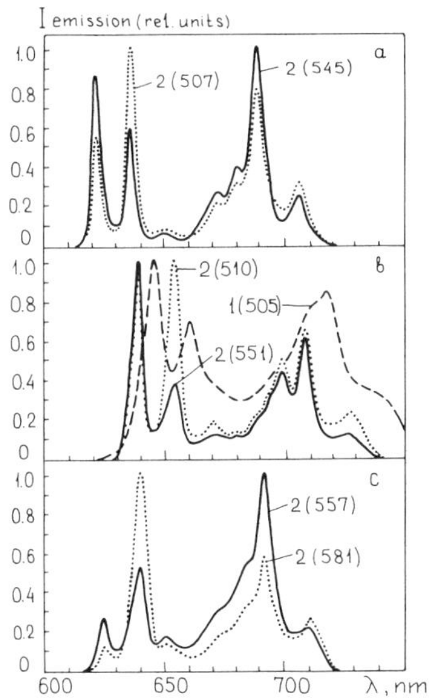

Figure 3. Fluorescence emission spectra of compounds 2(a) and 10(b) in EPIP and 6(c) in THF-DE at 293 K (1) and 77 K (2). $\Delta\lambda_{excit}$ = 4 nm, $\Delta\lambda_{monit}$ = 2 nm. Excitation wavelengths are shown in brackets. (Reproduced with permission from Ref. 12. Copyright 1984 North-Holland Physics Publishing Company.)

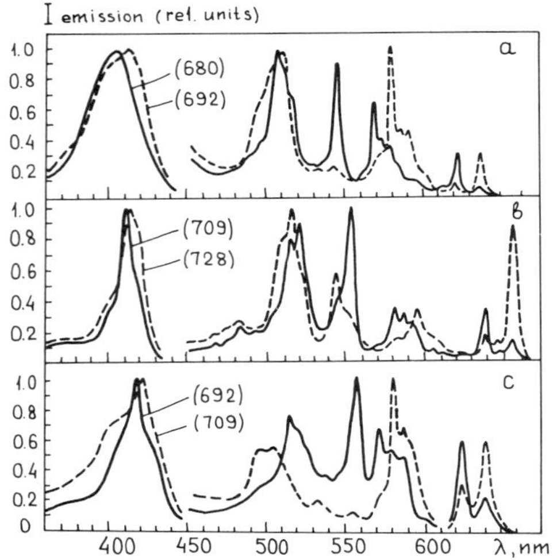

Figure 4. Fluorescence excitation spectra of compounds 2(a) and 10(b) in EPIP and 6(c) in THF-DE at 77 K. The recording wavelengths are shown in brackets. $\Delta\lambda_{excit}$ = 2 nm, $\Delta\lambda_{monit}$ = 4 nm. Correction of exciting light spectral distribution has been done up to 600 nm. (Reproduced with permission from Ref. 12. Copyright 1984 North-Holland Physics Publishing Company.)

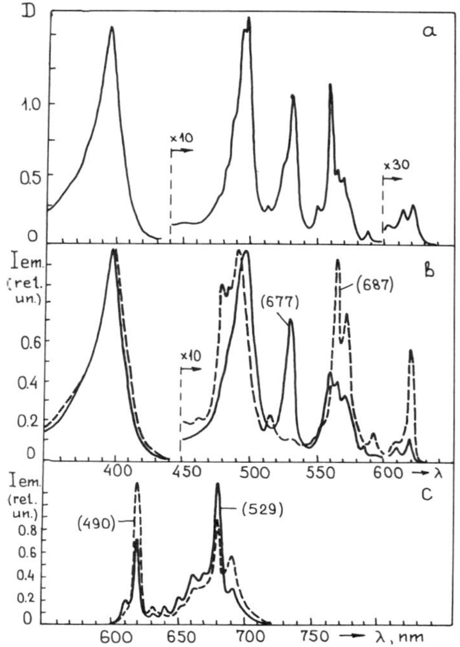

Figure 5. Absorption (a), fluorescence excitation (b) and fluorescence (c) spectra of compound 12 in EPIP at 77 K. Recording (b) and excitation (c) wavelengths are shown in brackets. Correction of exciting light spectral distribution for spectra c has been done up to 600 nm.

TABLE I. Maxima of absorption and fluorescence spectra (λ, nm), band relative intensities (in brackets), energy separation ΔE between S_0-S_1 transitions, quantum yields (B) and lifetimes (τ_s) of fluorescence and triplet state lifetimes (τ_T) of NH-tautomers in porphyrins with non-symmetrical substitution at 77 K.

		ABSORPTION					ΔE	FLUORESCENCE				
C	T	B(0.0)	Qy(1.0)	Qy(0.0)	Qx(1.0)	Qx(0.0)	cm^{-1}	I^{0-0}	I^{vib}	B[b)]	τ_s ns	τ_T ms
I	II	III	IV	V	VI	VII	VIII	IX	X	XI	XII	XIII
1	*a)	399	499(1.0)	532(0.26)	561(0.74)	612(1.0)	262	614(1.0)	679(0.48)	0.10	21.3	19
	**	403	498(1.0)	529(0.67)	569(0.74)	622(0.23)		624(1.0)	692(0.90)		18.9	20
2	*	407	507(1.0)	544(0.88)	568(0.67)	621(0.19)	379	622(0.94)	689(1.0)	0.09	21.9	31
	**	413	504(0.94)	535(0.36)	580(1.0)	636(0.72)		637(1.0)	707(0.28)		20.3	25
3	*	400	496(0.97)	533(0.58)	561(1.0)	612(0.81)	263	614(1.0)	680(0.51)	0.10	17.6	
	**	403	496(0.98)	528(1.0)	570(0.94)	622(0.25)		624(1.0)	692(0.87)		18.3	
4	*	419	518(0.56)	559(1.0)	581(0.67)	635(0.29)	123	636(0.5)	707(1.0)	0.065	9.9	
	**	421	516(0.95)	543(0.62)	583(1.0)	640(1.0)		641(1.0)	712(0.12)		10.4	
5	*	397	496(1.0)	532(0.35)	561(0.68)	613(0.91)	261	614(0.76)	681(1.0)			
	**	398	(499(1.0)	529(0.74)	570(0.71)	623(0.36)		624(0.29)	693(1.0)			
6	*	418	514(0.70)	557(1.0)	571(0.71)	623(0.06)	377	624(0.33)	692(1.0)	0.065	10.8	
	**	420	507(0.78)	533(0.2)	581(1.0)	638(0.9)		639(1.0)	709(0.18)		12.6	
7	*	421	521(0.59)	562(1.0)	578(0.75)	634(0.25)	197	635(0.59)	704(1.0)			
	**	426	516(0.79)	543(0.42)	543(0.42)	583(1.0)		643(1.0)	713(0.2)			
8	*	422	523(0.40)	563(1.0)	589(0.66)	637(0.19)	338	638(0.3)	707(1.0)	0.07	9.3	
	**	426	513(1.0)	546(0.52)	603(0.68)			652(1.0)	726(0.17)		12.5	

Continued on next page

Table I. Continued

		ABSORPTION					ΔE	FLUORESCENCE			τ_S	τ_T
C	T	B(0.0)	Qy(1.0)	Qy(0.0)	Qx(1.0)	Qx(0.0)	cm^{-1}	I^{0-0}	I^{vib}	B b)	ns	ms
I	II	III	IV	V	VI	VII	VIII	IX	X	XI	XII	XIII
9	*	429	530(0.69)	570(1.0)	588(0.76)	643(0.17)	238	644(0.25)	714(1.0)		9.7	
	**	431	515(0.83)	549(0.4)	605(1.0)	653(0.60)		654(1.0)	730(0.21)		9.2	
10	*	412	520(0.85)	552(1.0)	582(0.32)	638(0.38)	360	639(1.0)	709(0.49)	0.09	17.3	2.4
	**	413	513(0.81)	543(0.48)	594(0.28)	653(1.0)		654(1.0)	728(0.17)		17.9	2.3
11	*c)	400	500	534	561	613	380	--	--			
	*	412	509	549	569	622		624(0.9)	692(1.0)	0.10	20.0	
	**	423	501	534	580	637		638(1.0)	709(0.3)		19.2	
12	*	396	494(1.0)	529(0.68)	559(0.42)	610(0.03)	212	611(0.1)	677(1.0)	0.065	22.9	12
	**	397	490(1.0)	514(0.22)	564(0.95)	618(0.67)		619(1.0)	687(0.3)		22.7	14
13	*	399	493(1.0)	525(0.2)	560(0.6)	614(0.77)	210	615(1.0)	681(0.45)	0.10	24.3	
	**		497(1.0)	530(0.47)	569(0.5)	622(0.30)		623(1.0)	695(0.2)		22.8	
14	*	400	500(0.85)	535(1.0)	559(0.75)	612(0.04)	263	613(0.09)	678(1.0)	0.007	2.4	2.1
	**	401	479(1.0)	516(0.25)	568(0.93)	622(0.68)		623(1.0)	691(0.18)		2.5	1.8
15	*	400	499(1.0)	531(0.9)	559(0.73)	609(0.05)	265	610(0.07)	677(1.0)	0.06	21.8	10
	**		489(1.0)	515(0.12)	566(0.69)	619(0.9)		620(1.0)	688(0.28)		22.4	10

C = compound; T = tautomer; a)* tautomer 1 with shortwave S_0-S_1 transition; ** tautomer 2 with longwave S_0-S_1 transition; b) values of fluorescence quantum yield were measured for tautomers 1, which were the predominating component in solution at 77 K; c) fluorescence of compound 1 is strongly quenched in dimer at 77 K due to the irreversible energy transfer to compound 2.

solutions of the investigated compounds are not due to porphyrin molecule aggregation.

Photoinduced reversible conversion of the centers. Because of the intrinsic asymmetry caused by side substituents, the tautomers in the investigated compounds are chemically inequivalent and can absorb at rather different frequencies (see Table 1). Therefore, it is possible to realize transformation between the two tautomers, which can be achieved by irradiating either of them in the region of the priority absorption. As can be seen from Fig. 6, the fluorescence intensity of the first tautomer decreases in time upon excitation into its absorption band by constant (polarized or nonpolarized) light. At the same time the band of the first tautomer in the fluorescence excitation spectra decreases while that of the other increases. The reverse transformation can be realized by exciting the solution into the absorption band of the second tautomer. It is essential that we could observe in such experiments not only a decrease in the fluorescence intensity of the first tautomer, but an increase in the fluorescence intensity of the other, too (Fig. 6a, curve 3). Consequently, this photoprocess is exactly phototransformation and not photodestruction. The rate of transformation under the same excitation conditions is greater by 2-8 times for long-wave tautomers than for short-wave ones, which is due to the existence of a potential barrier at photoinduced NH-proton jumping at 77 K. As can be seen from Table 1, fluorescence lifetimes τ_S and triplet state lifetimes τ_T are almost equal for both tautomers of every compound. Therefore, the comparison of phototransformation rates and lifetime data evidently shows that the efficiency of NH-photoconversion process in excited states is very low against the other deactivation processes of these states.

We have also investigated the effect of deuterium substitution of the inner protons in the molecule center (12,13). For example, the addition of deuteroethanol to THF-DE solution of compound 2 causes a significant difference between the kinetic curves and, consequently, the quantum efficiency of photoprocesses as compared with a similar case where normal ethanol is added (Fig. 6a). This fact can readily be associated with the inhibition of ND-tautomerism as against NH-tautomerism due to the increase in the mass of the jumping particles. An additional evidence of the NH-tautomerism process is the observation of only one absorbing and luminescing form in solutions of the Zn-complex of compound 2 and its diprotonated derivative and the absence of any phototransformation processes under similar conditions. Moreover, we have not observed any photoinduced transformations of the two tautomers of compound 2 in rigid films of polyvinylbutyral at 293 K. This latter result is due to the quick thermal migration of inner protons, which inhibits the photoselection of individual tautomers.

Finally, the relative intensities of S_0-S_1 transitions of different tautomers of the same compound show significant dependence on the solution temperature. This temperature effect is caused by changing equilibrium tautomer concentrations due to the existence of a potential barrier for proton migration in the ground state. Because of large energy separation ΔE between S_0-S_1

transitions of NH-tautomers for compounds 2 and 10 (see Table 1) it is possible to calculate the enthalpy of NH-tautomerism activation (ΔH_{NH}) from temperature dependence of normal absorption spectra. For example, the experimental value ΔH_{NH} is equal to 510 cm^{-1} for compound 10 (Fig. 7). It has been found that in all cases at 77 K the predominating component in solution (~90 percent) was as a rule a tautomer with shortwavelength S_0-S_1 transition (tautomer 1).

The NH-tautomer structure. The solution of the question of internal protons location in each tautomer of the investigated porphyrins with non-symmetrical substitution will result in obtaining definite correlations between tautomer spectra and their structures. Here we present such information for compound 10. This compound has no alkyl substituents at pyrrole rings and therefore, one is able to measure the constant of spin-spin interaction of pyrrole ring protons with NH-protons using NMR^1H method (J = 1.8 Hz). Based on the complete assignment of signals in NMR^1H spectrum of compound 10 and using double resonance and Overhauser nuclear effect it has been found that in tautomer 1 NH-protons are located at the opposed pyrrole rings which are not bonded with isocycle (23). The experimental data obtained enable us to conclude that this is true for most isocycle containing porphyrins. In the case of compound 12, using low temperature NMR^1H spectra it has been found that in the shortwavelength tautomer 1 NH-protons are fixed at the opposed pyrrole rings having alkyl substituents. But at the same time for compound 13, which has additional isocycle ring NH-protons in tautomer 1 are fixed at the unsubstituted pyrrole rings. Then, according to the selective excitation spectroscopy data at 4.2 K (17) NH-protons in the shortwave tautomer 1 of compount 15 are fixed on rings with methyl groups, that is consistent with Gouterman considerations (31,18). Thus, we have now direct information as to which types of absorption spectra of individual tautomers correspond to real distribution of NH-protons in these forms.

Absolute orientation of longwave absorption oscillators in NH-tautomers. It has been known (28,18) that in free-base porphyrins due to D_{2h} symmetry the electric transition dipoles in the X and Y directions are not equivalent and Q_x and Q_y absorption electronic bands are modeled by mutually perpendicular linear oscillators X and Y passing through pyrrole rings. Moreover, for both symmetrical (28,32) and non-symmetrical (18,33) porphyrins the oscillator of longwave absorption Q_x band is believed to be oriented along the H-H axis, and to rotate through 90° following the migration of inner protons. The cyclic polyene model (34) also predicts band polarizations relative to the H-H axis. The only experimental work available to us (32) presents evidence for correspondence of the longwave absorption oscillator X to the H-H axis for the case of the symmetrical crystal tetraphenylporphine. Nevertheless, up to now there has been no straightforward independent information confirming the same situation for porphyrins with non-symmetrical substitution. Besides, the theoretical consideration of this problem for porphyrins with

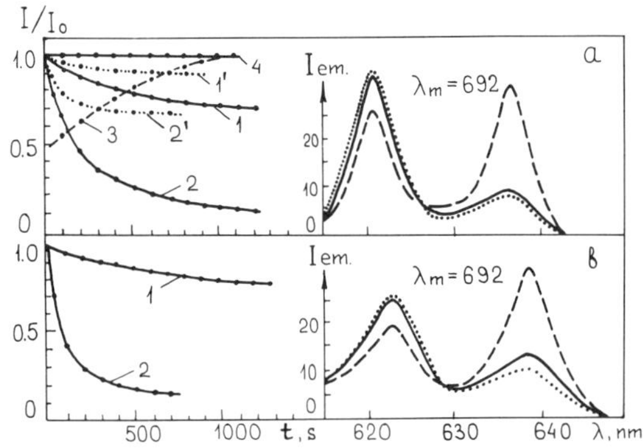

Figure 6. Changes of fluorescence intensity and fluorescence excitation spectra of compounds 2 in EPIP (a) and 6 in THF-DE (b) with irradiation time at 77 K. $\Delta\lambda_{excit}$ = 4 nm, $\Delta\lambda_{monit}$ = 4 nm. (a) 1,1' - λ_{ex} = 621 nm, λ_{mon} = 690 nm; 2,2' - λ_{ex} = 636 nm; λ_{mon} = 690 nm (1' and 2' - deuterated compound 2 in THF-DE); 3 - λ_{ex} = 622 nm, λ_{mon} = 707 nm; 4 - polyvinylbutyral film as a solvent, T = 293 K, λ_{ex} = 627 nm; λ_{mon} = 697 nm.
(b) 1 - λ_{ex} = 623 nm, λ_{mon} = 692 nm; 2 - λ_{ex} = 637 nm, λ_{mon} = 692 nm.
Excitation spectra: ——— initial; - - - - - - after irradiation into the absorption band of the tautomer 1; · · · · · after irradiation into the absorption band of the (longwavelength) tautomer 2. λ_m is the monitoring wavelength.

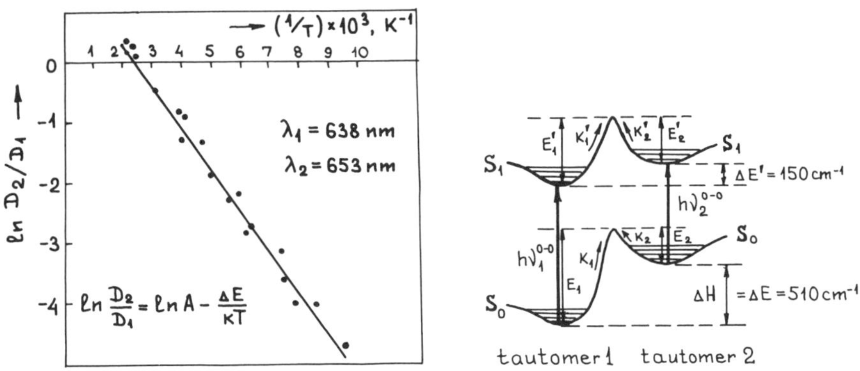

Figure 7. Temperature dependence of the Q(0,0) longwave bands optical densities of compound 10 NH-tautomers 1 and 2 (T = 77 - 500 K) and schematic diagrams of energies of the two lowest singlet (S_0, S_1) states of tautomers at 77 K according to experimental data.

external substituents (namely, 2,3,12,13-tetramethylporphyrin (18)) did not take into account the existence of two spectrally different NH-tautomers which have been initially discovered by us for such compounds (see Table 1).

In order to specify the behaviour of a system of molecular oscillators X and Y in non-symmetrical porphyrins as NH-protons move through 90°, i.e., when transition from tautomer 1 to tautomer 2 takes place, we experimentally investigated the dichroism d = (D - D)/(D + D) of photoinduced changes in absorption spectra of compound 10 at 77 K caused by polarized light irradiation of glassy solutions. It is common knowledge that at 77 K the above tautomerism may be induced only by selective irradiation of solutions into absorption bands of one of the NH-tautomers (5,12). As it can be seen from Table 1, the significant difference between S_0-S_1 transitions of two tautomers of compound 10 permits us to separately investigate changes in their longwave Q bands after irradiation. In our experiments, we always started from the initial situation when tautometer 2 concentration was extremely low relative to that of tautomer 1 (see spectrum A, Fig. 8). Then the solution being irradiated (for about 30 min) by linearly polarized light into the absorption Q_x band of tautomer 1 (638 nm, optical density D_1) we could measure the rising Q band of tautomer 2 (654 nm, optical density D_2) (see spectra B and C, Fig. 8).

When absorption spectra were recorded using a photometric beam polarized either parallel ($\parallel$) or perpendicular ($\perp$) to the excited beam the following basic differences were observed: $D_1^{\parallel} < D_1^{\perp}$ whereas $D_2^{\parallel} > D_2^{\perp}$, i.e., the dichroism of longwavelength Q bands of two tautomers is opposite in sign (d_1 = -9 percent, d_2 = +6 percent). This remarkable result, as seen from Fig. 8 seems to have only one explanation: the oscillator of longwave absorption band of tautomer 2, which appears after polarized irradiation in the Q band of tautomer 1, must be parallel to the exciting light vector E, and, correspondingly, to the oscillator of the longwavelength Q_x band of tautomer 1. Hence, under central proton displacement the H-H axis rotates through 90° whereas the longwavelength transition oscillator X remains presumably fixed at the same pyrrole rings for both tautomers. As far as for each tautomer of the same compound the electronic density distribution is not identical; therefore, it is not excluded that the mutual displacement of oscillators X and Y may be slightly different in them. But as a whole the system of oscillators X and Y does not rotate through 90° in going from tautomer 1 to tautomer 2 in the compounds considered. Such situation is consistent with MCD data. As seen from Fig. 9 there are two well distinguished bands in the red region of absorption spectra for compounds 6 and 10 even at 293 K. These bands belong to Q(0,0) transitions of individual NH-tautomers. But at the same time there is no sign inversion of MCD bands in this region; we observe only one band whose maximum is situated approximately between tautomer absorption maxima. This result shows that longwave transition oscillators in both tautomers are presumably fixed at the same position against the molecular structure of each compound. Finally, the changing of sign pattern in MCD spectra of compound 6 in comparison with that of compound 10

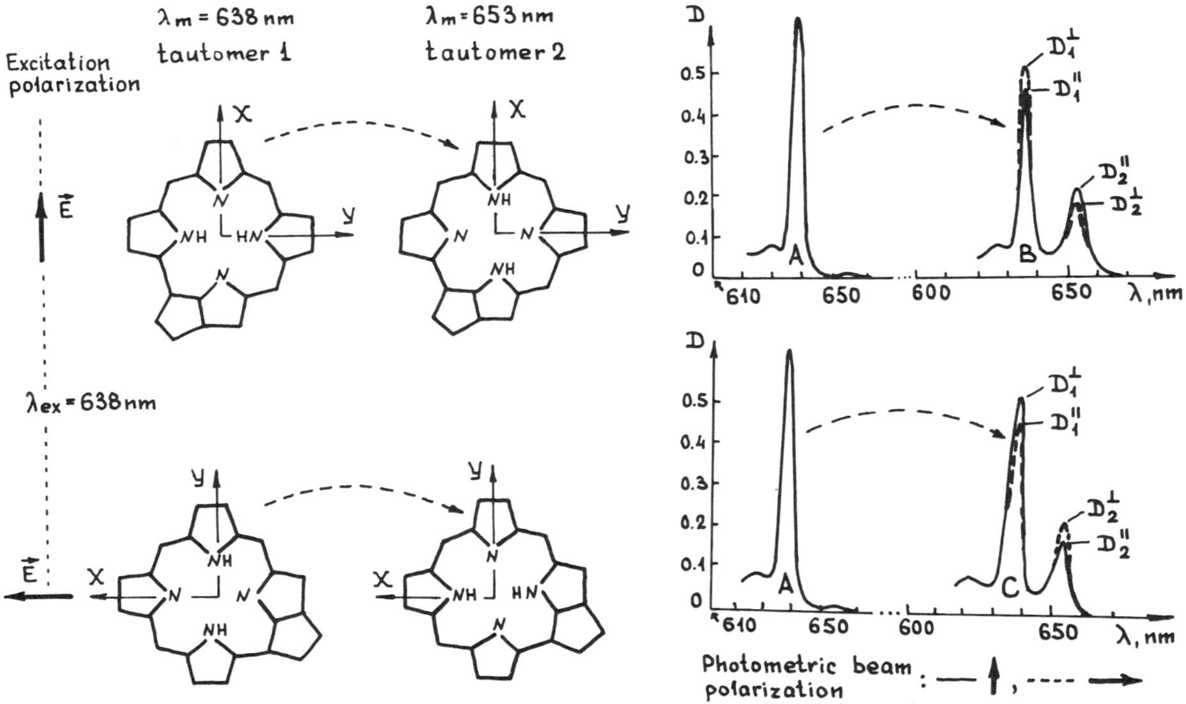

Figure 8. Absorption onset experiments for photoinduced conversion of compound 10 NH-tautomers in EPIP at 77 K: (A): Initial spectrum; (B,C): Monitored spectra after irradiation ($\Delta t \simeq 30$ min) by the polarized light (λ_{ex} = 638 nm). Conditions for polarized excitation and polarized monitoring are shown by arrows. The displacement of inner protons in tautomers conforms to experimental data. The direction of X-axis is chosen for tautomer 1 to be parallel to vector E of exciting light. (Reproduced with permission from Ref. 15. Copyright 1985 North-Holland Physics Publishing Company.)

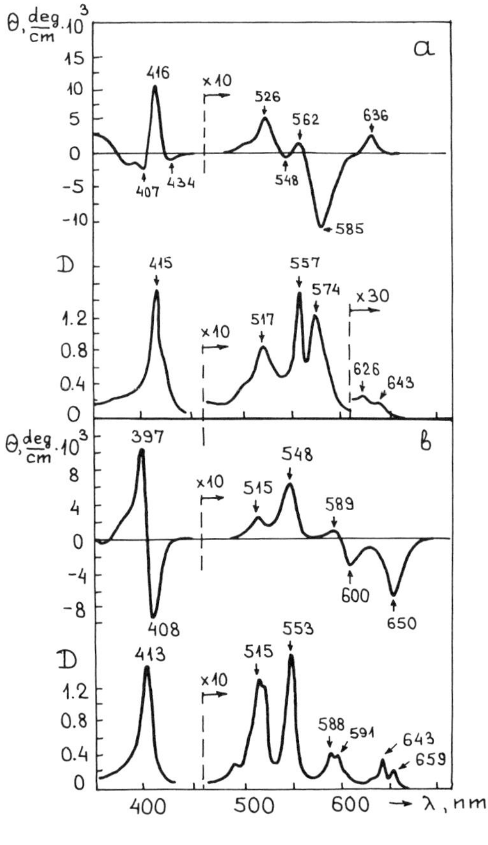

Figure 9. Absorption and magnetic circular dichroism spectra of compounds 6(a) and 10(b) in THF-DE at 293 K. Optical path is 2mm. (Reproduced with permission from Ref. 15. Copyright 1985 North-Holland Physics Publishing Company.)

may be explained as a result of the changing of the relative magnitudes of the orbital energy differences between the two highest occupied (ΔHOMO) and the two lowest unoccupied (ΔLUMO) molecular orbitals (35,36). Experiments and theoretical MO calculations in this field are performed in our laboratory.

Interpretation of NH-tautomer visible absorption spectra. The data summarized in Table 1 reveal that each of the NH-tautomers exhibits four visible bands which are generally characteristic of free-base porphyrins. But the observed intensity pattern dramatically changes when passing from one tautomer to another for each compound; while the situation $I(Q_x) < I(Q_y)$ is observed for one tautomer the spectral picture $I(Q_x) > I(Q_y)$ occurs for the second one and vice versa.

In the four-orbital model (18), low-lying π-π^* states of free-base porphyrins (symmetry D_{2h}) are considered as resulting from single electron excitation from a pair of nondegenerate highest occupied molecular orbitals (b_1, b_2) to a pair of nondegenerate lowest unoccupied molecular orbitals (c_1, c_2). In the case of symmetry D_{2h}, mutually perpendicular electric transition dipoles X and Y are not equivalent and, therefore, in the visible absorption spectra of free-base porphyrins two different electronic bands $Q_x(0,0)$ and $Q_y(0,0)$ are observed (Table 1 and Fig. 10). The question of relative intensities of Q_x and Q_y transitions which is a matter of focus in the present part of the paper has been solved by Gouterman (31) using perturbation theory. If it is assumed that states of different polarization (Q_x and Q_y) are not mixed, the Q states of an arbitrary free-base porphyrin are constructed as follows:

$$Q_x = Q_x + \lambda_x B_x, \quad \lambda_x = [\varepsilon(C_2) - \varepsilon(b_1)] - [\varepsilon(C_1) - \varepsilon(b_2)]/2\Delta$$
$$Q_y = Q_y + \lambda_y B_y, \quad \lambda_y = [\varepsilon(C_1) - \varepsilon(b_1)] - [\varepsilon(C_2) - \varepsilon(b_2)]/2\Delta \qquad (1)$$

where Δ is an initial energy gap between B^o and Q^o bands, $\varepsilon(b_i)$ and $\varepsilon(C_j)$ are orbital energies. The intensities of Q_x and Q_y states are proportional to λ_x^2 and λ_y^2. Hence, the amounts of absorption intensity gained in the initially forbidden Q_x^0 and Q_y^0 transitions of arbitrary free-base porphyriun by borrowing from appropriate B_x^0 and B_y^0 Soret states (by the so-called configuration interaction) are related to differences in orbital energies of states created by x- or y-polarized singlet electron excitation. That is, for each polarization, as the equality of the transition energies breaks down (or configuration interaction becomes weaker), the corresponding visible band becomes allowed.

In order to consider the inversion of $Q_x(0,0)$ and $Q_y(0,0)$ electronic transition intensities in NH-tautomers of non-symmetrical free-base porphyrins we calculated the ground-state orbital energies of the investigated molecules by a CNDO/2 method using the symmetrized crystal geometry of porphyrin molecule (37,38). On the basis of the above experimental results we must introduce a motionless system of molecular X and Y axes, identically fixed in both tautomers. Then using theoretical MO calculations and the analysis

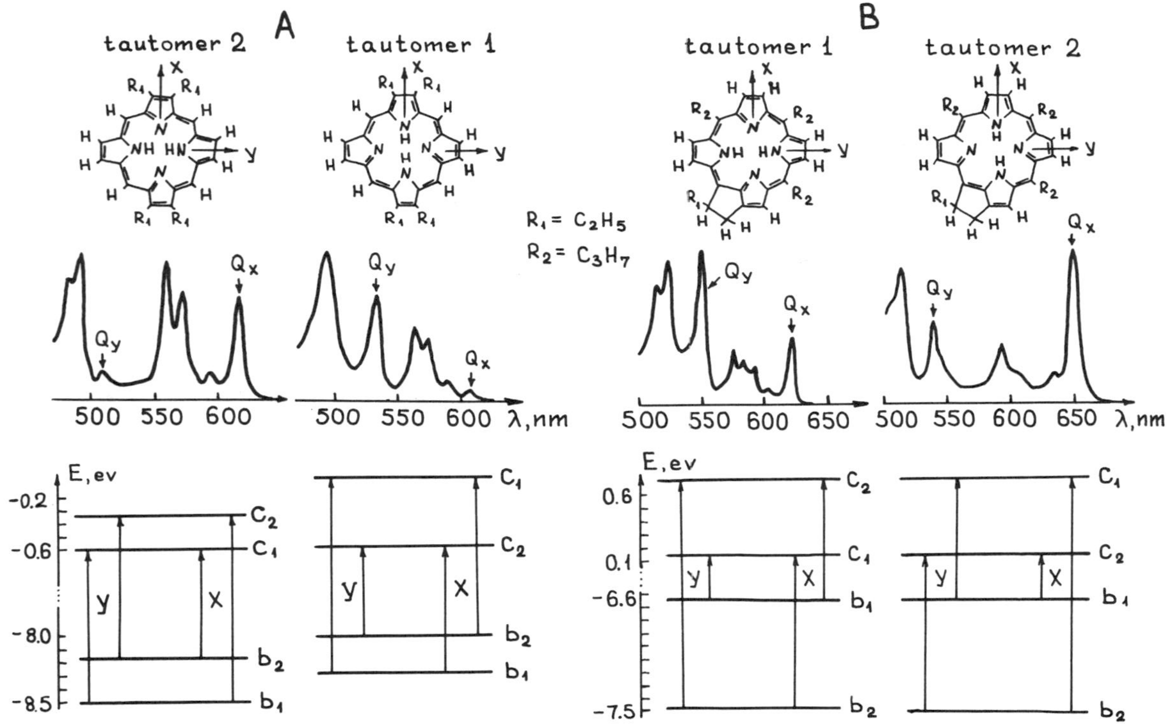

Figure 10: Structures, visible absorption spectra and four-orbital diagram of MO's for NH-tautomers of porphyrins with the non-symmetrical substitution. (A): compound 12; (B): compound 10. (Reproduced with permission from Ref. 15. Copyright 1985 North-Holland Physics Publishing Company.)

of their symmetry properties the ordering of orbitals b_1, b_2, c_1 and c_2 was determined. Figure 10 illustrates that, similarly to the case of symmetrical porphyrins (18), the ordering of c_1 and c_2 orbitals changes as NH-protons displace. It is essential that in symmetrical porphyrins the X axis would always keep the H-H axis direction (18,33). Therefore, as the H-H axis rotates through 90° (going from one tautomer to another) the orbital combinations (b_ic_j) determining X- and Y-polarized excitations must be changed. For example, this change for X-polarized excitations must be presented as follows: (b_1c_2), (b_2c_1) $\rightarrow$ (b_1c_1) (b_2c_2). As a result, the configuration interaction energies of X- and Y-polarized excitation do not change in going from one tautomer to another and consequently, their absorption spectra are indistinguishable from one another. However, for non-symmetrical porphyrins due to fixed X and Y axes the orbital combinations (b_ic_j) corresponding to X- and Y-polarized excitations remain unchanged under the NH-proton migration. Then, as seen from Fig. 10, the differences in orbital energies of the states reached by X and Y excitation really differ for each polarization when tautomer 1 $\rightarrow$ tautomer 2 transition occurs. Such variation of the configuration interaction energies for each polarization must result in principal changes in relative intensities of $Q_x(0,0)$ and $Q_y(0,0)$ absorption bands in two tautomers of the same molecule. Actually such a situation is experimentally observed for compounds 12 and 15 (Fig. 5 and Table 1): etio-type spectrum of tautomer 1 transforms into deoxophyllotype spectrum of tautomer 2. And our results, in fact, are in agreement with theoretical considerations of Gouterman on parallelism of H-H and methyl axes (18), but that is the case only for tautomer 1. Moreover, the lowest Q band of the compound 15 absorption spectrum (18) actually belongs to tautomer 2, as is seen from Table 1. Besides, at 293°C compounds 12 and 15 show just an etiotype spectrum in contrast to a rhodotype spectrum, used in paper (18). Our experimental data (NMR, mass-spectrometry and absorption spectra) permit us to confirm that a rhodo spectrum belongs to compound 14 which is the accompanying product during the synthesis of compounds 12 and 15. Table 1 shows that the Br-derivative has also two tautomers with different spectra.

These types of arguments have been used by us for explanation of characteristic peculiarities in spectra of compound 10 NH-tautomers (Fig. 10) and were applied to understand the principal changes in the NH-tautomer visible absorption of investigated porphyrins with non-symmetrical substitution. It may be concluded that difficulties in the theoretical treatment of spectral peculiarities of porphyrins with π-acceptor substituents (33) are caused by the absence of detailed absorption and fluorescence excitation spectra of individual tautomers of these compounds.

Conclusions

We believe that detailed analysis of absorption, fluorescence and fluorescence excitation spectra permits us to investigate the main spectroscopic features of individual NH-tautomers in porphyrins with non-symmetrical substitution. Then, obtained information together with NMR^{1}H data and selective excitation spectroscopy results at 4.2 K provide the possibility to establish which types of absorption spectra of individual tautomers correspond to real distribution of NH-protons in every compound. MO calculations based on the four-orbital model and carried out with account of the absolute orientation of longwave band oscillators in NH-tautomers explain the unexpected types of transformation of visible absorption spectra under NH-proton migration.

Literature Cited

1. Storm, C.B.; Teklu, Y. J. Am. Chem. Soc., 1972, 94, 1745-7.
2. Eaton, S.S.; Eaton, G.R. J. Am. Chem. Soc., 1977, 99, 1601-4.
3. Hennig, J.; Limbach, H.-H. J. Chem. Soc. Faraday Trans., 1979, 2, 752-66.
4. Abraham, R.J.; Hawkes, G.E.; Smith, K.M. Tetrahedron Lett., 1974, 1483-6.
5. Zalesski, E.I.; Kotlo, V.N.; Sevchenko, A.N.; Solov'ev, K.N.; Shkirman, S.F. Dokl. Akad. Nauk SSSR, 1972, 207, 1314-7.
6. Korotaev, O.N.; Personov, R.I. Opt. i Spektroskopiya, 1972, 32, 400-2.
7. Solov'ev, K.N.; Zalesski, I.E.; Kotlo, V.N.; Shkirman, S.F. Pis'ma v ZhoTPh, 1973, 17, 463-6.
8. Shkirman, S.F.; Arabey, S.M.; Egorova, G.D. Zh. Prikl. Spektroskopii, 1979, 31, 817-21.
9. Volker, S.; Macfarlane, R.M. Liquid Cryst., 1979, 50, 213-6.
10. Arabey, S.M.; Egorova, G.D.; Solov'ev, K.N.; Shkirman, S.F. Zh. Prikl. Spektroskopii, 1984, 40, 92-8.
11. Mauring, K.; Avarmaa, R. Chem. Phys. Lett., 1981, 81, 446-9.
12. Zenkevich, E.I.; Shulga, A.M.; Chernook, A.V.; Gurinovich, G.P. Chem. Phys. Lett., 1984, 109, 306-11.
13. Zenkevich, E.I.; Shulga, A.M.; Chernook, A.V.; Gurinovich, G.P. Dokl. Akad. Nauk SSSR, 1984, 28, 1082-5.
14. Zenkevich, E.I.; Shulga, A.M.; Chernook, A.V.; Gurinovich, G.P.; Sagun, E.I. Zh. Prikl. Spektroskopii, 1985, 42, 772-9.
15. Zenkevich, E.I.; Shulga, A.M.; Filatov, I.V.; Chernook, A.V.; Gurinovich, G.P. Chem. Phys. Lett., 1985, 120, 63-8.
16. Zenkevich, E.I.; Shulga, A.M.; Filatov, I.V.; Chernook, A.V.; Gurinovich, G.P. Dokl. Akad. Nauk SSSR, 1985, 285, No. 1.
17. Shulga, A.M.; Gladkov, L.1.; Stanishevski, I.V.; Starukhin, A.S. Teoreticheskaya i Experimentalnaya Khimiya, 1985, 5, 554-60.
18. Gouterman, M. J. Mol. Spectrosc., 1961, 6, 138-63.
19. Ponomarev, G.V.; Shulga, A.M. Khimija Heterocycl. soedinenii, 1984, 479-84.
20. Ponomarev, G.V.; Shulga, A.M. Khimija Heterocycl. soedinenii, 1984, 485-9.
21. Shulga, A.M.; Ponomarev, G.V. Khimija Heterocycl. soedinenii, 1984, 922-7.
22. Ponomarev, G.V.; Shulga, A.M. Dokl. Akad. Nauk SSSR, 1983, 271, 365-7.
23. Shulga, A.M. Khimija Heterocycl. soedinenii, 1985, 132-3.

24. Kenner, G.W.; McCombie, S.W.; Smith, K.M. J. Chem. Soc. Perkin Trans., 1973, 1, 2517-23.
25. Kenner, G.W.; McCombie, S.W.; Smith, K.M. J. Chem. Soc. Perkin Trans., 1974, 1, 527-30.
26. Housier, C.; Sauer, S. Biophys. Biochem. Acta, 1969, 172, 492-502.
27. Gurinovich, G.P.; Zenkevich, E.I., Sagun, E.I. Dokl. Akad. Nauk SSSR, 1981, 257, 70-4.
28. Gurinovich, G.P.; Sevchenko, A.N.; Solov'ev, K.N. "Spektroskopiya khlorofilla i rodstvennykh soedinenii", Nauka i tekhnika, Minsk, 1968.
29. Gouterman, M. In "The Porphyrins"; Dolphin, D., Ed.; New York, 1978, III, 1-165.
30. Zenkevich, E.I.; Shulga, A.M.; Sagun, E.I.; Chernook, A.V.; Gurinovich, G.P. Zh. Prikl. Spektroskopii, 1985, 43, 455-61.
31. Gouterman, M. J. Chem. Phys., 1959, 30, 1139-61.
32. Anex, B-G.; Umans, R.S. J. Am. Chem. Soc., 1964, 86, 5026-7.
33. Djerassi, C.; Lu, Y.; Waleh, A. et al. J. Am. Chem. Soc., 1984, 106, 4241-58.
34. Simpson, W.T. J. Chem. Phys., 1949, 17, 1218.
35. Michl, J. J. Am. Chem. Soc., 1978, 100, 6812-8.
36. Keegan, J.D.; Stolzenberg, A.M.; Lu, Y.-C.; Linder, R.E.; Barth, G.; Moscowitz, A.; Bunnenberg, E.; Djerassi, C. J. Am. Chem. Soc., 1982, 104, 4305-16.
37. Caughey, W.S.; Ibers, J.A. J. Am. Chem. Soc., 1977, 99, 6639-43.
38. Fischer, M.S.; Templeton, D.H.; Zalkin, A.; Calvin, M. J. Am. Chem. Soc., 1972, 94, 3613-9.

RECEIVED June 10, 1986

6

Optical Detection of the Lanthanoid Ion Contraction by Internal Charge-Transfer Absorption of Rare Earth Bisporphyrinate Double-deckers

J. W. Buchler, K. Elsässer, M. Kihn-Botulinski, B. Scharbert, and S. Tansil

Institut für Anorganische Chemie, Technische Hochschule Darmstadt, D–6100 Darmstadt, Federal Republic of Germany

The optical spectra of a series of lanthanoid bis-(octaethylporphyrinates), $Ln(OEP)_2$ (Ln = La...Nd, Sm...Lu) are reported. All compounds have a doubledecker structure. Apart from the Ce^{IV} derivative, all the other representatives contain Ln^{III} ions and a defect electron on one of the porphyrin rings and have the composition $Ln^{3+}(OEP)^{2-}(OEP^{\bullet})^{-}$. These "odd" species all have near infrared (NIR) internal charge transfer absorption bands between 1480 (La) and 1135 nm (Lu). The wavenumber of this band is inversely proportional to the radius of the rare-earth ion for the coordination number 8. The neutral $Ce(OEP)_2$ can be oxidized to a radical cation which also has a NIR band at 1270 nm.

A complete series of lanthanoid bisoctaethylporphyrinates $Ln(OEP)_2$ (Ln = La...Lu, with the exception of Pm) has been synthesized, together with a series of bislanthanoid trisporphyrinates $Ln_2(OEP)_3$ (Ln = La...Gd) (2,3,4). (Abbreviations used: $(OEP)^{2-}$, $(TPP)^{2-}$, $(TTP)^{2-}$, $(Pc)^{2-}$ and $(P)^{2-}$ are the dianions of 2,3,7,8,12,13,17,18-octaethylporphyrin, 5,10,15,20- tetraphenylporphyrin, 5,10,15,20-tetra(p-tolyl)porphyrin, phthalocyanine, and a general porphyrin, respectively. L, Ln, M are general ligands, lanthanoids, metals.)

The complexes $Ln_2(OEP)_3$ all contain Ln^{III} ions. Apart from the diamagnetic Ce^{IV} complex $Ce(OEP)_2$ (2,3), the complexes $Ln(OEP)_2$ are

NOTE: This is number 42 in the series Metal Complexes with Tetrapyrrole Ligands.

0097–6156/86/0321–0094$06.00/0

described as Ln^{III} complexes of a dinegative porphyrinate and a mononegative porphyrin radical with a hole in the π-electron system (2). The crystal and molecular structures of $Ce(OEP)_2$ and $Ce_2(OEP)_3$, the parent species of the two series, have been determined and prove the doubledecker and tripledecker shapes of the two series, respectively (3). Essential structural parameters in the doubledecker $Ce(OEP)_2$ (A) or in the tripledecker $Ce_2(OEP)_3$ (B) are shown in Figure 1, which gives the previously used bar graphs (5,6) to represent the configuration. Above the bars, the distances of the mean planes of the 20 C and 4 N core atoms of the individual porphyrinato ligands are quoted, at the right of the bars the mean Ce-N bond lengths are listed. The mean value of these bond lengths is 247.5 pm in A and 263 pm in B, the longer distance in the latter corresponding to the fact that B is a Ce^{III} complex and A a Ce^{IV} complex. The Ce-N distances in B differ for the two external rings on one hand and for the internal one on the other; the internal ring has to hold two cerium ions and therefore is connected through somewhat longer bonds.

Optical Absorption Spectra and Electronic Structure

The optical spectra of all the doubledeckers are listed in Table I. On first glance, $Ce(OEP)_2$ has a "normal" spectrum (7). However, the spectrum shows extra bands and therefore should be called "hyper". A small band appears at 467 nm (maybe a ligand-to-metal charge transfer band), and broad features extend far into the near infrared (NIR). The latter absorption may be due to exciton interactions. Contrary to the known rare earth monoporphyrins (7), it has been shown for the closely related cerium(IV) bis(tetra-p-tolylporphyrinate), $Ce(TTP)_2$ (8) that the doubledeckers do not emit, and their excitation is dissipated very rapidly (9).

The other doubledeckers containing the metals La, Pr, Nd, Sm - Lu have rather similar spectra that have less well distinct bands in the visible region. The extinction at 573 nm is much less than in A. Peaks occur at about 540 and 670 nm. The latter absorption is typical for π-cation radicals (10). Indeed, the composition $Ln(OEP)_2$ can only be accomodated with the normal trivalent state of nearly all of the lanthanoid ions if one assumes that the

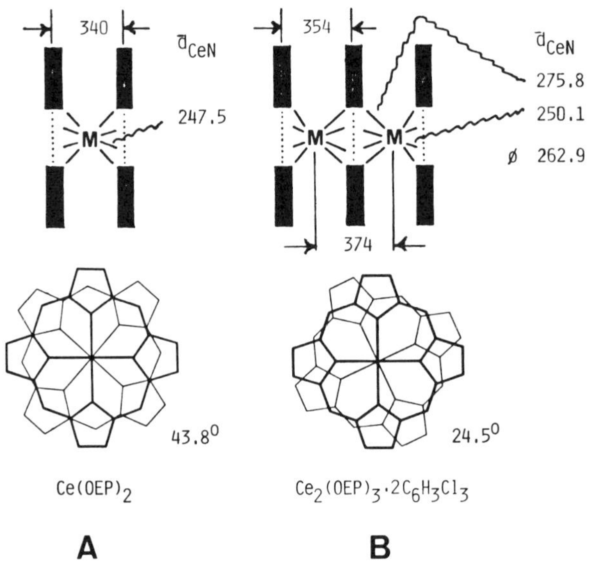

Figure 1. Synopsis of the structural features of $Ce(OEP)_2$ (A) and $Ce_2(OEP)_3$ (B). Essential distances are given in pm. Below the bar graphs of the doubledecker A and the tripledecker B, the relative orientations of the porphyrin rings in A and B are shown.

Table I. Electronic Absorption Spectra of Lanthanoid Bis(octaethylporphyrinates), $Ln(OEP)_2$, in cyclohexane (4)

[λ_{max} in nm, logε in brackets, Hewlett-Packard spectrophotometer HP 8451 A]

Complex	NIR band	UV/VIS bands				
$La(OEP)_2$	1480 (2.75)		576 (3.70)	540 (3.85)	494 (3.75)	394 (5.15)
$Ce(OEP)_2$	---- ----	661 (3.16)	573 (4.24)	530 (3.77)	467 (3.92)	378 (5.08)
$Pr(OEP)_2$	1400 (3.50)	670 (3.21)	576 (3.64)	540 (3.80)	490 (3.72)	391 (5.18)
$Nd(OEP)_2$	1360 (3.28)	668 (3.08)	577 (3.48)	542 (3.97)	488 (3.53)	390 (4.81)
$Sm(OEP)_2$	1300 (3.20)	674 (3.21)	578 (3.78)	544 (3.86)	489 (3.79)	390 (5.11)
$Eu(OEP)_2$	1280 (3.89)	676 (3.36)		543 (3.87)		376 (5.04)
$Gd(OEP)_2$	1250 (3.74)	674 (3.26)		542 (3.71)		380 (4.88)
$Tb(OEP)_2$	1237 (3.49)	680 (3.30)	580 (3.70)	544 (3.81)	492 (3.80)	378 (5.00)
$Dy(OEP)_2$	1205 (3.46)	679 (3.27)	582 (3.77)	546 (3.84)	494 (3.82)	386 (5.09)
$Ho(OEP)_2$	1182 (3.65)	684 (3.24)	582 (3.83)	546 (3.91)	494 (3.89)	376 (5.10)
$Er(OEP)_2$	1175 (3.68)	679 (3.24)	582 (3.67)	544 (3.79)	494 (3.78)	374 (5.06)
$Tm(OEP)_2$	1152 (3.24)	672 (3.31)	584 (3.78)	546 (3.76)	494 (3.77)	388 (5.08)
$Yb(OEP)_2$	1144 (3.83)	672 (3.21)		530 (3.76)	498 (3.75)	370 (4.98)
$Lu(OEP)_2$	1135 (3.85)	674 (3.36)		534 (3.77)	490 (3.76)	370 (5.02)

porphyrinato ligands occur in different oxidation states, i. e. a normal $(OEP)^{2-}$ dianion and a porphyrin radical monoanion, $(OEP^{\bullet})^{-}$ (2). The composition may then be described as $Ln^{+3}(OEP)^{-2}(OEP^{\bullet})^{-}$. This is easily seen from the paramagnetism of $Lu(OEP)_2$ which has the magnetic moment of a radical, i. e. 1.7 B. M. in the solid state (4). As lutetium ions other than diamagnetic f^{14} Lu^{III} are not known, only a porphyrin radical state is left to explain the properties. A similar radical state has been found for another doubledecker molecule, lutetium bis(phthalocyaninate), $Lu(Pc)_2$, which is hence to be described as $Lu^{3+}(Pc)^{-2}(Pc^{\bullet})^{-}$ (11,12). All the Ln^{III} doubledeckers furthermore show characteristic IR bands at about 1550 cm^{-1} in the IR spectra (2,4), bands that have been shown to be typical for metalloporphyrins with electron-deficient porphyrin ligands (13).

Apart from $Ce(OEP)_2$, all the other doubledeckers $Ln(OEP)_2$ each have an extra band of a molar extinction coefficient of about 10^3 in the near IR between 1200 and 1500 nm (see Table I and Figure 2). This band is typical for those bisporphyrins that are electron-deficient in one of the porphyrin rings. It is absent in $Ce(TTP)_2$, $PrH(TTP)_2$, $NdH(TTP)_2$, and the anions obtained by deprotonation of the latter two species (8,14) as well as in all the tripledeckers. We tentatively explain it as an internal charge transfer band: aromatic donor molecules form coplanar electron donor acceptor (EDA) complexes with aromatic acceptors; these EDA complexes have typical additional bands at longer wavelengths than the individual separated components (15). The geometry fixed by the sandwiched lanthanoid builds up a very efficient EDA complex from the donor, $(OEP)^{2-}$, and the acceptor, $(OEP^{\bullet})^{-}$, to which system then belongs the NIR absorption with the rather high extinction coefficient. (A factor decreasing the overlap within the EDA complex is certainly the staggered configuration of the porphyrin ligand as taken from Figure 1 (3).) The NIR band is therefore called an "internal charge transfer" (CTI) band.

The CTI band has nothing to do with the varying f-electron configuration of the rare earth ions. If this would be the case, the energy of this band should vary characteristically like the f-electron spectra or the magnetic moments of the rare earth ions. On the contrary, a quite systematic behavior with the lanthanoid ion contraction is observed. A plot of the energy of CTI band versus the ionic radius at coordination number 8 of the Ln^{III} ions (16) indicates a linear decrease of this energy as the ionic radius increases. This is seen in Figure 3, and illustrates a bare size effect of the energy variation. The phenomenon is called "optical detection of the lanthanoid ion contraction by internal charge transfer absorption".

The complexes $(P)^{2-}$-Ln-$(P^{\bullet})^{-}$ could be regarded as "inverse" mixed-valence complexes (17). The normal mixed-valence complexes, e.g. M^{2+}-L-M^{3+}, consist of two metals in different oxidation states connected by a certain bridging ligand. In the inverse situation, e.g. L^{2-}-M-L^{1-}, two ligands in different oxidation states are held together by a bridging metal. Mixed-valence complexes have characteristic long-wave absorption bands in the visible spectra. It is tempting to classify our systems according to Hush's theory of mixed-valence complexes. They would belong to Class II (optical spectra of constituent ions with minor modifications and additional NIR bands). However, the observed distance dependence of the NIR bands is opposite to what would be expected by Hush's theory. The NIR bands show a solvent dependence; a hypsochromic shift is observed as the polarity of the solvent increases. This means that the complex cannot have a totally symmetrical electron distribution, and hence, the electronic "hole" cannot be completely delocalized in the molecule. Possibly the hole is localized on the NIR timescale on one of the rings, and NIR excitation moves the hole somewhat more towards the other ring, but not completely onto it. Then the central metal ion would repel the moving charge the more the smaller the metal-porphyrin distances become, in accord with the observed situation.

Another explanation for the effect of the lanthanoid ion contraction on the energy of the NIR-CTI band may be found in the extent of doming of the porphinato ligands in the $Ln(P)_2$ structures. A comparison of the inclination angles of the pyrrole units in A and

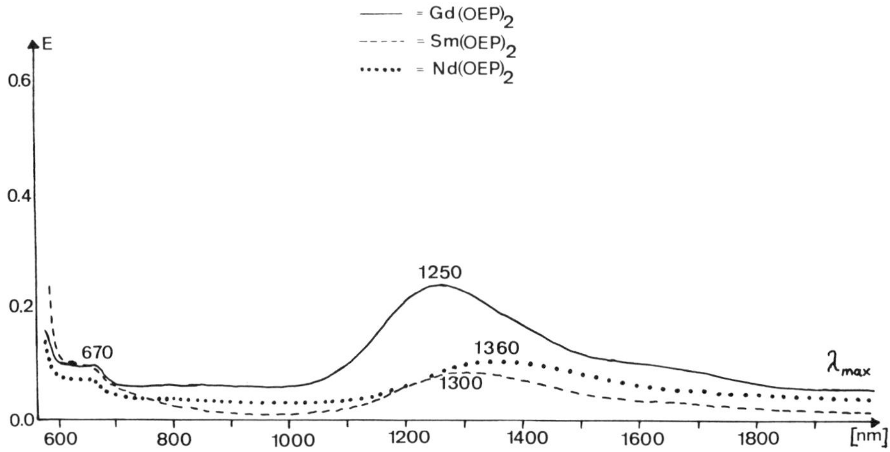

Figure 2. Near infrared spectra of $Nd(OEP)_2$ (········), $Sm(OEP)_2$ (--------), and $Gd(OEP)_2$ (————) in cyclohexane (see Table I).

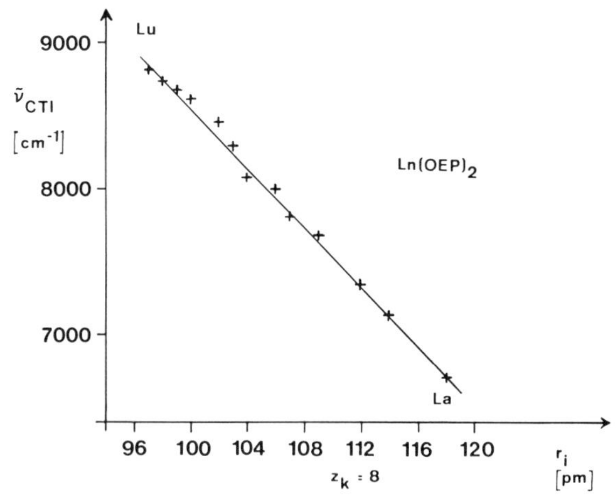

Figure 3. Plot of the wavenumber (cm^{-1}) of the NIR bands of $Ln(OEP)_2$ (Ln = Lu.....La) as a function of the ionic radius (pm) of the Ln^{III} ions at the coordination number 8 ("Optical detection of the lanthanoid ion contraction" by the CTI bands).

B (see Figure 1) with respect to the mean plane of the four porphinato nitrogen atoms leads to the conclusion that these angles increase with decreasing ionic radius of the lanthanoid ion and increasing strength of the Ln-(P) bonds (3,4). In a given series of $Ln(P)_2$ complexes, the most pronounced doming of the porphinato core should occur in the $Lu(P)_2$ system which, on the other hand, will have the smallest Ln-N bond distances. A stronger outwards doming of the porphinato ligands in a $Ln(P)_2$ doubledecker could then reduce the inter-ring charge transfer interactions despite a decrease in the distance of the mean planes of the porphinato core atoms and hence cause a hypsochromic shift to the CTI bands. This idea has to be tested by further crystallographic studies in the $Ln(P)_2$ series.

Analogies of Rare Earth Porphyrin Doubledeckers with the Special Pair of Bacteriochlorophylls in Bacterial Photosynthesis

The photosynthetic apparatus of purple bacteria, e. g. of Rhodopseumonas viridis, contains the so-called "special pair" of bacteriochlorophyll-b molecules, $(BChl\text{-}b)_2$, which consists of contiguous tetrapyrrole disks overlapping at the rings I of the bacteriochlorophyll b moieties which lie about 300 pm apart from each other (18). The photoionization of the special pair according Equation 1 is thought to be the origin of the charge separation driving the electron flow in bacterial photosynthesis (18, 19).

$$(BChl\text{-}b)_2 \quad + \quad h\nu \quad \longrightarrow \quad (BChl\text{-}b)_2^{+} \quad + \quad e^{-} \qquad (1)$$

A similar NIR band occurs at about 1300 nm in the oxidized form of the special pair, $(BChl\text{-}b)_2^{+}$, which can be generated either by illumination or chemical oxidation (19).

The similarities in NIR absorption between the series of lanthanoid doubledeckers and the oxidized special pairs suggested experiments which might produce oxidized cationic species from the neutral $Ce(P)_2$ sandwiches. Thus it would be possible to prove the idea that these NIR bands were typical for whatever kind of oxidized pair of porphyrin rings. Indeed, oxidation of the Ce^{IV} sandwiches according Equation 2 produces π-cation radical salts like $Ce(OEP)_2^+$ $SbCl_6^-$ (1,20), which also show a NIR-CTI band, as does $Pr(TTP)_2$, the oxidation product of $PrH(TTP)_2$. In fact, these species may have more than one NIR band; the spectra of the neutral sandwiches also show shoulders at the low energy tails of the main NIR band (see Figure 2). Examples and optical spectra are shown in Table II. The species have been characterized by elemental analyses, conductivity measurements, cyclic voltammetry, IR-, and NMR spectra.

$$Ce(P)_2 \longrightarrow Ce(P)_2^+ + e^- \qquad (2)$$

The NIR absorptions between 1100 and 1500 nm therefore are a common property of two closely associated porphyrin or hydroporphyrin rings one of which is oxidized to a radical. Especially noteworthy is the parallelism between the redox pairs defined in Equations 1 and 2 which is presently further investigated in our laboratories (1).

Table II. Electronic Absorption Spectra of A) Cerium Bisporphyrinate Salts (1), B) Praseodymium Bis(tetra-p-tolylporphyrinate) Sandwich Species (14) (λ_{max} in nm, log ε in brackets, Hewlett-Packard spectrophotometer 8451 A; NIR: Zeiss photometer DMR 21)

Compound	NIR band	UV/VIS bands
A) Salt		
$Ce(OEP)_2^+$ $SbCl_6^-$	1270(3.50)[a]	515(3.87), 354(4.94)
$Ce(TPP)_2^+$ $SbCl_6^-$	1350(3.93)	662(3.53), 384(5.04)
$Ce(TTP)_2^+$ PF_6^-	1350(3.42)	542(4.08), 396(5.27)
B) Complex		
$PrH(TTP)_2$ (8)	----	556(4.03), 510(3.94), 413(5.60)
$Pr(TTP)_2$ (20)	1140(3.57)[b]	672(3.69), 412(4.95)

a) Other bands: 1762(3.14), 808(4.50).-

b) Other band: 1450(3.36).-

Acknowledgments

Financial support of the Deutsche Forschungsgemeinschaft, the Fonds der Chemischen Industrie, and the Vereinigung von Freunden der Technischen Hochschule Darmstadt is gratefully acknowledged. The authors thank Professors A. Vogler (Regensburg) and M. Gouterman (Seattle) for illuminating discussions, Professors R. Weiss and J. Fischer (Strasbourg) for the excellent cooperation on the crystal structure determinations, and a reviewer for a useful comment.

Literature Cited

1. Buchler, J. W., Elsässer, K., Kihn-Botulinski, M., Scharbert, B. Angew. Chem. 1986, in press. Number 41 in the series Metal Complexes with Tetrapyrrole Ligands.
2. Buchler, J. W., Kihn-Botulinski (formerly Knoff), M. In "Optical Properties and Structure of Tetrapyrroles", G. Blauer, H. Sund, Eds., de Gruyter, Berlin (1985), pp. 91-105.
3. Buchler, J. W., De Cian, A., Fischer, J., Kihn-Botulinski, M., Paulus, H., Weiss, R. J. Am. Chem. Soc. 1986, in press.
4. Kihn-Botulinski, M. Doctoral Dissertation, Technische Hochschule Darmstadt, 1986.
5. Buchler, J. W. In "Porphyrins and Metalloporphyrins", Smith, K. M., Ed.; Elsevier: Amsterdam, 1975; pp. 157-231.
6. Buchler, J. W. In "The Porphyrins", Dolphin, D., Ed.; Academic: New York, 1978; Vol. I, pp.389-483.
7. Gouterman, M. In "The Porphyrins", D. Dolphin, Ed.; Academic: New York 1978; Vol. III, pp. 1-165.
8. Buchler, J. W., Kapellmann, H.-G., Kihn-Botulinski (formerly Knoff), M., Lay, K. L., Pfeifer, S. Z. Naturforsch. 1983, 38b, 1339-1345.
9. Holten, D., and Gouterman, M., personal communication (see also contribution to this volume).
10. Fuhrhop, J. H., Wasser, P., Riesner, D., Mauzerall, D. J. Am. Chem. Soc. 1972, 94, 7996-8001.
11. Chang, A. T., Marchon, J. C. Inorg. Chim. Acta 1981, 53, L 241-L243.
12. De Cian, A., Moussavi, M., Fischer, J., Weiss, R. Inorg. Chem. 1985, 24, 3162-3167.
13. Shimomura, E. T., Philippi, M. A., Goff, H. M., Scholz, W. F., Reed, C. A. J. Am. Chem. Soc. 1981, 103, 6778-6780.
14. Tansil, S. Diplomarbeit, Technische Hochschule Darmstadt, 1985.
15. Briegleb, G. "Elektronen-Donator-Acceptor-Komplexe". Springer: Berlin 1961.
16. Shannon, R. D., Prewitt, C. T. Acta Crystallogr. 1969, B25, 925-929.
17. Lever, A. B. P. "Inorganic Electronic Spectroscopy" (second edition). Elsevier: Amsterdam 1984, pp. 647-658.

18. Deisenhofer, J., Epp, O., Miki, K., Huber, R., Michel, H. J. Mol. Biol. 1984, 180, 385-398.
19. Davis, M. S., Forman, A., Hanson, L. K., Thornber, J. P., Fajer, J. J. Phys. Chem. 1979, 83, 3325-3332.
20. Elsässer, K. Diplomarbeit, Technische Hochschule Darmstadt, 1985.

RECEIVED April 3, 1986

7

The S_2 Emission of Metalloporphyrins

Hiroshi Kobayashi and Youkoh Kaizu

Department of Chemistry, Tokyo Institute of Technology, O-okayama, Meguro-ku, Tokyo 152, Japan

Upper excited-state emission was observed for a variety of diamagnetic metal porphyrins. Characteristic Q and B (Soret) bands are assigned as the S_1 and S_2 excited states, which are described as a 50-50 admixture of two common $^1(\pi,\pi^*)$ excited configurations. Almost parallel surfaces of the S_1 and S_2 excited states retard $S_2 \rightsquigarrow S_1$ relaxation and a greater transition moment granted to the $S_2 \rightarrow S_0$ transition yields emission from the Soret excited singlet.

Paramagnetic rare-earth porphyrins, TPPErOH (TPP: 5,10,15,20-tetraphenylporphin) and TPPEr(dpm) (dpm: 2,2,6,6-tetramethyl-3,5-heptanedionate), emit not only S_1 but also S_2 fluorescence even if a sequence of the sublevels of (f,f*) excited states are present below the S_2 state. Upon irradiation in the ultraviolet band around 300 nm, TPPEr(dpm) emits both from porphyrin S_2 and coordinated β-diketone T_1 excited states.

Exciton coupling in μ-oxo dimer $[TPPNb]_2O_3$ shifts the Soret band to the blue in comparison with monomeric $TPPNbO(CH_3COO)$. S_2 fluorescence was observed for $TPPNbO(CH_3COO)$ while it was not detected for $[TPPNb]_2O_3$. In the dimer, the exciton coupling results in a fast relaxation to the S_1 manifold.

Diacids emit S_2 fluorescence, while free bases do only weakly. However, S_2 emission of a covalently-linked free-base dimer is detected with a benzene solution containing a low concentration of CF_3COOH, in which $PH_2 \sim PH_2$, $PH_2 \sim PH_4^{2+}$ and $PH_4^{2+} \sim PH_4^{2+}$ (PH_2: free base; PH_4^{2+}: diacid) coexist. The excitation spectrum of the S_2 emission shows peaks not only in the Soret

0097-6156/86/0321-0105$06.00/0

band of $\sim PH_4^{2+}$ but also in that of $\sim PH_2$. An intramolecular energy transfer from the higher S_2 state of $\sim PH_2$ in $PH_2\sim PH_4^{2+}$ to the lower S_2 state of $\sim PH_4^{2+}$ yields S_2 emission of the $\sim PH_4^{2+}$.

Emission from the upper electronic excited states of polyatomic molecules, in violation of Kasha's rule which allows emission only from the lowest excited states (1), have been observed in a reasonable variety of molecules (2-11). With notable exceptions of azulene and thioketone, however, such emission is usually very weak, because the rates of nonradiative decay processes greatly exceed the rates of radiative processes when excited states other than the lowest excited states are involved.

Azulene displays the fluorescence originating from the second excited singlet ($S_2 \rightarrow S_0$) which is very much stronger than that from the first ($S_1 \rightarrow S_0$) (12). Quantum yields of the emission of gaseous azulene are about 0.2 for $S_2 \rightarrow S_0$ and 8×10^{6} for $S_1 \rightarrow S_0$ (13), while in solution, the yield of S_2 emission is around 0.03 (14). This "anomalous" behavior of azulene has been attributed, at least in part, to a large S_2-S_1 separation (the S_1 and S_2 are at 14300 and 28800 cm^{-1}, respectively), which reduces the rate of $S_2 \rightsquigarrow S_1$ relaxation, in accordance with the energy gap law (15).

In metalloporphyrins, there is a relatively large energy separation between the S_1 and S_2 excited states. Characteristic Q and B (Soret) bands of metalloporphyrins in the visible and near-ultraviolet region are assigned as the S_1 and S_2 excited states. These excited states are described as a 50-50 admixture of two common $^1(\pi,\pi^*)$ excited configurations $^1(a_{1u}e_g)$ and $^1(a_{2u}e_g)$ in accidental degeneracy and the energy surfaces of the S_1 and S_2 excited states are parallel. This retards $S_2 \rightsquigarrow S_1$ intramolecular electronic relaxation. In addition, as suggested by the high intensity of the B (Soret) band, the radiative process from S_2 excited state to the ground state is highly accelerated. For these reasons, diamagnetic metalloporphyrins, in which no triplet excited states are present between the S_1 and S_2 excited states, are anticipated to fluoresce from the S_2 excited state.

$S_2 \rightarrow S_0(S_2)$ fluorescence was first observed by Bajema et al.(16) for zinc tetrabenzoporphyrin (TBPZn) in octane containing 0.5 % pyridine. The yield was determined to be 1.6×10^{-3} and the lifetime was estimated to be about 2.4 ps from the quantum yield (16), which is in good agreement with the recent value obtained by Even et al. (17). Even et al. have applied the method of supersonic expansion to the study of interstate coupling and electronic relaxation in the porphyrin S_2 state (17-19). From line broadening of the B band in fluorescence excitation spectra, they determined the lifetimes of S_2 state of free-base porphyrins as well as metalloporphyrins.

Martarno et al. found S_2 fluorescence for diamagnetic Y(III), Lu(III) and Th(IV) complexes of tetraphenylporphin (TPP) and determined their yields as 1×10^{-3}, 1×10^{-3} and 4×10^{-4}, respectively (20). Kurabayashi et al. also observed S_2 fluorescence for TPPAlCl, TPPZn and TPPCd in CH_3CN and evaluated the rates of radiative ($k_f(S_2)$) and nonradiative ($k_d(S_2)$) processes of the S_2 state as Table I (21).

Table I. Fluorescence Properties of Some Metal Complexes of TPP in Acetonitrile at 296 K

	$\phi_f(S_1)$[a]	$\phi_f(S_2)$[a]	$k_f(S_2)\times10^{-9}$ /s^{-1}	$k_d(S_2)\times10^{-12}$ /s^{-1}	$\Delta E(S_2-S_1)$ /cm^{-1}
TPPAlCl	1.4×10^{-1}	6.7×10^{-4}	1.4	2.2	7100
TPPZn	3.3×10^{-2}	3.7×10^{-4}	1.5	4.1	6900
TPPCd	1.0×10^{-3}	1.1×10^{-4}	1.1	9.6	6700
TBPZn[b]	0.35	1.6×10^{-3}	0.67	0.42	7400

[a] $\phi_f(S_1)$ and $\phi_f(S_2)$ stand for the quantum yields of S_1 and S_2 fluorescence.

[b] Ref. 16

Tsvirko et al. studied the S_2 emissions with a series of rare-earth (Sm(III), Eu(III), Gd(III), Tb(III), Yb(III) and Lu(III)) complexes of tetratolylporphin (TTP) and the quenchings of TTPEu and TTPYb were attributed to the participations of a "ligand-to-metal" charge-transfer state lower than the S_2 state and low-lying 4f levels, respectively (22,23).

A further striking finding by Stelmakh and Tsvirko is the S_2 fluorescence following excitation to the S_1 state by a low-energy photon. They also proposed possible excitation mechanisms to explain their results (24-26). Pumping can be achieved by two different ways of unimolecular successive two-photon absorption;

$$S_0 \xrightarrow{h\nu_1} S_1(\text{Q state}) \xrightarrow{h\nu_2} S_n \rightsquigarrow S_2(\text{B state}) \quad (1)$$

$$S_0 \xrightarrow{h\nu_1} S_1(\text{Q state}) \rightsquigarrow T_1 \xrightarrow{h\nu_2} T_n \rightsquigarrow S_2(\text{B state}). \quad (2)$$

Another pumping path is also possible by T-T annihilation mechanism;

$$S_0 \xrightarrow{h\nu_1} S_1(\text{Q state}) \rightsquigarrow T_1,\ T_1 + T_1 \longrightarrow S_0 + S_2(\text{B state}). \quad (3)$$

S_2 emission of TBPCd is observed upon excitation of a mixture of TBPCd and TPPPd in pyridine by the 514.5 nm argon ion laser line. The line can excite TPPPd but not TBPCd. The pumping is performed by T-T annihilation mechanism as follows (26);

$$S_0(\text{TPPPd}) \xrightarrow{h\nu_1} S_1(\text{TPPPd}) \rightsquigarrow T_1(\text{TPPPd})$$

$$T_1(\text{TPPPd}) + S_0(\text{TBPCd}) \longrightarrow S_0(\text{TPPPd}) + T_1(\text{TBPCd})$$

$$T_1(\text{TBPCd}) + T_1(\text{TBPCd}) \longrightarrow S_0(\text{TBPCd}) + S_2(\text{TBPCd}).$$

The optical-optical double resonance technique was applied to determine which mechanism is predominant in the pumping of TPPZn (27, 28). Because of the large difference in the lifetimes of S_1 and T_1 states, TPPZn was excited by two-color laser pulses with varied time intervals. The first laser pulse was fixed at 540 nm corresponding to $S_0 \rightarrow S_1$ excitation of TPPZn and the second was fixed at 640 nm which corresponds to the $S_1 \rightarrow S_n$ and $T_1 \rightarrow T_n$ excitations. The intensity of S_2 emission was measured with a variety of time delay between the two pulses. The S_2 emission is appreciably diminished for such a long delay as 20 ns, during which $S_1 \rightsquigarrow T_1$ intersystem crossing is almost completed. This indicates that the pumping is achieved by the mechanism (1) (28). However, it should be noted that a high-density excitation to the S_1 excited state of TPPZn in nonpolar solvent yields a delayed S_2 fluorescence which is ascribed to the T-T annihilation mechanism (3) (29).

In this paper, we will present our recent observations on the upper excited-state emission of a variety of metalloporphyrins.

S_2 Emission of Diamagnetic Metalloporphyrins

Al(III), Ga(III) as well as In(III) porphyrin are typical porphyrins incorporated with a tervalent metal ion: Characteristic Q and B bands in the visible and near-ultraviolet region, respectively, arise from the (π,π^*) excitations in the porphyrin ring with only minor perturbation from the outershell electrons of the central metal ion. The Q band is of forbidden character, however, the Q band can borrow the intensity by vibronic couplings from the allowed B band (30). The intensity of the Q(1,0) band is much less sensitive to the peripheral substituents, the axial ligands and the central metal ions, while that of the Q(0,0) band without excitation in the skeletal vibrational modes is rather sensitive to various substituents.

The Q and B bands are assigned as the excitations to the S_1 and S_2 states. Since these excited states are described as a 50-50 admixture of two common excited configurations $^1(a_{2u}e_g)$ and $^1(a_{1u}e_g)$, the energy surfaces of the lowest excited singlet states are almost parallel. This retards $S_2 \rightsquigarrow S_1$ intramolecular electronic relaxation and thus the strong S_2 (Soret) oscillator can emit. There are no triplet excited states between the S_1 and S_2 excited states (31).

TPPAlCl, TPPGaCl, TPPInCl as well as other diamagnetic metal complexes and diacids of TPP and its analogues emit S_2 fluorescence, while the free-bases $TPPH_2$, $OEPH_2$ (OEP: 2,3,7,8,12,13,17,18-octaethylporphin) as well as OEP metal complexes do not emit (32). Figure 1 shows S_1 and S_2 fluorescence bands of TPPGaCl in ethanol.

Excitation spectra of the S_2 fluorescence are in good agreement with absorption spectra. This implies that the fluorescence is not originating from some impurities. Since the excitation spectra of S_1 fluorescence of the complexes which emit S_2 fluorescence are almost identical to respective absorption spectra up to the range of 250 nm, the internal conversion $S_2 \rightsquigarrow S_1$ must be faster than the intersystem crossing $S_2 \rightsquigarrow T_n$ and the internal conversion $S_2 \rightsquigarrow S_0$.

S_2 fluorescence of any OEP metal complexes except OEPPb can not be detected by the sensitivity used in the present work for which a quantum yield 0.03×10^{-4} is the detectable limit. A prominent role as accepting modes was assigned to the stretching modes of peripheral ethyl groups in OEP complexes (21). However, the quenching of S_2 emission in OEP complexes is to be attributed to a fast relaxation to the S_1 state.

TPPZn and PZn emit S_2 emission while OEPZn does not. The yield of S_2 emission decreases in a sequence

$$T(p\text{-}C_2H_5O)PPZn \sim TPPZn > PZn \sim TFPPZn > OEPZn$$

($T(p\text{-}C_2H_5O)PP$: tetra(p-ethoxyphenyl)porphin; TFPP: tetra(pentafluorophenyl)porphin). This is reverse to the ordering found for the intensity ratio Q(0,0)/Q(1,0). In OEPZn, the Q(0,0) band is granted intensity from the B band more than that of TPPZn. As stated earlier, the surfaces of the S_1 and S_2 states are parallel when the component $^1(\pi,\pi^*)$ excited configurations are in accidental degeneracy. The parallel excited-state surfaces, however, shift from the ground state surface. In fact, the Franck-Condon overlap is so small that only weak Q(0,0) band is observed in TPPZn and PZn. In OEPZn, on the other hand, the electron-donating peripheral substituents, which partly lift the accidental degeneracy, yield some difference in shift between the S_1 and S_2 surfaces. This results in an increase in Franck-Condon overlap, which enhances the intensity of Q(0,0) band, and also prompts the $S_2 \rightsquigarrow S_1$ internal conversion. The yields of S_2 emission of Zn porphyrins are not necessarily explained by the energy gap between the S_1 and S_2 states.

In Pb(II) porphyrins, the Q and B bands are shifted to the red and an intense new band appears in 28×10^3 cm^{-1} (33). In the vicinity of the B state, a "metal-to-porphyrin" charge-transfer excited state is present. Configuration interaction increases the intensity of Q(0,0) in TPPPb but decreases in OEPPb. The detectable S_2 fluorescence of OEPPb is attributable to a partial recovery of the parallel surfaces of S_1 and S_2 states (32).

The yields of S_2 fluorescence obtained in the present work are summarized in Table II.

Table II. Quantum Yields of S_2 Fluorescence

	Solvent	$\phi_f(S_2)\times10^4$ [a]	$\Delta E(S_2-S_1)/10^3\ cm^{-1}$
TPPAlCl	benzene	3.1	6.86
TPPGaCl	benzene	3.4	6.69
TPPInCl	benzene	4.0	6.81
TPPAlCl	ethanol	6.1	7.11
TPPGaCl	ethanol	5.1	7.00
TPPInCl	ethanol	5.2	7.06
TPPZn	benzene	3.9	6.69
T(p-C_2H_5O)PPZn	benzene	4.0	6.61
TFPPZn	benzene	0.51	6.57
PZn	benzene	0.80	7.23
OEPZn	benzene	<0.04	7.21
TPPPb	benzene	0.47	6.24
OEPPb	benzene	0.04	
$TPPH_4^{2+}$	benzene	0.65	7.51
PH_4^{2+}	benzene	0.62	7.90

[a] The yields were evaluated with reference to the yield of TTPZn in benzene ($\phi_f(S_1)$ = 0.033 ref. 34).

The Free-Base and Diacid Porphyrins

S_2 emissions of the porphyrin free-bases are too weak to be detected unless intense laser excitations are used. As indicated by enhanced Q_y band in the free-bases, a strong coupling of Q_y and B states give rise to a crossing of the S_1 and S_2 surfaces which is responsible for the fast deactivation of S_2 state. It should be noted that laser excitation yields a rather broad S_2 emission band for the free-bases in comparison with sharp S_2 emission of metalloporphyrins (28). A large Stokes shift and the broadness of S_2 emission as shown in Figure 2 indicate that a dynamic Jahn-Teller distortion happens in the B state of the free-base to greater extent than that of metalloporphyrins.

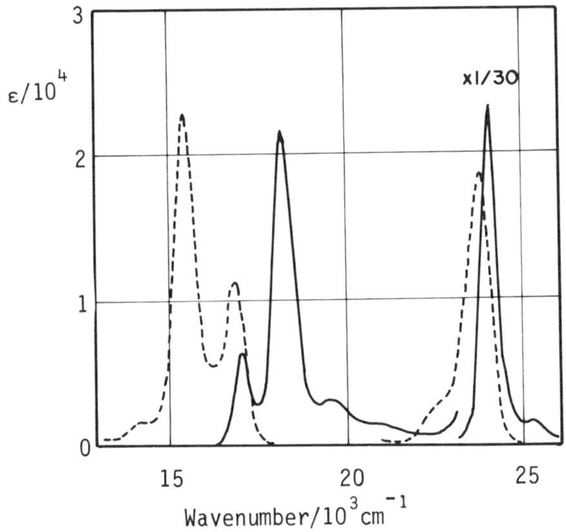

Figure 1. Absorption (——) and fluorescence (-----) spectra of TPPGaCl in ethanol.

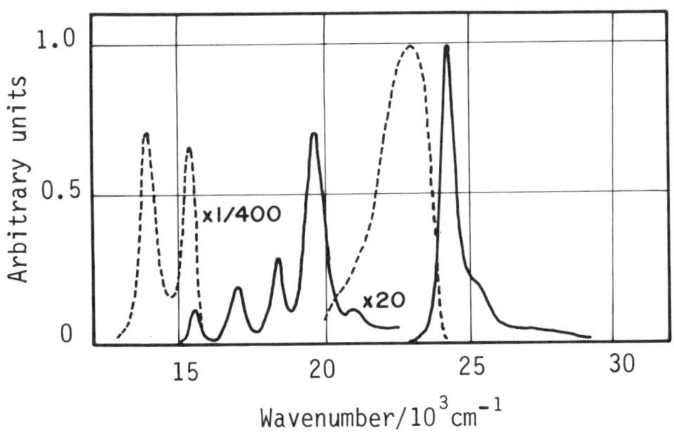

Figure 2. Absorption (——) and fluorescence (-----) spectra of $TPPH_2$ in EPA at room temperature.

In contrast with the free-bases, diacids $TPPH_4^{2+}$ and PH_4^{2+} (P: porphin) emit rather intense S_2 fluorescence. The quantum yield of PH_4^{2+} is as high as that of PZn, while the yield of $TPPH_4^{2+}$ is about 1/5 of TPPZn.

Paramagnetic Metalloporphyrins

The lowest excited states of paramagnetic metal complexes are described by configuration interactions of the porphyrin (π,π^*) excited singlet and triplet states and the "porphyrin-to-metal" or "metal-to-porphyrin" charge-transfer excited states (35,36). Thus T_1 (phosphorescence) emission of paramagnetic metal complexes decays in a very short period (37,38). When the charge-transfer excited states are sufficiently low, the S-T absorption bands are observed to the red of Q band instead of the T_1 emission (39). Ni(II) porphyrins, even in diamagnetic ground state, show neither S_1 nor T_1 emission (40,41). On the other hand, Pd(II) and Pt(II) porphyrins with isoelectronic outershells emit intense phosphorescence even at room temperature. The quenching of luminescence in Ni(II) porphyrins was attributed to a $^1(d,d^*)$ excited state at lower energy than the S_1 state (42). The low-lying $^{1,3}(d,d^*)$ excited states of Pd(II) and Pt(II) porphyrins are lifted in between the S_1 and S_2 states. This, however, results in quenching of the S_2 emission of these porphyrins.

Paramagnetic Er(III) porphyrins TPPErOH and TPPEr(dpm) (dpm: 2,2,6,6-tetramethyl-3,5-heptanedionate) show not only S_1 but also S_2 emission, even if there are a sequence of the f sublevels ^4I, ^4F, ^4S, ^2H below the S_2 state (43). OEPErOH emits only S_1 emission like many other OEP metal complexes. In CH_3OH/C_2H_5OH (7:3) glassy medium at 77 K, TPPEr(dpm) is present in two different conformations I and II. A spectral splitting is observed upon freezing but the change is reversible. At room temperature, a spectrum close to that of TPPErOH is revived. TPPErOH itself does not show any spectral change upon freezing of the medium.

The metal ion in conformer II of TPPEr(dpm) is shifted out of the porphyrin plane, as suggested by an increase in the Q(0,0) intensity and a comparable red shift of the Q and B bands, while that of conformer I is in the plane as TPPErOH in methanol/ethanol. However, the conformation change might be due to a change of coordination number by solvent coordination rather than the shift of the central metal ion. Excitation spectra of the S_2 emission of TPPEr(dpm) indicate that only conformer I with a weak Q(0,0) band exhibits S_2 emission, while conformer II with intense Q(0,0) band does not emit.

TPPEr(dpm) emits not only porphyrin S_2 emission but also coordinated β-diketonate T_1 emission, when it is excited in an absorption band of coordinated dpm origin around 300 nm (43). A fast relaxation within β-diketonate accelerated by the paramagnetic central metal ion yields a high population in the lowest dpm T_1 state which is slightly higher than porphyrin S_2 state. This results in dpm phosphorescence. The rate of inter-ligand energy transfer is not as fast as intersystem crossing within the β-diketonate. Emission spectrum (I) in Figure 3 is attributable to a superposition of porphyrin S_2 fluorescence, β-diketonate phosphorescence and intense Soret reabsorption. The figure also shows phosphorescence of $Er(dpm)_3$ (b) which is to be close to that of the dpm moiety of TPPEr(dpm) and also intense Soret reabsorption (a). The porphyrin S_2 emission can be excluded by use of a rotating sector. The spectrum (II) shows phosphorescence of the β-diketonate moiety superposed by intense Soret reabsorption.

Porphyrin Dimers

Soret band of μ-oxo dimer $[TPPNb]_2O_3$ shifts to the blue in comparison with $TPPNbO(CH_3COO)$ and exhibits a long tailing to the red of the Soret band. Exciton coupling in the "face-to-face" stacking results in an allowed state at higher energy than the monomer state and a forbidden state at lower energy (30,44). The μ-oxo dimer emits both S_1 fluorescence and phosphorescence. A rather intense S_2 fluorescence was observed for $TPPNbO(CH_3COO)$, while the fluorescence could not be detected for $[TPPNb]_2O_3$ (32). Exciton couplings are possible not only between the component S_2 states but also between the component S_1 and S_2 states and result in a fast relaxation to the S_1 manifold.

In the present work, we used a conformationally flexible covalently-linked porphyrin dimer (5,5-[1,3-Propanediyl bis(oxy-2,1-phenylene)]bis[10,15,20-tris(4-methylphenyl)-21H,23H-porphin](o,o'-$C_3(TTPH_2)_2$). The dimer is mainly in two different conformations. The red component of the dimer Soret band is ascribed to a twisted conformer "A" and the blue component to a stacked conformer "B". Computer calculations can predict two conformations A and B with dihedral angles of the porphyrin moieties of 80° and 56°,respectively.

S_2 emission was detected for o,o'-$C_3(TPPZn)_2$ and o,o'-$C_3(TPP\text{-}AlCl)_2$ and their corresponding monomers. Excitation spectra of the monomer S_2 emissions are in good agreement with respective absorption spectra even in the ultraviolet region. S_2 emission of the free-base dimers like many other free-bases was too weak to be detected.

Excitation spectra of the dimer S_2 emissions correspond well to the red components of Soret bands of the dimers and also to the Soret bands of the respective monomers. This indicates that S_2 emission comes only from the dimer in twisted conformation. The quantum yield of S_2 emission of the dimer is much less than that of the corresponding monomer, while those of the S_1 and T_1 emissions of the dimer coincide with those of the monomer. The dimer in a conformation like "A" emits since $S_2 \rightsquigarrow S_1$ internal conversion is retarded, while strong exciton coupling prompts the internal conversion in the dimer of conformation "B".

The fluorescence excitation polarization of the monomer is almost 1/7 regardless of the excitation wavelength. A value of 1/7 is typical when both the absorption and the emission oscillators are degenerate and polarized in the same plane. Since the dimer is regarded as a weakly coupled, three-dimensional, double-oscillator, energy transfer between the dimer partners will randomize the excitation between the two porphyrin planes oriented in a tilt angle. In fact, the observed polarization of the dimer is less than 1/7.

Free-base porphyrins are easily protonated in acidic media. The diacid, a doubly protonated species, emits S_2 fluorescence, while the free-base does not. The second-derivative spectra indicate that o,o'-$C_3(TTPH_4)_2^{4+}$ is in two different conformations as o,o'-C_3-$(TTPH_2)_2$ regardless of the positive charges accumlated on the two porphyrin moieties. The quantum yield and the lifetime of S_1 emission of o,o'-$C_3(TTPH_4)_2^{4+}$ conform with those of $TTPH_4^{2+}$, while the yield of S_2 emission is reduced in the dimer.

Absorption spectrum of the free-base monomer varies with increasing $[H^+]$, exhibiting isosbestic points. The spectrum of the free-base dimer, on the other hand, varies with increasing $[H^+]$ showing a shift of the isosbestic point. A two-step equilibrium must be considered for protonation of the free-base dimer:

$$PH_2 \sim PH_2 + 2H^+ \rightleftharpoons PH_2 \sim PH_4^{2+}$$

$$PH_2 \sim PH_4^{2+} + 2H^+ \rightleftharpoons PH_4^{2+} \sim PH_4^{2+}.$$

The excitation spectra of S_2 emission of $TTPH_4^{2+}$ and o,o'-$C_3(TTPH_4)_2^{4+}$ are in good agreement with the Soret band of the monomer and the red component of Soret band of the dimer, respectively. Figure 4 shows the excitation spectra of S_2 emission of the free-base monomer and dimer in benzene containing a low concentration of CF_3COOH. The free-base dimer shows S_2 emission, which is close to that of $TTPH_4^{2+}$, in a dilute acidic solution, where $PH_2{\sim}PH_2$, $PH_2{\sim}PH_4^{2+}$ and $PH_4^{2+}{\sim}PH_4^{2+}$ coexist. The excitation spectrum of the dimer shows peaks not only in the Soret band of the diacid but also in the Soret band of the free base, whereas that of the monomer coincides with absorption spectrum of $TTPH_4^{2+}$ but not with the spectrum of the $TTPH_4^{2+}$ and $TTPH_2$ mixture. This indicates that part of the S_2

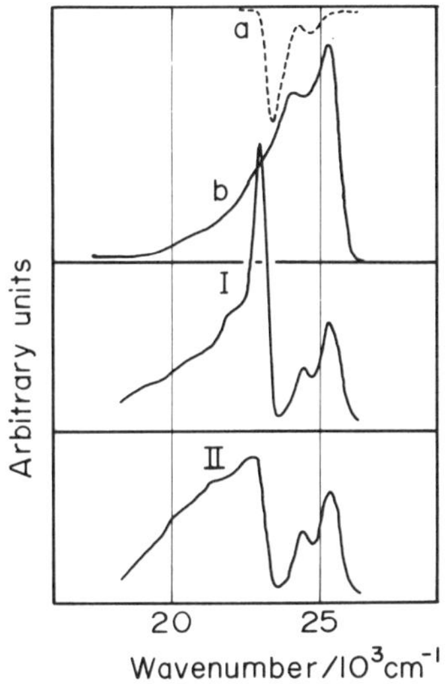

Figure 3. Emission spectra of TPPEr(dpm) observed upon irradiation in the $^1(\pi,\pi^*)$ band of coordinated dpm around 300 nm. In (II), the porphyrin S_2 component emission was cut by a chopper interposed in the detection beam.

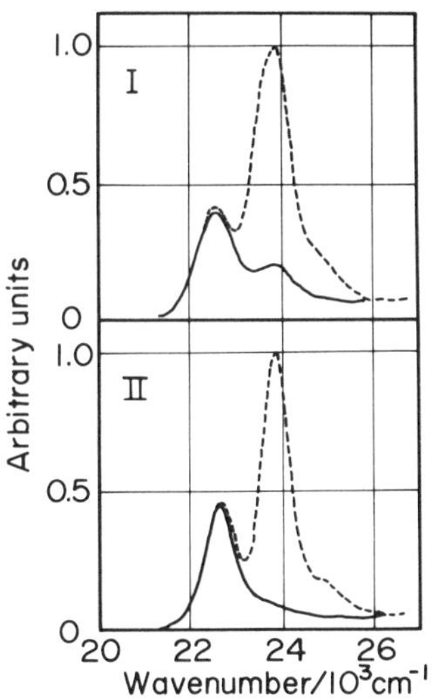

Figure 4. Excitation spectra of S_2 emission (——) and absorption spectra (-----) in the Soret region of 1.4×10^{-7}M o,o'-$C_3(TTPH_2)_2$ (I) and 3.2×10^{-7}M $TTPH_2$ (II) in benzene containing 1.0×10^{-5}M CF_3COOH.

emission of the diacid moiety in PH_2~PH_4^{2+} arises from intramolecular energy transfer from the higher S_2 state of the free-base moiety in PH_2~PH_4^{2+} to the lower S_2 state of the diacid counterpart (45).

Acknowledgments

The present work has been achieved in collaboration with our students, Dr. Osamu Ohno, Mr. Hideaki Maekawa and Miss Motoko Asano. The authors are grateful to Professor Ikuzo Tanaka for his continuous interest and encouragement.

Literature Cited

1. Kasha, M. Discuss. Faraday Soc. 1950, 9, 14-9.
2. Hui, M.H.; DeMayo, P.; Suau, R.; Ware, W.R. Chem. Phys. Lett. 1975, 31, 257-63, and references therein.
3. Levine, S.Z.; Knight, A.R.; Steer, R.P. Chem. Phys. Lett. 1974, 29, 73-6.
4. Oka, T.; Knight, A.R.; Steer, R.P. J. Chem. Phys. 1975, 63, 2414-20; ibid. 1977, 66, 699-706.
5. Eber, G.; Gruneis, F.; Schneider, S.; Dorr, F. Chem. Phys. Lett. 1974, 29, 397-404.
6. Huber, J.R.; Mahaney, M. Chem. Phys. Lett. 1975, 30, 410-2.
7. Margulies, L.; Yogev, A. Chem. Phys. Lett. 1976, 37, 291-4.
8. Eber, G.; Schneider, S.; Dorr, F. Chem. Phys. Lett. 1977, 52, 59-62.
9. Lin, H.-B.; Topp, M.R. Chem. Phys. Lett. 1977, 48, 251-5.
10. Morgante, C.G.; Struve, W.S. Chem. Phys. Lett. 1979, 68, 267-71.
11. Plummer, B.F.; Al-Saigh, Z.Y. Chem. Phys. Lett. 1982, 91, 427-32.
12. Beer, M.; Longuet-Higgins, H.C. J. Chem. Phys. 1955, 23, 1390-1.
13. Huppert, D.; Jortner, J.; Rentzepis, P.M. J. Chem. Phys. 1972, 56, 4826-33.
14. Murata, S.; Iwanaga, C.; Toda, T.; Kokubun, H. Chem. Phys. Lett. 1972, 15, 152.
15. Englman, R.; Jortner, J. Mol. Phys. 1970, 18, 145-64.
16. Bajema, L.; Gouterman, M.; Rose, C.B. J. Mol. Spectrosc. 1971, 39, 421-31.
17. Even, U.; Magen, J.; Jortner, J.; Friedman, J. J. Chem. Phys. 1982, 77, 4384-90.
18. Even, U.; Magen, J.; Jortner, J.; Friedman, J.; Levanon, H. J. Chem. Phys. 1982, 77, 4374-83.
19. Even, U.; Jortner, J. J. Chem. Phys. 1982, 77, 4391-9.
20. Martarano, L.A.; Wong, C.-P.; Horrocks, W.DeW. Jr.; Goncalves, A.M.P. J. Phys. Chem. 1976, 80, 2389-93.
21. Kurabayashi, Y.; Kikuchi, K.; Kokubun, H.; Kaizu, Y.; Kobayashi, H. J. Phys. Chem. 1984, 88, 1308-10.
22. Tsvirko, M.P.; Stelmakh, G.F.; Pyatosin, V.E.; Solovyov, K.N.; Kachura, T.F. Chem. Phys. Lett. 1980, 73, 80-3.
23. Tsvirko, M.P.; Solovev, K.N.; Stelmakh, G.F.; Pyatosin, V.E.; Kachura, T.F. Opt. Spektrosk. 1981, 50, 555-60 (Opt. Spectrosc. 1981, 50, 300-3).

24. Stelmakh, G.F.; Tsvirko, M.P. Opt. Spektrosk. 1980, 48, 185-8 (Opt. Spectrosc. 1980, 48, 105-6).
25. Stelmakh, G.F.; Tsvirko, M.P. Opt. Spektrosk. 1981, 49, 511-6 (Opt. Spectrosc. 1980, 49, 278-80).
26. Stelmakh, G.F.; Tsvirko, M.P. Opt. Spektrosk. 1981, 50, 998-1000 (Opt. Spectrosc. 1981, 50, 547-8).
27. Tobita, S.; Tanaka, I. Chem. Phys. Lett. 1983, 96, 517-21.
28. Tobita, S.; Kaizu, Y.; Kobayashi, H.; Tanaka, I. J. Chem. Phys. 1984, 81, 2962-9.
29. Kurabayashi, Y.; Kikuchi, K.; Kokubun, H.; Kaizu, Y.; Kobayashi, H. XIIth International Conference on Photochemistry, Tokyo, Japan, 1985, Abstracts p206.
30. Kaizu, Y.; Misu, N.; Tsuji, K.; Kaneko, Y.; Kobayashi, H. Bull. Chem. Soc. Jpn. 1985, 58, 103-8.
31. Sekino, H.; Kobayashi, H. to be published.
32. Ohno, O.; Kaizu, Y.; Kobayashi, H. J. Chem. Phys. 1985, 82, 1779-87.
33. Sayer, P.; Gouterman, M.; Connell, C.R. J. Am. Chem. Soc. 1977, 99, 1082-7.
34. Quimby, D.J.; Longo, F.R. J. Am. Chem. Soc. 1975, 97, 5111-7.
35. Kobayashi, H.; Yanagawa, Y.; Osada, H.; Minami, S.; Shimizu, M. Bull. Chem. Soc. Jpn. 1973, 46, 1471-9.
36. Kobayashi, H.; Higuchi, T.; Kaizu, Y.; Osada, H.; Aoki, M. Bull. Chem. Soc. Jpn. 1975, 48, 3137-41.
37. Gouterman, M.; Hanson, L.K.; Khalil, G.E.; Leenstra, W.R.; Buchler, J.W. J. Chem. Phys. 1975, 62, 2343-53.
38. Harriman, A. J. Chem. Soc. Faraday Trans. 1 1981, 77, 369-77.
39. Kobayashi, H.; Tsuji, K.; Kaizu, Y.; Misu, N. to be published.
40. Ake, R.L.; Gouterman, M. Theoret. Chim. Acta 1970, 17, 408-16.
41. Kim, D.; Kirmaier, C.; Holten, D. Chem. Phys. 1983, 75, 305-22.
42. Kobayashi, T.; Straub, K.D.; Rentzepis, P.M. Photochem. Photobiol. 1979, 29, 925-31.
43. Kaizu, Y.; Asano, M.; Kobayashi, H. J. Phys. Chem. in press.
44. Gouterman, M.; Holten, D.; Lieberman, E. Chem. Phys. 1977, 25, 139-53.
45. Kaizu, Y.; Maekawa, H.; Kobayashi, H. J. Phys. Chem. in press.

RECEIVED March 28, 1986

8

Metalloporphyrin Dimeric S_2 Emission Resulting from Triplet–Triplet Annihilation

G. F. Stelmakh and M. P. Tsvirko

A. N. Sevchenko Research Institute of Applied Physical Problems, Minsk, 220106, USSR

S_2 emission resulting from triplet-triplet annihilation is reported. It differs from prompt S_2 emission in being red shifted by 80 to 500 cm^{-1}. It also appears slightly narrower in halfwidth. The difference is ascribed to the fact that for triplet-triplet annihilation delayed fluorescence comes from a dimeric species before it has time to dissociate.

Triplet-triplet annihilation (TTA), i.e., deactivation of triplet molecules as a result of their interaction, is one of the main pathways of triplet state decay in solutions, crystals and the gas-phase. TTA may become the determining pathway of triplet state deactivation under conditions of high concentrations of triplet-state molecules resulting in particular from powerful laser photoexcitation.

The existence of this TTA bimolecular process makes it possible to carry out cooperative excitation of energy rich molecular states by low energy quanta. In particular, formation of upper excited singlet states as a result of TTA has been observed for several aromatic hydrocarbons (1) and metalloporphyrins (2) by the demonstration of annihilation-induced delayed fluorescence (ADF) from upper singlet states.

Systematic research of two-quantum excitation of tetrapyrrolic pigments (2,3) has made it evident that cooperative triplet-triplet excitation of metalloporphyrin upper electronic states successfully competes with the processes of stepwise two-quantum excitation. This made us pay greater attention to porphyrin TTA since high quantum yields of intersystem crossing to triplet T_1 states is characteristic of these types of compounds. (It should be mentioned that interest in photoprocesses in excited states of tetrapyrrolic pigments has been greatly increased by development of solar energy converters based on metalloporphyrins (4,5)).

0097–6156/86/0321–0118$06.00/0

For several compounds probabilities of excited singlet state formation as a result of TTA processes have been determined in our previous paper (6) with the help of a method based on measurement of the ratio of the integrated intensities of ADF and prompt fluorescence (PF) under the conditions of powerful photoexcitation (see Table I).

Table I
Probabilities of Cooperative S_1 States (P_a) and S_2 State (P_2) Formation as a Result of TTA in Solutions at Room Temperature (6)

Compound	Solvent	P_a	P_2
Tetraphenylporphyrin	Toluene	0.16 ± 0.02	
Mg-tetraphenylporphyrin	Toluene	0.175 ± 0.02	0.01
Zn-tetraphenylporphyrin	Toluene	0.24 ± 0.02	0.04
Pd-tetraphenylporphyrin	Toluene	0.27 ± 0.04	
Cd-tetrabenzporhyrin	Pyridine	0.08 ± 0.01	8.10^{-4}
Zn-tetrabenzporhyrin	Pyridine	0.10 ± 0.01	$< 10^{-3}$
Chlorophyll a	Ethanol	0.014 ± 0.002	
Anthracene	Ethanol	0.045 ± 0.005	

Notes: 1) P_a and P_2 have been calculated for one annihilation act of two triplet molecules. 2) For the P_2 calculation it was assumed that probabilities of radiative $S_2 \rightarrow S_0$ transition are the same for one-photon or cooperative mechanism of S_2 state excitation.

Probabilities of cooperative S_1 - excited molecule formation (P_a) for metalloporphyrins range from 0.1 to 0.25 per one act of TTA. At the same time probabilities of S_2 state formation from TTA (P_2) which have been determined with the assumption that the ADF and the PF are emitted from identical states turned out to be considerably less. This difference between P_a and P_2 remained unexplained since for two T_1 molecules the total excitation energy is similar to the S_2 state energy as far as metalloporphyrins are concerned. Thus one may expect that some quasi-resonant upper singlet state is being predominantly generated, i.e., cooperative excitation of singlet states takes place according to the scheme $2T_1 \rightarrow S_2 \rightarrow S_1$ and correspondingly $P_a \simeq P_2$.

In this report the TTA process has been studied taking into account new experimental data indicating that the ADF emitted from metalloporphyrin upper electronic states is of dimeric nature. Moreover low values of P_2 obtained in (6) can be explained by the lesser probability of dimer singlet states emitting in comparison with the monomer. But true probabilities of S_n state formation from TTA processes may be quite high. We have been able to establish simultaneous relations between diffusion parameters and annihilation characteristics obtained from spectral kinetic measurements.

We have carried out a comparative study of S_2 state fluorescence emitted from several metalloporphyrins: Zn - tetraphenylporphin (Zn TPP), Zn - tetrabenzporphin (Zn TBP) and Cd TBP for direct one-quantum excitation and for annihilating two-quantum excitation. Low concentration ($C = 1 - 2 \times 10^{-6}$M) solutions have been studied at room temperature. Spectral luminescence measurements have been carried out with the Raman spectrometer "Ramalog 4" and spectrofluorimeter "Spex Fluorolog" (more detailed procedure of measurements is described in (2)).

The ADF has been measured in deoxygenated solutions with continuous excitation of either argon, krypton or helium-neon lasers. The exciting wavelengths corresponded to the bands of $S_0 \rightarrow S_1$ metalloporphyrin absorption (excitation wavelengths: λ_{ex} = 514.5; 530.9; 568.2 nm for ZnTPP and λ_{ex} = 568.2; 632.8 nm for ZnTBP and CdTBP).

Application of continuous anti-Stokes excitation makes it possible to record the emission from the S_2 state when the PF $S_2 \rightarrow S_0$ is absent (2). Prompt $S_2 \rightarrow S_0$ fluorescence has been measured at direct $S_0 \rightarrow S_2$ excitation (krypton laser: λ_{ex} = 406.7; 413.1 nm) of oxygenated solutions (in this case the ADF is absent due to strong quenching of triplet states by oxygen (2)).

We have found that emission spectra of the cooperatively excited ADF from upper electronic states and of the one quantum excited PF are different. ADF spectrum is shifted in the longwave direction with respect to PF spectrum (see Table II). The shift effect depends on solvent type. ADF spectral shift is most considerable and comprises 500 cm^{-1} (see Fig. 1) for the case of ZnTPP in decane.

In principle, fluorescence spectra from shortlived upper metalloporphyrin states in solution can be shifted in the longwave direction with vibrational energy relaxation at the cost of short-wave excitation (7). However, in our case the energy difference of annihilating and direct excitation was slight ($\leq 10^3$ cm-1) and could not lead to such considerable difference of ADF and PF spectra. The observed difference can be explained in the following way: ADF takes place in biomolecular complexes while the PF is ordinary monomer fluorescence.

At present it is universally acknowledged that TTA as triplet-triplet energy transfer is caused by exchange interaction of electrons in bimolecular complexes which takes place during molecular diffusion encounters in solution (in gas phase - molecular collisions are examined; in crystals - triplet exciton diffusion is the responsible annihilation process (8-10)). No doubt, interaction of molecular partners in a diffusion complex may lead to the change of probabilities of fluorescent state radiative and nonradiative deactivation. Nevertheless, it is normally considered that as a result of TTA the energy of two triplet partners is accumulated in one molecule which emits the ADF (11). Interaction with the second deactivated partner is not taken into account, i.e. it is assumed that the ADF is of monomer nature and its spectrum coincides with the PF spectrum. Apparently the latter may be true when the ADF takes place from S_1 state the lifetime of which (τ_{S_1} ~ 10-8 - 10-9 s) is much longer than the lifetime of diffusion encounter complex (~ 10-10 - 10-11 s in liquid solutions). As a matter of fact we have not observed considerable ADF and PF spectral difference when S_1 metallo-

TABLE II

Spectral Characteristics of the Fluorescence from Metalloporphyrin S_2 States at Direct $S_0 \rightarrow S_2$ Excitation (PF) and Cooperative Triplet Excitation (ADF)

Compound	Solvent	PF		ADF		shift
		ν maxcm^{-1}	$\Delta\nu$ 1/2cm^{-1}	ν maxcm^{-1}	$\Delta\nu$ 1/2cm^{-1}	cm^{-1}
Zn TPP	decane	23800 ± 20	750 ± 20	23300 ± 20	640 ± 20	500 ± 20
Zn TPP	toluene	23330 ± 20	830 ± 20	23030 ± 20	620 ± 20	300 ± 20
Zn TPP	ethanol	23460 ± 20	650 ± 20	23270 ± 20	500 ± 20	190 ± 30
Zn TPP	pyridine	22990 ± 20	610 ± 20	22850 ± 20	600 ± 20	140 ± 30
Zn TBP	dimethylformamide	23220 ± 20	590 ± 20	23140 ± 20	590 ± 20	80 ± 20
Cd TBP	pyridine	22620 ± 20	730 ± 20	22500 ± 20	770 ± 20	120 ± 30

Note: ν_{max} - maximum fluorescence wavenumber, $\Delta\nu_{1/2}$ - spectrum half-width.

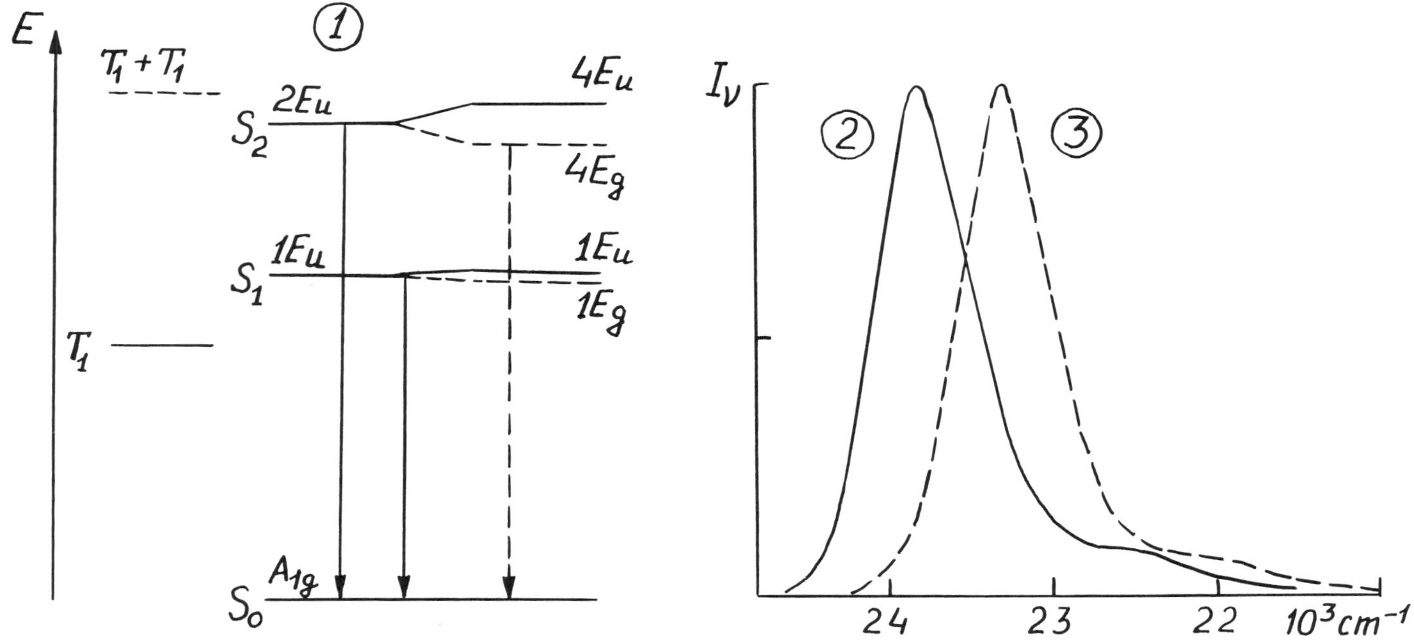

Figure 1
Left Side:
(1) Splitting of excited singlet states in metalloporphyrin sandwich dimer of D_{4h} symmetry (according to the data (16).

Right Side:
(2) $S_2 \rightarrow S_0$ fluorescence spectrum (λ_{ex} = 413.1 nm);
(3) spectrum of annihilating fluorescence (λ_{ex} = 530.9 nm).

porphyrin fluorescence is studied (τ_{S1} = 2 ns for Zn TPP and Zn TBP (7,12)). For Zn TPP in toluene the shift of ~ 30 cm^{-1} was within the range of experimental error. S_2 state lifetime of the observed metalloporphyrins (τ_{S2} = 2 - 5 ps (7,12)) is considerably shorter than encounter complex lifetime (τ_c). Time of τ_c which may be treated as the time of coexistance of two molecules in a solvent cage can be estimated proceeding from the theory of diffusion in solutions (13):

$$\tau_c = \frac{d^2}{6\,D_{AB}} \qquad (1)$$

where d = solvent molecular diameter, D_{AB} = coefficient of the molecule-partner relative diffusion. In our case, D_{AB} = 2D, where D = coefficient of metalloporphyrin molecular diffusion which may be estimated by molecular radius r and solvent viscosity η using the Stokes-Einstein formula:

$$D = \frac{kT}{6D\,\eta\,r} \qquad (2)$$

Proceeding from the molecular volume V = 400 cm^3/mole (12) and X-ray data (14) and assuming that r = 0.55 nm we get D = 4 X 10^{-6} cm^2/sec, the coefficient of metalloporphyrin diffusion at room temperature in solvents with typical viscosity η = 10^{-3} gm cm^{-1} sec^{-1}. Assuming that d = 0.5 nm (typical value for typical solvents) we get from (1) τ_c = 5.10^{-11}s $\gg \tau_{S2}$.

Thus we may consider that ADF from S_2 states takes place in a dimeric complex which dissociates after deactivation of the cooperatively excited S_2 state.

As is known, the triplet and singlet tetrapyrrolic pigment states are caused by excitation of delocalized π electrons of the heteroaromatic macrocycle (15). The most favourable structure of dimeric complexes for metalloporphyrin TTA seems to be a sandwich dimer as in the case when molecular orientation of plane to plane π electronic shells reaches the maximum overlap. Quantum chemistry calculations of metalloporphyrin dimers indicates that for the dimeric emission process the S_2 state is split into states of higher energy (S_2^+) and lower energy (S_2^-). In the case of a sandwich dimer of D_{4h} symmetry, the following states are seen:

$$S_2^+ \equiv 4\,E_u,\; S_2^- \equiv 4\,E_g$$

At room temperature the splitting energy value ΔE (S_2^+ - S_2^-) is considerably higher than that of thermal energy kT and fluorescence from the lower S_2^- state will take place in the dimeric complex. In this case for ADF the spectral shift with respect to the PF spectrum is determined by difference of S_2 state energy in the monomer and S_2^- state energy in the dimer. However, for dimeric emission processes the S_1 state is only slightly split (ΔE (S_1^+ S_1^-) << kT except for very short distances between the partners (16)). This may be also a reason for the insignificance of spectral difference between the ADF and the PF from S_1 metalloporphyrin states.

The following conclusion of the theory (16) is extremely important. The radiative transition $S_2 \rightarrow S_0$ in a sandwich dimer is forbidden. In case of a dimer of D_{4h} symmetry, the transition S_2^- ($4E_g$) $\rightarrow$ S_0 (A_{1g}) is forbidden because of parity. There is no principle difference in the splitting nature of S_2 and S_1 states for sandwich type dimers with lesser than D_{4h} symmetry and the $S_2 \rightarrow S_0$ transition remains quasiforbidden. This makes it possible to explain low P_2 values obtained in (16) by a decrease of the $S_2 \rightarrow S_0$ transition radiative probability, i.e., by decreasing of $S_2 \rightarrow S_0$ fluorescence quantum yield in dimeric TTA complexes. In the case of non-sandwich dimer structures with location of subunits in one plane, the S_2 state also is split into two states (high $2B_{2u}$ and low $2B_{3u}$). However, two radiative transitions $S_2^+(B_{2u})$ $\rightarrow S_0$ and $S_2^-(B_{3u}) \rightarrow S_0$ are allowed. In this case the observed decrease in ADF intensity cannot be explained. Dependence of ADF spectral shift on solvent type (see Table II) also confirms the correctness of a TTA complex sandwich structure: the shift decreases when solvents which are able to coordinate with metal central atoms are used. Here, an extra-ligand is being created perpendicular to the plane of porphyrin ring (pyridine, dimethylformamide). Extra- ligand presence creates steric obstacles for porphyrin macrocycles to draw together and as a result of this interaction between the subunits the corresponding splitting energy $\Delta E(S_2^+ - S_2^-)$ decreases.

Along with the experimentally determined ADF spectral shift it has been found that for ZnTPP some ADF spectral narrowing takes place as compared to the PF spectra (see Table II). It seems that the inhomogenous broadening caused by different locations of the phenyl rings with respect to the porphyrin macrocycle plane (17) has its impact on the width of tetra-meso-phenylporphyrin spectral bands. It is evident that when two ZnTPP molecules converge in a sandwich dimer the freedom of phenyl ring rotation is restricted and this decreases inhomogeneous broadening. With extra- ligand presence which obstructs close convergence of macrocycle planes the rotational freedom does not decrease and that is why for ZnTPP in pyridine spectral narrowing is not observed.

In general deactivation of triplet excited molecules is determined by the kinetic equation:

$$\frac{d\,[T_I]}{dt} = k_m\,[T_I] + k_a\,[T_I]^2 \qquad (3)$$

where k_m is the total rate constant of monomolecular decay and impurity quenching, where $k_m = k_a$ - TTA. Data obtained about the nature of excited singlet state creation during TTA processes and ideas concerning TTA mechanisms which have been derived on the basis of analyses of magnetic field effects on the ADF (see (8,9) and references to them) make it possible to imagine a scheme of porphyrin TTA in the following way:

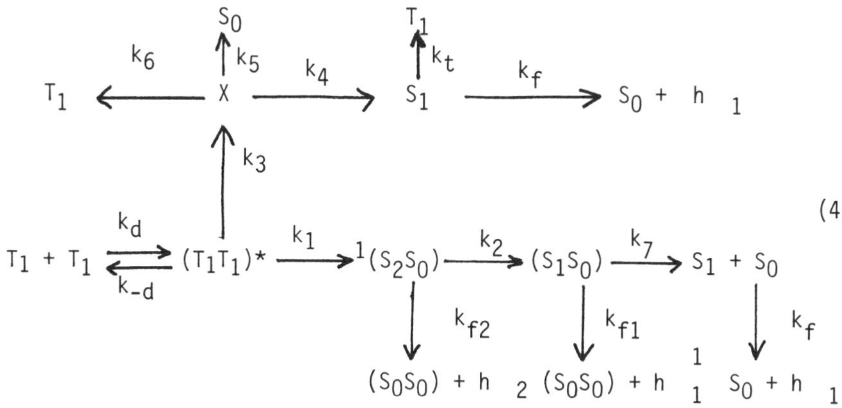

(4)

Here $(T_1T_1)^*$ is the encounter complex of two triplet molecules or, in other words, a triplet pair with energy $2E_{T_1}$ capable of dissociating back into two triplet molecules or which may undergo intracomplex relaxation (spin exchange, vibrational relaxation, electron transfer) creating, for instance, either singlet, or triplet, or quintet dimers (8,9), or ion-radical pairs (18).

$^1(S_2S_0)$ in the scheme (4) denotes singlet excited dimer and X is any other energetically allowed state of the relaxated triplet pair. Intermediate stages of triplet pair relaxation may be complicated but the only important thing to be mentioned is that one of its final stages may be once again a molecule in the T_1 state (both through S_1 state and bypassing it (8)).

Analysis of kinetic equations for the scheme (4) taking into account steady-state concentrations leads to the following relation between the TTA rate constant k_a and constant k_d, the diffusion encounter rate:

$$k_a = k_d \left[1 - \frac{1}{2} (Q\,\phi_t - R) \right] \qquad (5)$$

where

$$Q = \tau_p \left(k_1 + k_3 \frac{k_4}{k_4 + k_5 + k_6}\right. \qquad (6)$$

is the probability of S_1 state creation from the triplet pair $(T_1T_1)^*$, i.e., probability of S_1 state creation per single encounter of two triplet molecules:

$$R = \tau_p \left(2k_{-d} + k_3 \frac{k_6}{k_4 + k_5 + k_6} \right) \quad (7)$$

is probability of T_1 state creation from the triplet pair bypassing the S_1 state:

$$\tau_p = \frac{I}{k_{-d} + k_I + k_3} \quad (8)$$

is lifetime of non-relaxated triplet pair $(T_1T_1)^*$:

ϕ_t = quantum yield of $S_1 \rightsquigarrow T_1$ intersystem crossing

Equation (5) reflects a well-known experimental fact that TTA rate constant in liquid solutions is usually less than that of diffusion rate (for tetrapyrrolic pigments, for instance, $k_a \simeq (0.2 - 0.4)\ k_d$ [19].

Q probability is linked with P_a probability of S_1 excited molecule creation at annihilation of two triplet molecules (6) by the simple relation:

$$P_a k_a = Q\ k_d \quad (9)$$

Some quantitative conclusions on TTA processes can be made after examination of the concrete case of ZnTPP - a porphyrin which is widely used in photochemical research. For ZnTPP in toluene $k_a = 3.10^9 \cdot l \cdot sec \cdot mole^{-1}$ (19), $P_a = 0.24$ (6), $k_d = 1.1 \cdot 10^{10} \cdot l \cdot sec \cdot mole^{-1}$ (calculated according to the standard equation $k_d = 8RT/3\ \eta$ (10,13)). Then from (5) and (9) we get $Q = 0.065$ and $R = 1.4$. Combined analysis of the equations (6) and (7) taking into account concrete probability values for Q and R and obvious inequalities:

$$k_I > 0 \quad \text{and } 0 \leq k_4 + k_6 \leq k_4 + k_5 + k_6 \quad (10)$$

leads to the following relations between k_1 and constants and the reverse diffusion rate constant $k_{-d} = \tau_c^{-1}$:

$$k_3 \geq 3.5\ k_I \quad (11)$$

$$0.43\ k_{-d} \leq k_I + k_3 \leq 1.1\ k_d \quad (12)$$

and also

$$\frac{k_4}{k_4 + k_5 + k_6} < \frac{2}{9} \quad (13)$$

The equations (11) - (13) indicate that relaxation of the triplet pair $(T_1T_1)^*$ through an intermediate state X, i.e. bypassing $^1(S_2S_0)$, is the important process but the efficiency of S_1 molecule creation from X is low. The equation (12) makes it possible to estimate the $(T_1T_1)^*$ pair lifetime. For ZnTPP in toluene (when T = 293 K; η = 0.6 X 10^{-3} gm cm^{-1} sec^{-1}; r = 0.55 nm; d = 0.57 nm) we get from (1) and (2) k_{-d} = 2.5 • 10^{10} s^{-1} and correspondingly from (8) and (12), 20 ps $\leq \tau_p \leq$ 30 ps.

Assuming that k_1 and k_3 are linked with each other by spin-statistical probabilities 1/9 and (3+5)/9, i.e., $8k_1 = k_3$, it is easy to get $k_1 \geq k_3 k_4/(k_4 + k_5 + k_6)$ from (6), (7) and (10). It means that the ZnTPP S_1 excited molecule is created by TTA process predominantly by relaxation of the triplet pair through an upper excited state of the dimeric complex, i.e., P_2 values may be quite high ($0.5 P_a \leq P_2 \leq P_a$).

Acknowledgments

We would like to express our gratitude to G.D. Egorova and T.F. Kachura for synthesizing and purifying the observed compounds and to N.M. Ksenofontova for the help in taking measurements.

Literature Cited

1. Nickel, B., Helv. Chim. Acta, 1978, 61, 198.
2. Stelmakh, G.F., Tsvirko, M.P., Opt. i Spectrosk., 1980, 49, 511.
3. Stelmakh, G.F., Tsvirko, M.P., Opt. i Spectrosk., 1980, 48, 185.
4. Bagdasaryan, Kh.S., Chem. Phys. (Russian), 1982, 3, 391.
5. Darwent, J.R., Douglas, P., Harriman, A., Porter, G., Richoux, M.-C., Coord. Chem. Rev., 1982, 44, 83.
6. Stelmakh, G.F., Tsvirko, M.P., Zh. Prikl. Spectrosk., 1982, 36, 609.
7. Stelmakh, G.F., Tsvirko, M.P., Opt. i Spectrosk., 1983, 55, 858.
8. Lendi, K., Gerber, P., Labhart, H., Chem. Phys., 1976, 18, 449.
9. Birks, J.B., Repts. Progr. Phys., 1975, 38, 903.
10. Ermolaev, V.L., Bodunov, E.V., Svechnikova, E.B., Shakhverdov, F.A., "Non-Radiative Transfer of the Energy of Electronic Excitation"; Nauka, Leningrad, 1977, in Russian.
11. Parker, C.A., Photoluminescence of Solutions, Elsevier, Amsterdam, 1968.
12. Aaviksoo, J., Freiberg, A., Savikhih, S., Stelmakh, G.F., Tsvirko, M.P., Chem. Phys. Lett., 1984, III, 275.
13. North, A.M., The Collision Theory of Chemical Reactions in Liquids. London: Methuen, New York: Wiley, 1964.
14. Collins, D.M., Hoard, J.L., J. Amer. Chem. Soc., 1970, 92, 3761.
15. Gouterman, M., In "The Porphyrins", Vol. 3; Dolphin, D., Ed.; Academic Press; New York, 1978, p. 1.
16. Kuzmitsky, V.A., Kravchuk, O.V., Solovjov, K.N., Zh. Prikl. Spectrosk., 1980, 33, 326.
17. Eaton, S.S., Eaton, G.R., J. Amer. Chem. Soc., 1975, 97, 3760.
18. Ballard, S.G., Mauzerall, D.C., J. Chem. Phys., 1980, 72, 933.
19. Pekkarinen, L., Linschitz, H., J. Amer. Chem. Soc., 1960, 82, 2407.

RECEIVED June 3, 1986

9

Triplet-State Decay Rate Constants of Porphyrin in the Presence of Carotenoids

An Electron Paramagnetic Resonance and Electron Spin Echo Study

L. D. Kispert[1], J. Joseph[1,3], C. Lin[2], and J. R. Norris[2]

[1]Department of Chemistry, University of Alabama, Tuscaloosa, AL 35486
[2]Chemistry Division, Argonne National Laboratory, Argonne, IL 60439

EPR and ESE study of the triplet-state decay of 1 mM tetraphenylporphyrin (TPP) in a 1:1 ethanol/toluene or micellar frozen glass at 4 K containing 1 mM of either trans β-carotene, β-8'-apocarotenal or canthaxanthin show that carotenoids with carbonyl functional groups increase the triplet-state decay rate constant of the porphyrin. No effect is observed in the presence of trans-β-carotene which does not contain a carbonyl functional group. Similar measurements on tetraphenylporphyrin tetrasodium sulfonate (TPPS) in a water/glycerol glass containing crocetin shows the same trend. The triplet-state decay rate constants, were determined by EPR methods by noting the phase angle of the modulated Hg/Xe light source required for maximum EPR signal relative to a standard reference phase. More accurate decay rate constants were determined by ESE measurements using the 530 nm light from the output of a frequency doubled Nd:YAG laser.

Cartenoids are one of the most widely distributed natural pigments. These polyenes are considered to be essential for photosynthesis (1,2) and in all green plants and algae carotenoids occur along with chlorophyll, a magnesium complex of porphyrin. It is believed that in the photosynthetic apparatus, carotenoids play a dual role, on the one hand as a light harvesting antenna (1), and on the other as a photoprotect device (2) to shield the plant tissues from the lethal effects of singlet oxygen formed during photosynthesis. Therefore it is obvious that studying the nature of the interactions between porphyrins and carotenoids would not only shed more light on the intricacies of photosynthesis but may also lead to the design of better solar energy converters.

[3]Current address: National Biomedical ESR Center, The Medical College of Wisconsin, 8701 Watertown Plank Road, Milwaukee, WI 53226

0097-6156/86/0321-0128$06.00/0

Recently it has been reported (3) that in a triad molecule where a porphyrin is juxtaposed between a carotenoid and a quinone, a charge transfer donor-acceptor pair with a lifetime similar to that found experimentally in biological systems was produced on light irradiation. It was suggested that an electrical potential similar to the type developed in this donor-acceptor pair may be important in driving the chemical reactions in natural photosynthesis.

The triplet-state of carotenoids has been very difficult to observe by EPR in artificial systems (in vitro) since the singlet excited states are found to be extremely short lived (4,5), making it not possible to populate a triplet-state by intersystem crossing (ISC). The only instance where the triplet state of β-carotene has been observed in vitro by EPR has been in detergent micelles and phospholipid vesicles by Frank et al (6). Unfortunately, the EPR spectra are extremely weak, making it difficult to study in detail. Nonetheless, carotenoid triplet states can be formed and are being studied by EPR (in vivo) in photosynthetic bacteria where the carotenoid triplets are possibly formed by triplet-triplet transfer (6) or by other processes (7). The carotenoid triplet state lifetime is rather short (e.g. 9 μs for β-carotene in benzene solution (8)) compared to most molecules studied by EPR. Flexibility and twisting of the long molecules may provide the pathway for the radiationless deactivation of their triplets (9). Although the triplet state of carotenoids are difficult to study by EPR, the triplet state of porphyrins have been studied in great detail by EPR (10). Using a modulated light source, one can eliminate the spurious radical signals from the porphyrin triplet-state EPR spectrum (11). By taking note of the phase angle of the modulated light required for the most intense EPR triplet-state line for the x, y and z components at high modulation frequency and low microwave power, the k_x, k_y and k_z triplet decay rate constants can be determined (12). Normally, the phase angle is calibrated by knowing the value for one of the levels. The triplet state decay rate contants for the x, y and z levels and the population rates have been determined for TPP in several solvents by ODMR measurements (13). It has been shown that typically the 6 line EPR spectrum of a frozen glass at short observing times has line intensities that alternate ea^weae^wa, where "e" means emission, "a" means absorption and w means weak. The relative emission or absorption characteristic is very dependent on phase angle at a given modulation frequency. At a modulation frequency of 16 Hz and at a temperature near the melting point of the glass, the EPR spectrum exhibits all transitions in the absorption mode while at a modulation frequency of 1000 Hz the six line EPR spectrum exhibits the ea^weae^wa pattern. At low modulation frequency, the long time interval between pulses allows the spin system to relax to that given by a Boltzmann distribution (14). Despite the fact that the carotenoid-porphyrin interaction is very significant in photosynthesis, there has been no report on the effect of carotenoid or on the porphyrin triplet state in frozen glasses. Here we report results of EPR and electron spin echo (ESE) studies on how the triplet state of porphyrin is affected by the presence of a cartenoid in binary systems.

Experimental

The carotenoids, trans-β-carotene (Sigma), I; canthaxanthin (Fluka), II; β-8'-apocarotenal (Fluka), III; and crocetin (Sigma), IV were used in this study (see structure). Of the above four carotenoids, only crocetin is water soluble. The porphyrins; 5,10,15,20-tetraphenyl-21H,23H-porphine (TPP) and 5,10,15,20-tetra-(4-sodiumsulfonatophenyl)-21H,23H-porphine, TPPS were obtained from Aldrich. TPPS is a water soluble porphyrin. Carotenoids I-III were used as supplied, however crocetin was purified according to published methods (15) just prior to use.

Samples of a 1:1 ethanol/toluene solvent mixture containing 1 mM each of porphyrin and carotenoid were made up in 4 mm pyrex EPR tubes. Care was taken to prepare the samples in the absence of oxygen and direct light. Crocetin and TPPS were dissolved in a 1:1 water/glycerol mixture to a concentration of 1 mM each. Carotenoids were also dissolved in a detergent micelle using IGEPAL CO 630 (GAF) following the procedure of Frank et al (6). All these samples formed glassy matrices on freezing. The same samples made up in suprasil quartz tubes were used for ESE measurements.

EPR spectra were recorded on an X-band Varian E-12 spectrometer. The field was calibrated using NMR markers and the microwave frequency was measured with a H/P 5246L frequency counter. EPR measurements were carried out at 110 K using a Varian variable temperature accessory. A 1 kW Hg/Xe lamp housed in a model LH 151/2 Kratos universal housing was used as the light source and the beam was passed through several inches of distilled water in a pyrex container to eliminate most of the UV and infrared radiation before reaching the sample in the EPR cavity. The light was modulated electronically by modulating the PRA model 301 lamp power supply with a Wavetek model 111 voltage controlled frequency oscillator. The EPR signal was double frequency detected, first at 100 kHz (EPR field modulation) and secondly, at the light modulation frequency using an EG and G lock-in amplifier model 128 A. ESE measurements were carried out at 4 K using 530 nm laser light and details of the spectrometer are given elsewhere (16). Nuclear Overhauser effect (NOE) studies of the carotenoid on the porphyrin in a binary mixture were carried out using a Nicolet NT-200 multi-nuclear 200 MHz superconducting NMR spectrometer.

Results

The phase angle for maximum EPR intensity of the triplet-state at a modulation frequency of 16 Hz and at 4 K, was observed to change upon the addition of carotenoids II or III to a sample containing TPP or carotenoid IV to a sample containing TPPS. According to Equations 1 and 2 the variation in phase angle (14), θ, indicates

$$N_i = \frac{P_i}{(k_i^2 + \omega^2)^{0.5}} \sin(\omega t + \tan^{-1} \frac{\omega}{k_i}) \tag{1}$$

$$\tan \theta = \frac{\omega}{k_i} \tag{2}$$

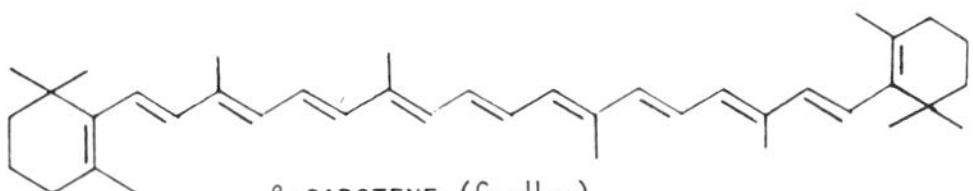

I

β-CAROTENE ($C_{40}H_{56}$)

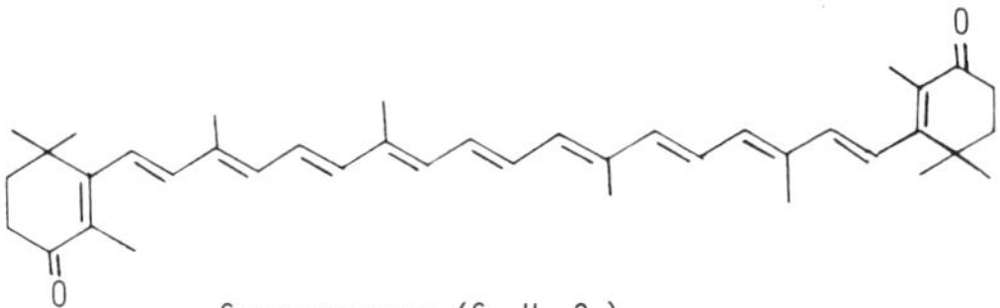

II

CANTHAXANTHIN ($C_{40}H_{52}O_2$)

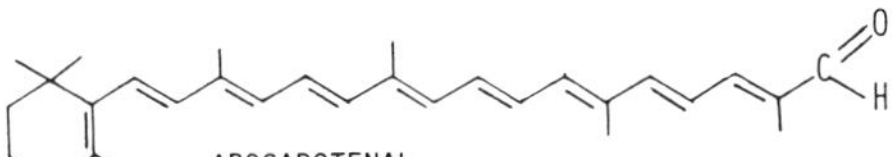

III

APOCAROTENAL

(β- APO-8′-CAROTENAL); ($C_{30}H_{40}O$)

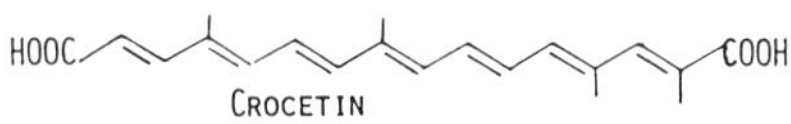

IV

CROCETIN

(8,8′-DIAPOCAROTENEDIOIC ACID) $C_{20}H_{24}O_4$

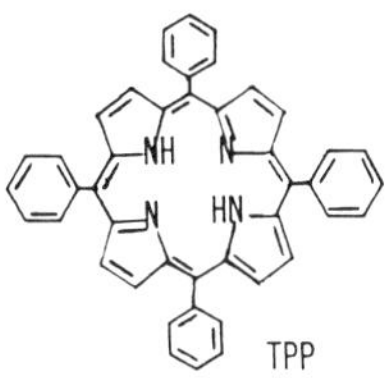

TPP

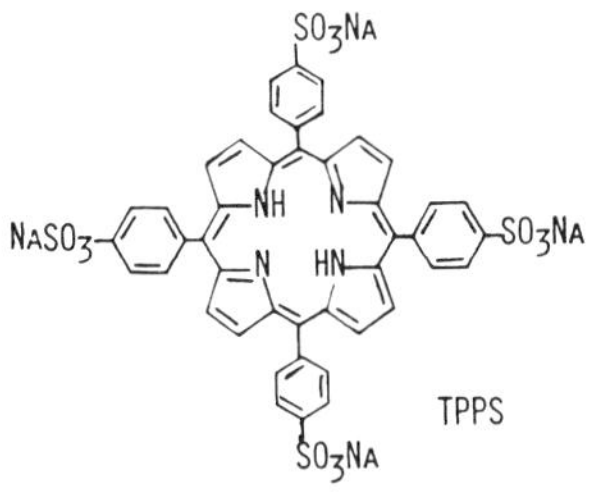

TPPS

Structure

that a change has occurred in k_i, the kinetic decay rate constant for the ith triplet-state level. Equations 1 and 2 assume that spin-lattice relaxation time, T_1, is not competitive; ie. $1/T_1 << \omega$. The intensities of the EPR transitions are of course dependent on the differences in the spin population N_i of the ith triplet-state level. N_i in turn depends on P_i, the relative populating rates for the ith level, k_i, the kinetic decay rate constant for the ith triplet state level; and ω, the light modulation frequency at time t. Since frequency ω is the known audio oscillator frequency, the rate constants k_i can be measured by noting the phase angle relative to a reference light signal at which the maximum EPR signal occurs (12,14). This change in phase is depicted in Figure 1 as a variation in the relative intensity of the high and low field x, y, and z components of the porphyrin triplet-state as a function of carotenoid since the spectra were recorded using the phase angle that gave the most intense components at a modulation frequency of 16 Hz for TPP at 4 K. The relative intensity ratio of the low field to the high field z component varied from 2.3 for I to 1.8 for II to 1.1 for III and for the y component from 1.5 for I to 1.2 for II and III. The emission character of the high field x component appears to decrease from I to III and II. The $ea^{w}eae^{w}a$ spectra intensity pattern at low (16 Hz) modulation frequencies normally expected has an asymmetry between the height of the low field emission (e) line (z) to that of the high field absorption (a) line (z). At low modulation frequencies, T_1 causes the (a) lines to have some (e) character and the (e) lines to be more intense. At low modulation frequencies, small changes in k_i's will produce variations in all line intensities when the phase angle for the maximum intensity of one line is used as a reference point.

This change in the triplet decay rate constant was verified by ESE measurements where spin lattice relaxation effects can be minimized from the decay rate constant measurements (14). Observed decay rate constants are given in Table I for the y triplet level

Table I. The triplet-state decay rate constants*, k_y, (s^{-1}), for TPP and TPPS in the presence of carotenoids

Porphyrin Medium	Carotenoids None	β-carotene	Apocarotenal	Crocetin
TPP EtOH/Tol	957(3)	760(3)	2090(115)	--------
TPP Micelle	608(1)	581(18)	873(36)	--------
TPPS H_2O/glycerol	432(12)	--------	401(9)	532(72)

*ESE measurement at 4 K (Argonne National Laboratory). Numbers in parenthesis are the computed errors [(01) = ±1] from the ESE decay curves. The rate constants were derived from the reciprocal of the ESE measured lifetimes for the downfield y-component.

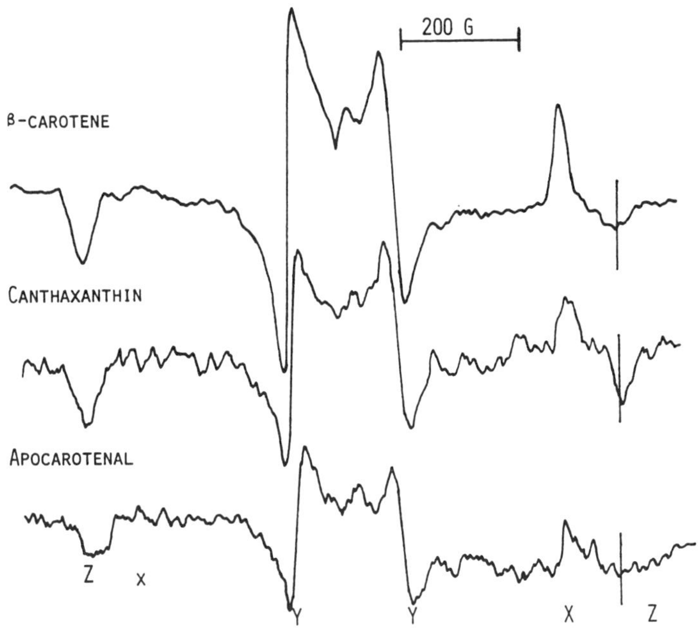

Figure 1. EPR triplet-state spectra of TPP in a toluene-ethanol glass at 10 K containing the indicated carotenoid.

(center lines). A determination of the decay rate constant for the x and z levels were not measured for any samples due to the poor ESE signal to noise for samples containing apocarotenal and canthaxanthin. It is to be noted that little change occurs in k_y for TPP upon addition of β-carotene. However addition of apocarotenal significantly increased k_y for TPP in either an EtOH/toluene or a micelle medium. k_y for TPPS in a H_2O/glycerol solution was not changed upon addition of the water insoluble apocarotenal probably due to solubility problems, however a significant increase in the decay rate constant did occur upon the addition of water soluble crocetin.

It is to be noted that k_y for TPP or TPPS increases upon the addition of a carotenoid containing a carbonyl functional group. One possible mechanism for the increase in k_y could be the presence of a triplet-triplet energy transfer. Therefore, attempts were also made to detect the carotenoid triplet-state. Unfortunately none was observed for a mixture of apocartenal and TPP in a 1:1 toluene/ethanol frozen glass where the large change in k_y was observed nor upon replacing the apocartenal by β-carotene. Further attempts were tried using various single crystals and powdered mixtures containing approximately 1% β-carotene using various laser sources as the excitation source (Table II). It was thought that better signal to noise might be possible in a single crystal. Crystals containing triplet sensitizers such as naphthalene or

Table II. Attempts to Detect Carotenoid Triplet-state by ESE*

Sample	Laser Source	Results
Frozen 1:1 ethanol/toluene glass containing 1 mM β-carotene and TPP	Nd:YAG-530 nm 4 K	No transient carotene triplet-state
Frozen 1:1 ethanol/toluene containing apocarotenal and TPP	Nd:YAG-530 nm 4 K	No transient carotene triplet-state
Pellet - 2% β-carotene in rubrene (CH_2Cl_2)	Nd:YAG-355 nm 3.7 K, 26 J	No transient carotene triplet-state
Crystal of C_6H_5-$(CH)_6$-C_6H_5 containing β-carotene ($\leq$1%)	488 nm Argon ion 0.4 watt CW	No transient carotene triplet-state
Crystal of C_6H_5-$(CH)_6$-C_6H_5 containing β-carotene ($\leq$1%)	Nd:YAG 355 nm 3.7 K, 26 J	No transient carotene triplet-state
β-Carotene pellet	Nd:YAG-355 nm 3.7 K, 26 J	No transient carotene triplet-state
Crystal of Benzophenone containing β-carotene ($\leq$1%)	Nd:YAG-355 nm 3.7 K, 26 J	No transient carotene triplet-state

*Spin Echo spectrometer facilities at Argonne National Lab.

pellets of rubrene and β-carotene failed to produce any carotene triplet signals. Efforts using either EPR or ESE techniques were also tried (Tables II and III) using hosts that did not contain

Table III. Attempts to Observe Triplet-States of β-carotene Doped in Various Hosts by EPR Methods

Host	Irradiation	Observed
Durene Crystal	X-ray	No triplet due to β-carotene
Durene Crystal	UV(CW) Hg Arc.	No triplet due to β-carotene
Glycerol	UV	No triplet due to β-carotene
$(C_6H_5)_2(CH)_6$ Crystal 3 K	UV	No triplet due to β-carotene
$(C_6H_5)_2(CH)_8$ Pellet 3 K	UV	No triplet due to β-carotene
Naphthalene Crystal	UV	(Only naphthalene triplet state)
Naphthalene + Durene crystal	UV	(Only naphthalene triplet state)
Benzophenone Crystal	UV(CW)	No triplet
β-Carotene Crystal (soln CH_2Cl_2)	UV	No triplet

triplet sensitizers. However, in all cases no triplet EPR spectrum such as reported by Frank et al. (6) in photosynthetic bacteria was observed from β-carotene. This failure (at least for III in the presence of TPP) is presumably due to the concentration of carotenoid triplet being too low, a possibility if the triplet-state lifetime is on the order of 10 μs being populated at a 1 ms rate. On the other hand, the lifetime of the carotenoid triplet in the presence of porphyrin might be too short (< 50 ns) to be observed by ESE techniques. The observed ESE signal to noise ratio for the triplet-state y-level of TPP and III equaled 20/1.

To determine whether carotenoids containing carbonyl groups associate with porphyrin in solution, a nuclear Overhauser effect (NOE) experiment was carried out. A solution containing 1mM each of III and TPP in a 1:1 mixture of 99% toluene-d_8 and 99% ethanol-d_6 gave rise to the NMR spectrum depicted in Figure 2A before saturation of the methyl protons of III. Applying the saturating rf pulse gave rise to the spectrum in Figure 2B. The carotenoid methyl peaks at 1.8-1.9 ppm are no longer observable, however the difference spectrum in Figure 2C shows that the height of the phenyl protons in TPP located between 7.6 to 8.2 ppm are not changed. Unfortunately the absence of an NOE effect can be attributed to many different conditions, so the presence of carotenoid - porphyrin association in solution can neither be proved nor disproved.

Discussion

The triplet state decay rate constant (k_y) of TPP is increased in the presence of the carotenoids apocarotenal and canthaxanthin but the addition of β-carotene seems to have little effect. In the case of TPPS it is seen that the water insoluble apocarotenal has no effect whereas the water soluble crocetin increases k_y as was expected.

The triplet state of II in benzene solutions has been measured (17) optically to have a lifetime of 3.8 μs. The triplet state of II plus those of other carotenoids were formed by pulse radiolysis of a benzene solution containing a carotenoid (30 μM) and naphthalene (0.01 M) as a triplet sensitizer. The naphthalene intercepted most of the excited states of benzene and yielded after a few ns a longer lived naphthalene triplet that underwent energy transfer to acceptors of lower energy. The triplet lifetimes of the various carotenoids formed in solution fall in the range of 2-10 μs, a time easily observable by the ESE measurements. The failure to observe carotene triplet states by ESE in the present study suggests that the triplet-state concentration is below the limit of detectability. β-Carotene is used (18) in energy transfer experiments because of its low triplet energy level, small triplet yield and intense triplet absorption band. Because of the intense absorption band, the presence of a triplet has been usually detected optically (8,19,20). Assuming the yield of carotene triplet is below the sensitivity of our ESE measurements the most probable mechanism by which the triplet state lifetime of porphyrin was decreased is by a triplet-triplet energy transfer from the triplet state of porphyrin to the carotenoid. For this to happen the molecule in the frozen glasses must lie within close association to each other, (since diffusion processes would be too slow) a possibility in frozen glasses if the carbonyl group of the carotenoid is hydrogen bonded to the N-H group of the porphyrin. The possibility of hydrogen bonding is consistent with the absence of any interaction with β-carotene which has no carbonyl functional groups. Unfortunately, the NOE experiments failed to confirm this assumption. Furthermore the carotenoids with carbonyl groups allow the transition $S \longrightarrow T_{n,\pi*}$ which is more favored than the corresponding $S \longrightarrow T_{\pi,\pi*}$ pathway possible in β-carotene having no non-bonding electrons.

It has also been observed (19) that retinal triplet-triplet absorption in hexane can be observed in the absence of a sensitizer because the presence of a carbonyl group that gives rise to a pair of singlet and triplet n,π* states which lie intermediate in energy between the π,π* singlet and π,π* triplet states. It is assumed that the π,π* singlet decays to the n,π* singlet state followed by intersystem crossing to the n,π* triplet followed by rapid internal conversion to the π,π* triplet. However a similar study (8) with II and III in solution found no transient absorption in the absence

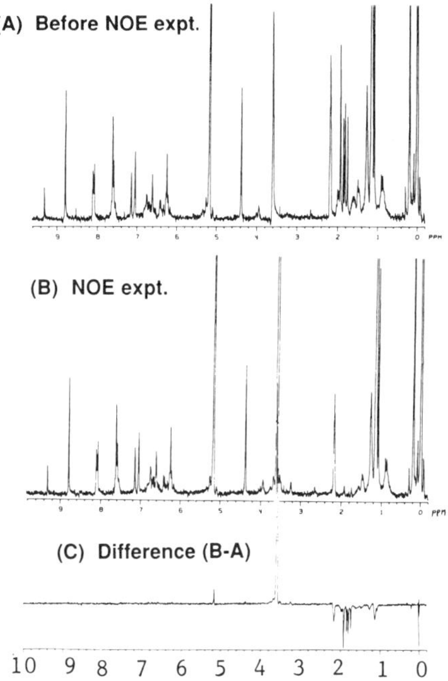

Figure 2. The proton NMR spectra of a 1 mM mixture of apocarotenal and TPP in a toluene-d_8/ethanol-d_6 1:1 solvent, (A) before the NOE experiment, (B) NOE spectrum, and (C) difference of (B) and (C).

of a sensitizer suggesting that the longer polyeneals had the n,π^* singlet states above the π,π^* states. It is known (8) that normal singlet and triplet absorption spectra are shifted to longer wavelengths with increasing chain length. Unfortunately, it is not known what effect the presence of ethanol/toluene as a solvent has on the energy levels in the carotenoid/TPP or TPPS system. However, it would appear that n,π^* singlet and triplet states do play a role in the triplet-triplet energy transfer in the carotenoid/porphyrin binary system.

The kinetic decay rate constant of the y-triplet level determined by ESE measurements and reported in Table I can be converted to the equivalent lifetime. This lifetime equals 1.05 ms for TPP in a 1:1 mixture of toluene and ethanol as the solvent. This is consistent with ODMR measurements (13) of the same level for TPP in n-octane where the lifetime equaled 1.45 ms.

Conclusion

It is found that carotenoids with carbonyl functional groups increase the triplet decay rate constants of the porphyrins TPP or TPPS in frozen toluene/ethanol glasses possibly by triplet-triplet transfer. The triplet-triplet transfer could be facilitated by the presence of hydrogen bonding between carotenoids containing carbonyl groups and the prophyrins. Carotenoids with no polar functional groups like β-carotene do not interact with porphyrin so as to affect the triplet life time of the latter.

Acknowledgments

We wish to thank V. Farr for running the NOE experiments and J. Dechter for extensive discussions regarding the NOE experiments. This work was supported by the Department of Energy (Office of Basic Energy Sciences) under contracts DE-AS05ER04062 and W-31-109-ENG-38. This is DOE document ORO-4062-95.

Literature Cited

1. Sauer, K. "Bioenergeties of Photosynthesis"; Govindjee Edn.; Acad. Press: New York, 1975; pp. 115-181.
2. Renger, G.; Wolff, Ch. Biochim. Biophys. Acta 1977, **460**, 47.
3. Moore, T. A.; Gust, D.; Mathis, P.; Mialocq, J-C.; Chachaty, C.; Bensasson, R. V.; Land, E. J.; Doizi, D.; Liddell, P. A.; Lehman, W. R.; Nemeth, G. A.; Moore, A. L., Nature 1984, **307**, 630.
4. Wasielewski, M. R.; Kispert, L. D. unpublished picosecond absorption results.
5. Dallinger, R. F.; Woodruff, R. F.; J. Rodgers, M. A. Photochem. Photobiol. 1980, **33**, 275.
6. Frank, H. A.; Bolt, J. D.; Costa, S. M. de B.; Sauer, K., J. Am. Chem. Soc. 1980, **102**, 4893.
7. Njijs, A. M.; Van Grondelle, R.; Joppe, H. L. P.; Van Bochore, A. C.; Duysens, L. N. M. Biochem. Biophys. Acta 1985, **810**, 94-105.
8. Truscott, T. G.; Land, E. J.; Sykes, A., Photochem. Photobiol. 1973, **17**, 43.

9. Jensen, N-H.; Wilbrandt, R.; Paysberg, P. B.; Sillesen; A. H.; Hansen, K. B. J. Am. Chem. Soc. 1980, **102**, 7441.
10. Van der Waals, J. H.; Van Drop, W-G.; Schaafsma, T. S. "The Porphyrins"; D. Dolphin Ed.; Academic: New York, 1979: pp. 257-309.
11. Lhoste, J. M.; Grivet, J. Ph. Adv. Radiat. Res. Phys. Chem. 1973, **1**, 327.
12. Levanon, H.; Weissman, S. L., J. Am. Chem. Soc., 1971, **93**, 4309. Paul, H. Chem. Phys. 1976, **15**, 115.
13. Van der Brent, S. J.; Schaafsma, T. J. Chem. Phys. Lett. 1975, **35**, 45.
14. Thurnauer, M. C. "Reviews of Chemical Intermediates"; Verlag Chemie International, Inc., 1979; p. 197.
15. Craw, M.; Lambet, C. Photochem. Photobiol. 1983, **38**, 241.
16. Norris, J. R.; Thurnauer, M. C.; Bowman, M. K., 'Advances in Biological and Medical Physics," Lawrence, J. H.; Optman, J. W.; Hayes, T. L., Ed.; Academic: New York, 1980; Vol. 17, p. 365.
17. Rodgers, M. A. J.; Bates A. L., Photochem. Photobiol. 1980, **31**, 533.
18. Beaumont, P. O.; Parsons, B. J.; Phillips, G. O.; Allan, J. C., Biochem. Biophys. Acta 1979, **562**, 214.
19. Abrahamson, E. W.; Ostroy, S. E. Prog. Biophys. Molec. Biol. 1967, **17**, 181.
20. Mathis, P.; Kleo, J. Photochem. Photobiol. 1973, **18**, 343.

RECEIVED July 2, 1986

10

Electron Paramagnetic Resonance Study of Dimerization Effects on Porphyrins in the Photoexcited Triplet State

Hans van Willigen, T. K. Chandrashekar[1], U. Das, and Marie H. Ebersole

Department of Chemistry, University of Massachusetts at Boston, Boston, MA 02125

An EPR study is made of the effect of dimerization on the photoexcited triplet state of a series of porphyrins. The systems of interest are tetra(benzo-15crown-5)porphyrin (TCP), tetra(4-sulfonatophenyl)porphyrin (TPPS), tetra(4-trimethyl-ammoniumphenyl)porphyrin (TTAP), and Zn-substituted derivatives of these porphyrins. Dimerization of TCP and ZnTCP can be induced by addition of cations. In aqueous solution TPPS and ZnTPPS are present in the form of monomers and dimers. Finally, admixture of TPPS (ZnTPPS) and TTAP (ZnTTAP) leads to dimer formation.

The study of these systems is concerned with the question of whether or not dimerization effects on zero field splitting (zfs) and kinetic parameters of the photoexcited triplet state can be related to dimer geometry. In the case of ZnTCP dimerization effects can be interpreted in terms of fast triplet excitation energy transfer between dimer constituents. This means that there is a well-understood relationship between dimerization effects and dimer geometry. The data on the other systems best fit a model in which the excitation energy is localized on one of the porphyrin rings. Dimerization induced changes in zfs and kinetic parameters in this case must be attributed to a perturbation of porphyrin structure and do not give an insight into dimer geometry.

Zero field splitting (zfs) values in photoexcited triplets of primary donor bacteriochlorophyll _a_ in photosynthetic bacteria are much lower than those found for _in vitro_ BChl_a_ triplets. There is a pronounced difference in kinetics of population and depopulation of the triplet sublevels as well. The differences have been attributed to the effect of BChl_a_ dimerization and it is now generally accepted that the primary electron donor in photosynthetic bacteria consists of a BChl_a_ dimer (special pair)(_1-6_).

The data on the photoexcited triplets are a potential source of structural information. Specifically, dimerization effects on zfs's and kinetics may be of aid in the characterization of special pair

[1]Current address: Department of Chemistry, Michigan State University, East Lansing, MI 48824

0097-6156/86/0321-0140$06.00/0

geometry and electronic structure (2-6). To obtain an insight into the relation between spectroscopic data and structure, studies are made of photoexcited triplets of model dimer systems. The studies focus on covalently-linked chlorophylls(7) and porphyrins(8) as well as dimers formed by spontaneous aggregation of tetrapyrroles (9-13). This paper is concerned with results of studies of model systems belonging to the latter category.

The molecules of interest are tetra(benzo-15-crown-5)porphyrin (TCP, I), tetra(4-sulfonatophenyl)porphyrin (TPPS, II), and tetra(4-trimethylammoniumphenyl)porphyrin (TTAP, III). (See Structure.)

TCP and its metal-substituted derivatives can be made to dimerize quantitatively by the addition of cations (NH_4^+, K^+, Cs^+, Ba^{2+}) to solutions of these porphyrins (14). The stoichiometry of the dimerization reaction establishes that two porphyrins are linked by four crown-cation-crown bridges (14). This forces the porphyrins in a parallel configuration. For steric reasons, one ring system must be rotated relative to the other. EPR (14) and ENDOR (15) studies of the ground state triplet $[CuTCP]_2$ and $[VOTCP]_2$ dimers fit a model of two parallel porphyrins with the transition metal ions positioned on a common axis perpendicular to the planes, with a center-to-center distance of 0.43 nm.

UV-VIS spectroscopic studies (9-11,16) establish that in aqueous solution TPPS (and its metal-substituted derivatives) is in equilibrium with $[TPPS]_2$. Dimerization can be promoted by increasing the ionic strength of the solution or by addition of cation-18crown6 (9-11) The EPR spectrum of CuTPPS in frozen H_2O-glycerol establishes the presence of a mixture of monomers and dimers. The EPR spectrum of $[CuTPPS]_2$ shows that the Cu-Cu distance is ~ 0.56 nm.

The combination of cationic (TTAP) and anionic (TPPS) water soluble porphyrins also leads to the formation of a stable 1:1 complex [TTAP/TPPS] (13,17). It is evident that dimerization is promoted by electrostatic interaction between the oppositely charged substituents. It is likely, therefore, that a face-to-face dimer is formed. EPR spectra of the [CuTTAP/CuTPPS] dimer indicate that the Cu-Cu distance is larger than 0.6 nm.

The study of these face-to-face dimers is concerned with the question of whether or not dimerization effects on zfs and kinetic parameters of the photoexcited triplet state can be related to dimer structure.

Experimental

TPPS (sodium salt) was obtained from Strem Chemicals and used as received. TTAP and some of its metal-substituted derivatives were kindly by Dr. E. Ojadi of Brandeis University. The other porphyrins used in this study were synthesized following published procedures (14,18).

EPR spectra were recorded with a Varian E9 X-band spectrometer using field (100 kHz) and light (13 or 83 Hz) modulation with phase-sensitive detection at the modulation frequencies (19). Typically, the field modulation amplitude employed ranged from 20 to 40 gauss, the microwave power from 0.1 to 0.5 mW. Measurements were performed on frozen solutions of the porphyrins at about 100 K using the standard Varian variable temperature accessory or at about 10 K with an Oxford Instruments helium gas cryostat. Light sources used for photoexcitation were a 1000 W Xe arc source powered by a Photochemical Research Associates Supply with electronic modulation capability and Spectra Physics argon (514.5 nm, 0.5 W) or krypton (647.1 nm, 0.5 W) lasers. Optical absorption spectra were recorded with a Cary 14 and fluorescence spectra with a Perkin-Elmer 650-40 spectrometer. A Nicolet 1180E computer interfaced to the Varian and Cary spectrometers was used for data acquisition and analysis. The kinetics of formation and decay of triplet ESR signals was recorded by exciting with square-wave-modulated light. The computer, used in the time-averaging mode, was triggered at the light-on edge, and the EPR signal amplitude vs. time profile was stored in a 1K memory block.

Results

Optical spectra

Dimerization effects on absorption and emission spectra of TCP are shown in Figures 1 and 2. Similar effects are observed upon dimerization of ZnTCP (12), formation of [TPPS/TTAP] and [ZnTPPS/ZnTTAP] (17). It is found that dimerization is accompanied by (a) a blue shift of the Soret band, (b) red shifts of the Q bands, (c) red shift of the fluorescence maximum, and (d) strong quenching of fluorescence. Optical spectra of face-to-face covalently-linked diporphyrins show the same trends (20,21). However, in one instance formation of a covalently-linked diporphyrin does not lead to a Soret band shift (22). Of the systems considered here, $[TPPS]_2$ and $[ZnTPPS]_2$ also fail to show a Soret shift even though the visible absorption bands and fluorescence show clear evidence of dimer formation (10). A summary of observed dimerization induced absorption band shifts is given in Table I.

EPR spectra

The EPR spectrum of the photoexcited triplet of TCP randomly oriented in frozen solution, recorded with 83-Hz light modulation at 10 K, is shown in Figure 3a. The effect of addition of Na^+ or K^+ to the TCP solution is illustrated in Figure 3, b and c, respectively. In general, EPR spectra of randomly oriented triplets will show three pairs of peaks (23) (labeled x, y, z in Figure 3). One pair for each of the three orientations for which the magnetic field is parallel to a principal axis (x, y, z) of the zfs tensor. The separations between the pairs of lines are given by 2D and D $\pm$ 3E, where D and E are the zfs parameters (23). The spectra in Figure 3 show absorption as well as emission peaks as a result of the spin selectivity of the intersystem crossing and the decay from the triplet sublevels (19,24). The observed pattern of absorption (A) and emission (E) peaks (marked in the figure) corresponds to that found for other tetraphenylporphyrin (TPP) triplets (9,10,25). It can be attributed to preferential population of the zero-field T_y sublevel (24) (following convention the order of the energy levels is given as $T_x > T_y > T_z$). It is evident that the introduction of sodium ions (which does not cause dimerization (14)) does not affect the values of the zfs parameters. It merely causes some enhancement in signal amplitude which may be the result of a slight reduction in line width. On the other hand, upon addition of K^+ (i.e. dimerization

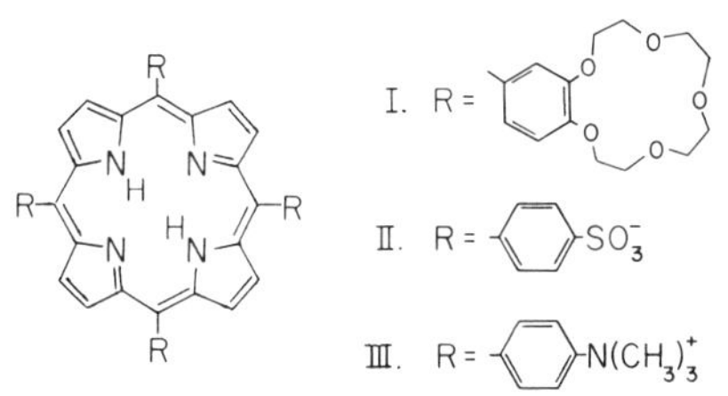

Structure

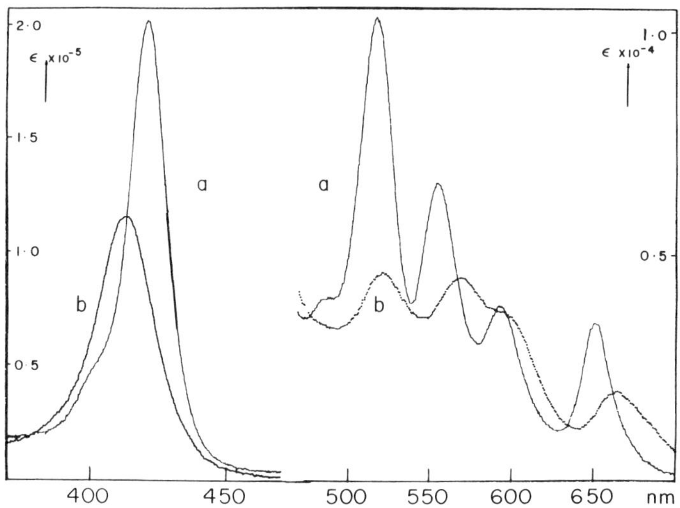

Figure 1. Optical absorption spectra of TCP (Soret, 3 x 10^{-6} M; Q bands, 6 x 10^{-5} M) in CH_3Cl-CH_3OH (1:1): (a) before and (b) after addition of KCl (~ 10^{-4}M). The vertical scale is based on monomer concentration.

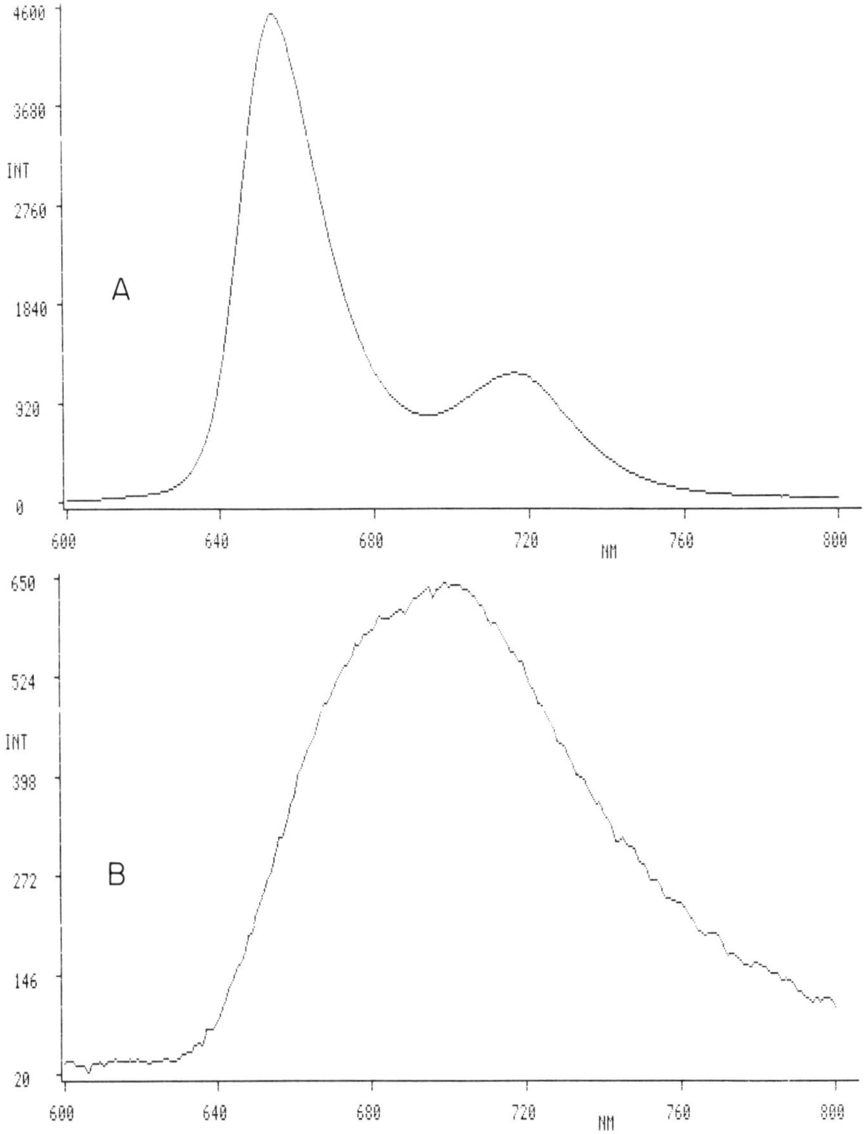

Figure 2. Fluorescence spectra of TCP (10^{-6} M) in CH_3Cl-CH_3OH (1:1) (a) before addition of KCl, excitation wavelength λ_{ex} = 426 nm, and (b) after addition of KCl (~ 10^{-4} M) λ_{ex} = 412 nm. Slitwidth 5 nm.

TABLE I: Dimerization induced absorption band shifts (cm^{-1}) for some porphyrins

System	Soret	Q bands				distance,[a] nm
$[TCP]_2$	458	-148	-411	-57	-300	0.43(14)
$[TPPS]_2$	-58	-296	-417	-519	-527	0.56
$[TPP]_2$[b]	0	-147	-192	+260	-590	
DP-7[c]	916	-200	-105	-155	-156	
DP-6[c]	1121					0.38-0.39[d]
DP-5[c]	1684					

[a]Center-to-center distance derived from EPR spectra of Cu-substituted derivatives. [b]Strati-bisporphyrin,[22] two TPP ring systems linked by four 5-atom-long bridges. [c]Diporphyrins linked by two 7-, 6-, or 5-atom-long bridges.(20) [d]Recent EPR measurements show that the distance between the porphyrin planes in these three dimers in virtually constant.

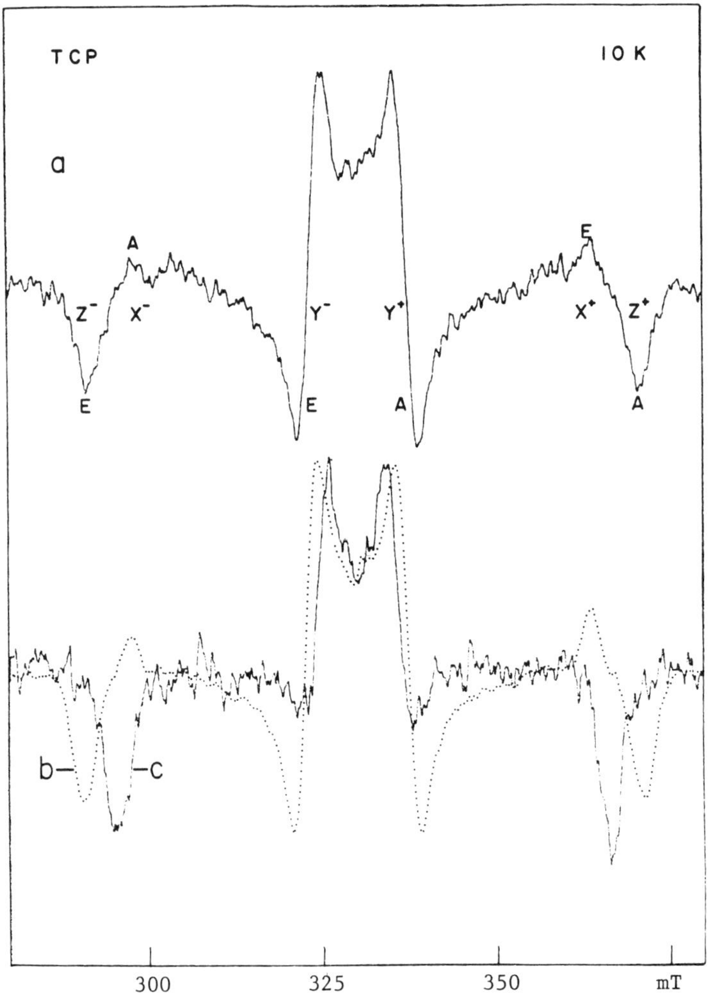

Figure 3. Triplet EPR spectra of TCP (~ 10^{-4} M) in CH_3Cl-CH_3OH recorded at about 10 K. Microwave power 0.5 mW, field modulation 20 G (100 kHz), excitation with square wave modulated (83 Hz) light of an argon laser (514.5 nm, 0.5 W). (a) No addition, (b) with NaCl (~ 5 x 10^{-3} M), and (c) with KCl (~ 5 x 10^{-3} M). Absorption and emission peaks have been labeled A and E, respectively.

(14)) the original triplet EPR spectrum is replaced by a new spectrum, reflecting a reduction in value of both D and E. The addition also leads to a marked reduction in signal amplitude. The effect of dimerization of TPPS (9,10) and formation of [TPPS/TTAP](13) on the triplet EPR spectra is very similar to that associated with $[TCP]_2$ formation.

Figure 4 shows the triplet EPR spectra of ZnTCP and $[ZnTCP]_2$ recorded at 10 K (12). The spectra reflect the absence of axial symmetry (D ≈ 3E) which is attributed to a Jahn-Teller effect (24). $[ZnTCP]_2$ formation has little or no effect on D whereas the value of E is strongly reduced (cf. Table II). By contrast, triplet EPR spectra of [ZnTPPS/ZnTTAP] (13) and $[ZnTPPS]_2$ (11) show no evidence of a dimerization effect on E. In the case of ZnTPPS dimerization leads to a reduction in D (11). As is illustrated in Figure 5, the triplet EPR spectrum of [ZnTPPS/ZnTTAP] is virtually indistinguishable from the spectra of the monomer precursors.

The triplet EPR spectra of the mixed dimers [ZnTPPS/TTAP] and [TPPS/ZnTTAP] also displayed in Figure 5 have the same characteristics as the free base dimer spectra (cf. Figures 3 and 5). A summary of the values of triplet parameters derived from the spectra is given in Table II. Included are literature values on face-to-face covalently-linked diporphyrins (8).

Discussion

The interpretation of the effect of dimerization on the values of the triplet parameters depends on the relative magnitudes of the exchange interaction (J) between the dimer constituents and the zfs parameters. The following three cases can be distinguished (5-8,24).

1. J < D, E. The triplet excitation energy is localized (on the time scale of the ESR measurement) on one porphyrin ring. If this condition applies, changes in D and E values, and triplet kinetics must stem from a dimerization induced change in structure of dimer constituents.

2. J ≥ D. There is rapid excitation energy transfer between the two porphyrins (exciton model). The values of the zfs and kinetic parameters for the dimer are determined by the monomer values and relative orientation of the porphyrin planes (5-8,24). (Assuming that the porphyrin structure is not modified by dimerization). For instance, with parallel porphyrin planes, the value of D is not affected by dimerization, whereas the effect on E depends on the relative orientation of the in-plane axes of the two rings (8).

3. J >> D. The triplet characteristics will be affected by a charge transfer contribution to the triplet electronic state. It is predicted that this will reduce the values of D and E, and increase the rate constants of decay from the triplet sublevels (5,8,24).

Dimerization causes shifts of optical absorption and emission bands of the order of several hundred cm^{-1} (cf. Table I). This suggests that the triplet EPR data must be interpreted in terms of exciton and charge transfer effects. In the case of ZnTCP the effect of dimerization on zfs values can be accounted for on the basis of rapid triplet excitation transfer between essentially unperturbed porphyrin moieties. If the exciton model applies the principal components of the zfs tensor in the dimer (X*, Y*, Z*) can be related

Table II. Zero-field Splitting parameters (in 10^{-4} cm^{-1}) and average triplet lifetimes of porphyrin monomers and dimers[a]

System	D	E	X	Y	Z	τ_{av} (msec)
TCP	377	79	205	46	-251	5
$[TCP]_2$	332	73	184	37	-221	5
TPPS or TTA	400	75	208	59	-267	9
$[TPPS]_2$[b]	334	80	191	31	-223	2[c]
[TPPS/TTAP]	343	80	195	34	-229	4
ZnTCP	298	99	199	0	-199	
$[ZnTCP]_2$	291	69	166	28	-194	
ZnTPPS or ZnTTAP	323	95	203	12	-215	
$[ZnTPPS]_2$[b]	274	100	192	-9	-183	
[ZnTTAP/ZnTPPS]	316	91	196	15	-211	
ZnTTAP/TPPS	361	75	195	46	-241	7
ZnTPPS/TTAP	380	75	201	52	-253	
mZnP4[d]	356	45	164	73	-237	145
dZnP4[d]	355	20	139	98	-237	153
mZnP5[d]	347	40	155	76	-231	
dZnP5[d]	347	25	141	90	-231	
mZnP6[d]	389	117	247	12	-259	
dZnP6[d]	365	110	231	12	-243	

[a]Measured at 10 K unless noted otherwise, estimated uncertainty $\pm$ 5 x 10^{-4} cm^{-1}.
[b]Dimerization induced by addition of K^+-18crown6 complex.[9-11]
[c]Average triplet lifetime at 100 K, monomer triplet lifetime at this temperature 5 msec.
[d]Monomer (mZnPn) and dimer (dZnPn) Zn porphyrins. Porphyrins in dimers covalently linked by two 4-, 5-, or 6-atom long chains. Measurements performed at 1.3 K. (8)

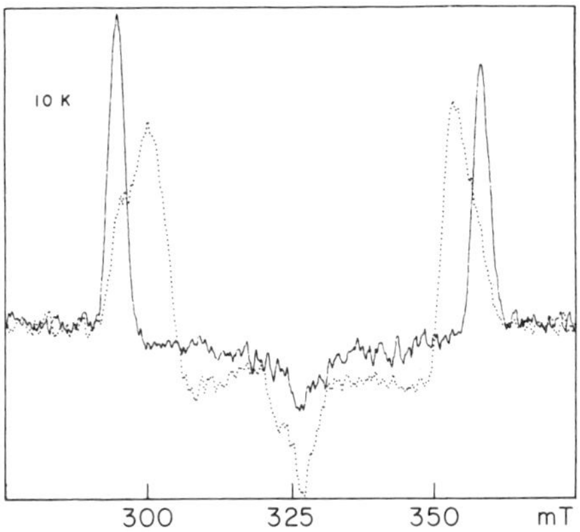

Figure 4. Triplet EPR spectra of ZnTCP (~10^{-4}) in CH_3Cl-CH_3OH recorded at 10 K before (solid line) and after (dotted line) addition of KCl (~5 x 10^{-3}). Microwave power 0.5 mW, field modulation 20 gauss (100 kHz), excitation with square wave modulated (13 Hz) light from an Xe high pressure arc (1000 W) passed through a $CuSO_4$ heat filter.

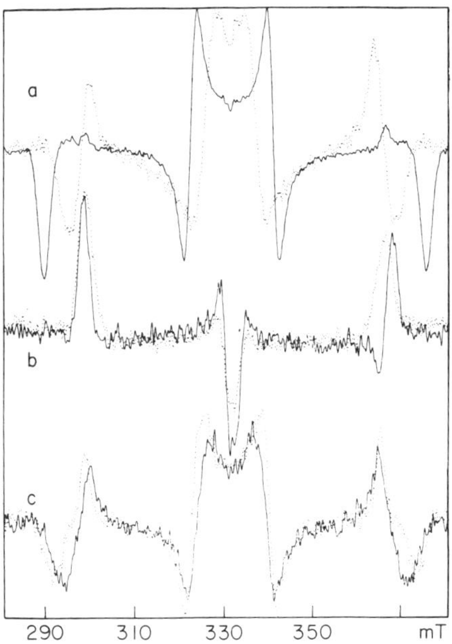

Figure 5. Triplet EPR spectra of porphyrins in frozen 1:1 H_2O-glycerol recorded at 10 K. (a)Solid line TTAP, dotted line TTAP/TPPS; (b)Solid line ZnTTAP, dotted line ZnTTAP/ZnTPPS; (c)solid line ZnTTAP/TPPS, dotted line ZnTPPS/TTAP.

to those in the monomer (X, Y, Z) by the expression (26)

$$\begin{vmatrix} X^* \\ Y^* \\ Z^* \end{vmatrix} = \begin{vmatrix} l_x^2 & l_y^2 & l_z^2 \\ m_x^2 & m_y^2 & m_z^2 \\ n_x^2 & n_y^2 & n_z^2 \end{vmatrix} \begin{vmatrix} X \\ Y \\ Z \end{vmatrix}$$

In this expression l, m, n denote the direction cosines specifying the relative orientations of the principal axes in the monomer and dimer. (The expressions defining the values D and E are D = -3/2 Z and E = 1/2 (Y - X).) For the face-to-face structure proposed for $[ZnTCP]_2$ (14) exciton theory predicts that dimerization should not affect the out-of-plane component (Z) of the tensor. The in-plane component, and therefore E, depends on the angle of rotation of one porphyrin plane relative to the other. According to the exciton model the observed reduction in E (cf. Table II) corresponds to an angle of rotation of about 23°. This is reasonably close to the value predicted by molecular models (14).

On the other hand, the intermediate- and strong-coupling models fail to explain the experimental results obtained for the other dimers. In systems derived from TPP, steric hindrance between the phenyls, which are not co-planar with the porphyrin rings (27), forces a rotation of the porphyrin rings relative to each other. According to exciton theory, non-coincidence of in-plane axes will cause a reduction in E. The data on $[ZnTCP]_2$ and covalently-linked diporphyrin (8) (cf. Table II) show that E is a sensitive function of rotation angle. As pointed out earlier, if charge transfer plays a role, D and E will both be reduced by dimerization. In view of these considerations it is noteworthy that, with the exception of $[ZnTCP]_2$, none of the dimer systems exhibits a significant reduction in E value. Instead, in cases where dimerization does affect zfs values it is reflected in a reduction in D rather than E.

The following interpretation of the experimental results is offered. The finding that the formation of [ZnTPPS/ZnTTAP] is not accompanied by a change in D and E (cf. Figure 5 and Table II) leads to the conclusion that the triplet energy must be localized on one porphyrin ring. Contrary to an earlier interpretation (11), the same is probably true for $[ZnTPPS]_2$ since the value of E is not reduced by dimerization. In the heterodimers [ZnTPPS/TTAP] and [TPPS/ZnTTAP] exciton transfer between unperturbed porphyrins is blocked because the free base triplet lies well below the triplet level of the Zn-substituted porphyrin (28). Photoexcitation could generate a charge-transfer triplet state in these systems (29). However, this is ruled out because the D and E values in the heterodimers are not that different from the values found for the free base monomers. Also, the average triplet lifetime is virtually the same as the triplet lifetime of the free base porphyrin (cf. Table II). It is concluded that the experimental data best fit a localized triplet state model with the excitation energy localized on the free base half of the dimer. The dimerization effect on zfs's is attributed to a perturbation of the structure of the free base porphyrin.

Formation of the free base porphyrin dimers $[TCP]_2$, $[TPPS]_2$ and

[TPPS/TTAP] is associated with a pronounced reduction in D, whereas E remains virtually unaffected. Furthermore, in the case of the last two dimers, dimerization causes about a factor of two reduction in average triplet lifetime. No lifetime change accompanies $[TCP]_2$ formation. Since the effect of dimerization on D and E cannot be reconciled with a shared triplet state, it follows that the triplet state must be localized on the time scale of the EPR experiment. In free base TPP two of the pyrrole rings are turned out of the least-squares plane defined by the porphyrin core (27). The observed D value reductions may be due to a change in the degree of nonplanarity induced by dimerization. Alternatively, it can reflect a dimerization induced change in phenyl ring orientation. The triplet lifetime is very sensitive to changes in structure of the porphyrin core (30), this may account for dimerization induced changes in triplet lifetime.

In the case of the 4- and 5-atom-amide-chain-linked diporphyrins studied by Gückel et al.(8) dimerization induced changes in D, E and triplet lifetime values (cf. Table II) can be interpreted satisfactorily in terms of the exciton formalism. With regards to separation and relative orientation of the porphyrin rings, the structure of the dimers considered here is very similar to that of the covalently-linked systems. Hence, one would expect the porphyrin-porphyrin interactions to be of the same order of magnitude. In fact, dimerization induced shifts in optical absorption and emission bands are similar, supporting this view. The triplet energies of TTAP and TPPS (ZnTTAP and ZnTPPS) may differ enough to account for a localized triplet state in [TTAP/TPPS] ([ZnTTAP/ZnTPPS]) even in the presence of a moderately strong porphyrin-porphyrin interaction ($J \geq D$). However, the finding of localized triplets in $[TCP]_2$, $[TPPS]_2$, and $[ZnTPPS]_2$ establishes that in these systems the exchange interaction between the rings must be much smaller than that in the covalently-linked systems. At this time no definitive explanation can be offered for this finding. Possibly it reflects an effect of the covalent links or subtle differences in the relative positions of dimer constituents. It is noteworthy that an EPR study (7) of photoexcited triplets of a series of chlorophyll-phanes with covalent links of different lengths established that the excitation energy in these systems is localized on one ring system as well.

As pointed out, dimerization induced absorption band shifts are of the order of several hundred cm^{-1} for the systems considered here. This finding may seem incompatible with the conclusion that the triplet state is localized. However, it should be noted that observed shifts may be due in part to dimerization induced changes in porphyrin geometry and solvation. At this time a quantitative analysis of the factors contributing to the shifts cannot be given. For this reason, they do not provide a measure of the strength of the electronic interaction between the porphyrins. Furthermore, it has been proposed that dimerization induced perturbations of the excited singlet electronic states can be interpreted in terms of the interaction between the transition dipole moments in the porphyrin moieties (31). According to this interpretation the interaction between ground state and triplet excited state porphyrins must be many orders of magnitude smaller than that found for singlet state molecules. For these two reasons it is concluded that a localized triplet state need not be incompatible with pronounced absorption and fluorescence peak shifts.

The results of EPR studies of photoexcited triplets of model systems show that it is not possible to give generally applicable rules for the interpretation of the spectroscopic data. In a number of cases there appears to be a well-understood relationship between dimerization effects and dimer geometry. In most of the systems considered here that is not the case. It is not clear to what difference in make-up of the dimers this discrepancy must be attributed and this is an interesting point of further investigation. Evidently, as long as the data on fairly well characterized model systems are not fully understood it will be impossible to derive definitive conclusions concerning the structure of the special pair from data on its photoexcited triplet state.

Acknowledgments

We thank Drs. H. Linschitz and E. Ojadi of Brandeis University for helpful discussions and a gift of some porphyrins. Support of this work by the U. S. Department of Energy (Division of Chemical Sciences, Office of Basic Energy Sciences), under grant No. FG02-84ER13242 is gratefully acknowledged.

References

1. Dutton, D. L.; Leigh, J. S.; Seibert, M. Biochem. Res. Commun. 1972, 46, 406. Leigh, J. S. Dutton, D. L. Biochim. Biophys. Acta 1974, 357, 67.
2. Clarke, R. H.; Connors, R. E.; Frank, H. A. Biochem. Biophys. Res. Commun. 1976, 71, 671. Clarke, R. H.; Conors, R. E.; Frank, H. A.; Hoch, J. C. Chem. Phys. Lett. 1977, 45, 523.
3. Hägele, W.; Schmid, D.; Wolf, H. C. Z. Naturforsch. A 1978, 33, 83.
4. Hoff, A. J.; Gorter de Vries, H. Biochim. Biophys. Acta 1978, 503, 94.
5. Levanon, H.; Norris, J. R. Chem. Rev. 1978, 78, 185.
6. Clarke, R. H., in "Light Reactions in Photosynthesis"; Fong, F., Ed.; Springer-Verlag: Berlin, 1982.
7. Litteken, S. R. Ph. D. Thesis, University of Illinois, 1983.
8. Gückel, F.; Schweitzer, D.; Collman, J. P. Bencosme, S.; Evitt, E.; Sessler, J. Chem. Phys. 1984, 86, 161.
9. Chandrashekar, T. K.; van Willigen, H. J. Am. Chem. Soc. 1983, 105, 6323.
10. Chandrashekar, T. K.; van Willigen, H.; Ebersole, M. H. J. Phys. Chem. 1984, 88, 4326.
11. Chandrashekar, T. K.; van Willigen, H. Chem. Phys. Lett. 1984, 106, 237.
12. Chandrashekar, T. K.; van Willigen, H.; Ebersole, M. H. J. Phys. Chem. 1985, 89, 3453.

13. van Willigen, H.; Das, U.; Ojadi, E.; Linschitz, H. J. Am. Chem. Soc. 1985, 107, 7784
14. (a) Thanabal, V.; Krishnan, V. J. Am. Chem. Soc. 1982, 104, 3643. (b) Thanabal, V.; Krishnan, V. Inorg. Chem. 1982, 21, 3606.
15. van Willigen, H.; Chandrashekar, T.K. J. Am. Chem. Soc. 19865, 108,709.
16. Krishnamurthy, M.; Sutter, J.R.; Hambright, P. J. Chem. Soc. Chem. Commun. 1975, 13
17. Ojadi, E.; Selzer, R.; Linschitz, H. J. Am. Chem. Soc. 1985,
18. Adler, A. D.; Longo, F. R.; Kampas, F.; Kim, J. J. Inorg. Nucl. Chem. 1970, 30, 2443.
19. Levanon, H.; Weissman, S. I. J. Am. Chem. Soc. 1971, 93, 4309.
20. (a) Chang, C. K.; Kuo, M.-S.; Wang, C.-B. J. Heterocycl. Chem. 1977, 14, 943. (b) Chang, C. K., ibid. 1977, 14, 1285. (c) Chang, C. K. Adv. Chem. Ser. 1979, No. 173, 162. (d) Eaton, S. S.; Eaton, G. R.; Chang, C. K. J. Am. Chem. Soc. 1985, 107, 3177.
21. Collman, J. P.; Chong, A. O.; Jameson, G. B.; Oakley, R. T.; Rose, E.; Schmittou, E. R.; Ibers, J. A. J. Am. Chem. Soc. 1981, 103, 516.
22. Kagan, N. E.; Mauzerall, D.; Merrifield, R. B. J. Am. Chem. Soc. 1977, 99, 5484.
23. Wasserman, E.; Snyder, L. C.; Yager, W. A. J. Chem. Phys. 1964, 41, 1763.
24. Van der Waals, J. H.; Van Dorp, W. G.; Schaafsma, T. J. in "The Porphyrins"; Dolphin, D., Ed.; Academic Press: New York, 1979; Vol. IV, Chapter 5.
25. Levanon, H.; Wolberg, A. Chem. Phys. Lett. 1974, 24, 96.
26. Schwoerer, M.; Wolf, H. C. Mol. Cryst. 1967, 3, 177.
27. (a) Fleischer, E. B. Acc. Chem. Research 1970, 105, 216. (b) Silvers, S. J.; Tulinsky, A. J. Am. Chem. Soc. 1967, 89, 3331.
28. Darwent, J. R.; Doublas, P.; Harriman, A.; Porter, G.; Richoux, M.-C. Coord. Chem. Rev. 1982, 44, 83.
29. Petke, J. D.; Maggiora, G. M. Chem. Phys. Lett. 1983, 97, 231.
30. Burgner, R. P.; Ponte Goncalves, A. M. Chem. Phys. Lett. 1977, 46, 275.
31. Gouterman M.; Holten, D.; Leiberman, E. Chem. Phys. 1977, 25, 139

RECEIVED April 3, 1986

11

Distance-Dependent Rates of Photoinduced Charge Separation and Dark Charge Recombination in Fixed-Distance Porphyrin–Quinone Molecules

Michael R. Wasielewski and Mark P. Niemczyk

Chemistry Division, Argonne National Laboratory, Argonne, IL 60439

Three zinc tetraphenylporphyrin-anthraquinone derivatives were prepared in which the edge-to-edge distances between the porphyrin and quinone π systems are fixed by a rigid hydrocarbon spacer molecule. Triptycene, trans-1,2-diphenylcyclopentane, and adamantane were used to fix the porphyrin-anthraquinone distance at 2.5, 3.7, and 4.9 Å, respectively. These molecules possess 1,2, and 3 saturated carbon atoms, respectively, between the porphyrin donor and the quinone acceptor. Rate constants for photoinduced electron transfer from the lowest excited singlet state of the zinc tetraphenylporphyrin donor to the anthraquinone acceptor were measured. In addition, the corresponding radical ion pair recombination rate constants for each of these molecules were also determined. The rate constants for both photoinduced charge separation and subsequent radical ion pair recombination decrease by approximately a factor of 10 for each saturated carbon atom intervening between the porphyrin donor and the quinone acceptor. These results are consistent with a model in which the rate of electron transfer is determined by weak mixing of the σ orbitals of the saturated hydrocarbon spacer with the π orbitals of the donor and acceptor.

In photosynthetic reaction centers the distances between the various electron donors and acceptors are restricted by the surrounding protein.(1) Studies of model systems possessing well-defined donor-acceptor distances and geometries are necessary to fully understand the critical role of these parameters in determining the efficiency of photoinduced charge separation. Porphyrins possessing covalent linkages to quinones have become increasingly important in the study of photoinduced electron transfer reactions.(2-4) Yet, few of these models possess linkages between the porphyrin electron donor and the quinone electron acceptor which restrict both the distance and the orientation between the donor and the acceptor.(2u-v,3,4)

We recently prepared a series of restricted distance porphyrin-quinone donor-acceptor molecules designed to study the dependence of the rate of electron transfer proceeding from the lowest excited singlet state of the porphyrin on the free energy of reaction.(3-4) We now report measurement

0097-6156/86/0321-0154$06.00/0

of the rate constants for both radical ion pair formation and recombination in a series of three molecules in which a zinc tetraphenylporphyrin is positioned at three different fixed edge-to-edge distances relative to an anthraquinone. The structures of these molecules, 1-3 are depicted in Figure 1. Compounds 1, 2, and 3 possess 1, 2, and 3 saturated carbon atoms, respectively between the π system of the zinc tetraphenylporphyrin electron donor and that of the anthraquinone electron acceptor. This results in an edge-to-edge distance of 2.5, 3.7, and 4.9 Å, for 1, 2, and 3, respectively.(5)

Experimental

The synthesis of compound 1 has already been outlined.(3) Compound 2 was prepared as follows: trans-1,2-diphenylcyclopentane, 4 was prepared in 65% yield by bis-alkylation of trans-stilbene with 1,3-dichloropropane.(6) Compound 4 was alkylated with $CHCl_2OCH_3$ in CH_2Cl_2 using $TiCl_4$ catalyst.(7) Hydrolysis of the resulting chloroether during work-up yielded trans-1-(4-formylphenyl)-2-phenylcyclopentane, 5, 51%. Aldehyde 5 was acylated with phthalic anhydride in 1,2-dichloroethane using an $AlCl_3$ catalyst to yield keto-acid 6, 83%. Keto-acid 6 was treated with polyphosphoric acid at 100° for 1 hr to give trans-1-(2-anthraquinonyl)-2-phenylcyclopentane, 7 in 10% yield. Treatment of 1 molar equivalent of 7 with 4 equivalents of pyrrole and 2 equivalents of benzaldehyde in refluxing propionic acid(8) for 30 min gave an 8% yield of the free base of 2 after chromatography. Treatment of the free base with excess $ZnOAc_2$ in refluxing $CHCl_3$ for 10 min gave 2 in quantitative yield: mass spectrum, m/z calcd 951.4, found 951.3.

Compound 3 was prepared as follows: 1,3-diphenyladamantane, 8 was prepared from 1-bromoadamantane in 47% yield by a one step literature route.(9) Compound 8 was alkylated with $CHCl_2OCH_3$ as described above to give after work-up and chromatography 1-(4-formylphenyl)-3-phenyladamantane, 9 in 63% yield. Aldehyde 9 was acylated as described above with phthalic anhydride to give keto-acid 10 in 79% yield. Ring closure of keto-acid 10 in polyphosphoric acid at 100° for 1 hr gave 1-(2-anthraquinonyl)-3-phenyladamantane, 11 in 40% yield. One step Adler porphyrin synthesis(8) using 11, benzaldehyde, and pyrrole gave the free base of 3 in 4% yield after chromatography. Treatment of free base with $ZnOAc_2$ as above gave 3 in quantitative yield: mass spectrum, m/z calcd 1017.4, found 1017.5.

Redox potentials for 1-3 were determined in butyronitrile containing 0.1M tetra-n-butylammonium perchlorate using a Pt disc electrode at 21°. These potentials were measured relative to a saturated calomel electrode using ac voltammetry.(10) Both the one electron oxidations and reductions of 1-3 exhibited good reversibility. The half-wave potentials for the one-electron oxidation and reduction of 1-3, ZnTPP, and two model quinones are given in Table I.

Fluorescence quantum yields for 1-3 and ZnTPP in butyronitrile were determined using 10^{-7} M solutions in 1 cm cuvettes at the 90° geometry. The samples were excited at 549 nm and the resultant fluorescence emission spectra were digitized. The emission spectra of 1-3 differed only in relative intensity. A direct comparison between the integrated and normalized fluorescence spectra of 1-3 was made with that of ZnTPP. The data normalized to the known fluorescence quantum yield of ZnTPP(11) are listed in Table II. Fluorescence lifetimes were determined on 10^{-7} M solutions of

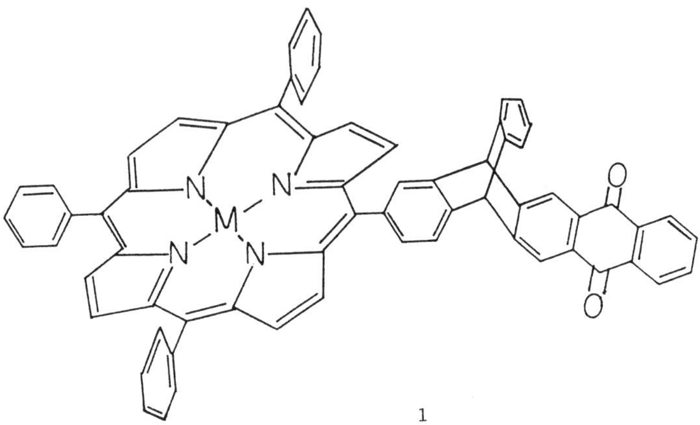

1

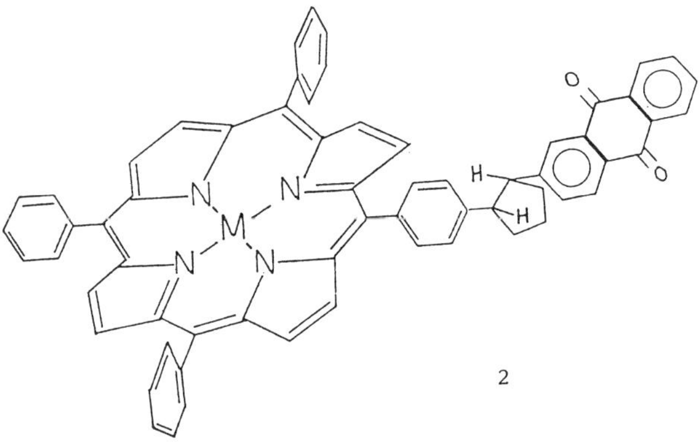

2

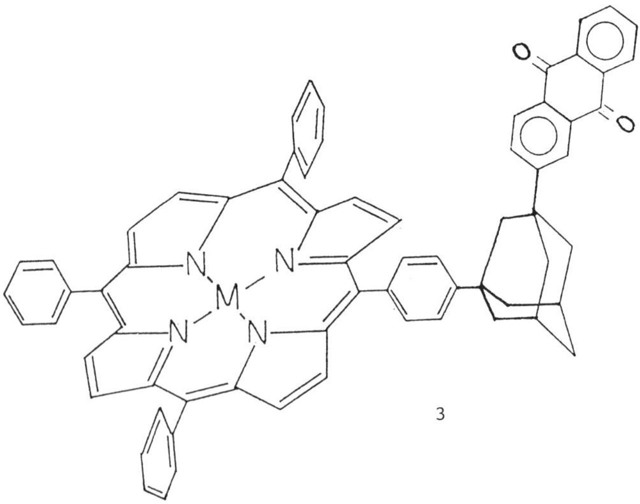

3

Figure 1. Structures of compounds 1-3.

Table I. Energetics

Compound	$E^{-}_{1/2}$	$E^{+}_{1/2}$	$-\Delta G_{cs}$	$-\Delta G_{cr}$
ZnTPP	-1.30	0.82		
triptycene-anthraquinone	-0.82			
2-methyl-anthraquinone	-0.82			
1	-0.82	0.82	0.43	1.64
2	-0.82	0.82	0.43	1.64
3	-0.82	0.82	0.43	1.64

Table II. Fluorescence Quantum Yields and Lifetimes

Compound	ϕ_F	τ_F
ZnTPP	0.030	2.0 ns
1	0.001	0.048
2	0.003	0.21
3	0.020	1.3

1-3 and ZnTPP by time-correlated photon counting(12) and are also listed in Table II.

Picosecond time-resolved transient absorption measurements were obtained as follows: Solutions of 2 x 10^{-4} M 1-3 were each prepared in butyronitrile contained in 2 mm pathlength cuvettes. The samples were degassed by three freeze-pump-thaw cycles. A 2 mm diameter spot on the sample cell was illuminated with the pump and probe beams of the transient absorption apparatus. The 600 nm, 0.4 ps, 0.5 nJ output of a mode-locked Ar^+ - synchronously pumped R6G/DQOCI dye laser was amplified to 1.5 mJ using a 4-stage R640 dye amplifier pumped by a frequency-doubled Nd-YAG laser operating at 10 Hz. The amplified laser pulse was split with a dichroic beam splitter. A 600 nm, 0.5 ps, 1.2 mJ pulse was used to generate a 0.5 ps white light continuum probe pulse. The remaining 0.3 mJ, 600 nm, 0.5 ps pulse was used to excite the sample. Absorbance measurements were made with a double beam probe configuration which employed optical multichannel detection. Time delays between pump and probe pulses were accomplished with an optical delay line. Analyses of the kinetics were carried out using the method of Provencher.(13)

Results and Discussion

The one electron redox potentials of 1-3 along with those of the appropriate reference compounds are presented in Table I. The redox potentials of both the porphyrins and the quinones are not altered by linking the two molecules. These potentials were used to obtain the exothermicity of the charge separation, $-\Delta G_{cs}$ and that of the charge recombination, $-\Delta G_{cr}$ from Equations 1 and 2, respectively:

$$-\Delta G_{cs} = E(S_1) - E^{ox}_{D} + E^{red}_{A} + e_o^2/\epsilon r \quad (1)$$

$$-\Delta G_{cr} = E^{ox}_{D} - E^{red}_{A} - e_o^2/\epsilon r \quad (2)$$

where $E(S_1)$ is the energy of the lowest excited singlet state of the molecule, in this case $E(S_1) = 2.07$ eV for zinc tetraphenylporphyrin(11), E^{ox}_{D} and E^{red}_{A} are the measured $E_{1/2}$ values for the one electron oxidation and reduction of the porphyrin donor and the quinone acceptor, respectively, ϵ is the solvent dielectric constant, r is the average center-to-center distance between the ions within the radical ion pair, and e_o is the charge of the electron. The term in equations 1 and 2 that depends on ϵ is the solvent dependent coulomb energy change upon ion pair formation or recombination. For compounds 1-3 r is about 11.5 - 15 Å, thus, in butyronitrile (ϵ= 20) the term $e_o^2/\epsilon r$ varies from 0.05 - 0.06 eV. Thus, for compounds 1-3 the exothermicities of both the charge separation and radical ion recombination reactions are approximately constant.

The degree of electronic interaction between the electron donor and acceptor in both the ground and lowest excited singlet states of 1-3 can be determined by an examination of their ground state optical absorption spectra and their fluorescence emission spectra. The ground state optical absorption spectra and fluorescence emission spectra of the porphyrins in 1-3 are not perturbed by the presence of the appended quinones and are typical of ZnTPP. This is illustrated for 1 in Figure 2. However, the fluorescence quantum yields of these compounds are a strong function of the distance between the porphyrin and the quinone, Table II. We have shown previously(3) that the fluorescence quenching mechanism for 1 in butyronitrile is electron transfer from the lowest excited singlet state of the ZnTPP to the anthraquinone. The fluorescence data in Table II suggest that electron transfer competes effectively with the intrinsic decay pathways of the $^1ZnTPP^*$ state in 2, but competes much less effectively with these pathways in 3.

We have obtained additional evidence supporting the electron transfer mechanism of fluorescence quenching in 2 and 3 from picosecond transient absorption and fluorescence measurements. The fluorescence lifetimes of 1-3 in butyronitrile are reported in Table II. These lifetimes are proportional to the observed fluorescence quantum yields of these compounds and therefore indicate that the observed fluorescence quenching is not due simply to a change in the radiative rate for emission.

The transient absorption spectrum observed 100 psec after excitation of 1 in butyronitrile with a 0.5 psec, 600 nm laser pulse is shown in Figure 3. We assign this spectrum to the $ZnTPP^+AQ^-$ radical ion pair, since it is very

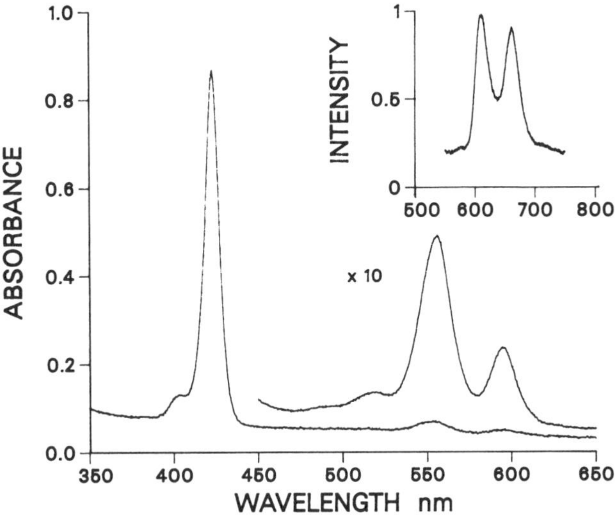

Figure 2. Optical absorption and emission (inset) spectra of 1 in butyronitrile.

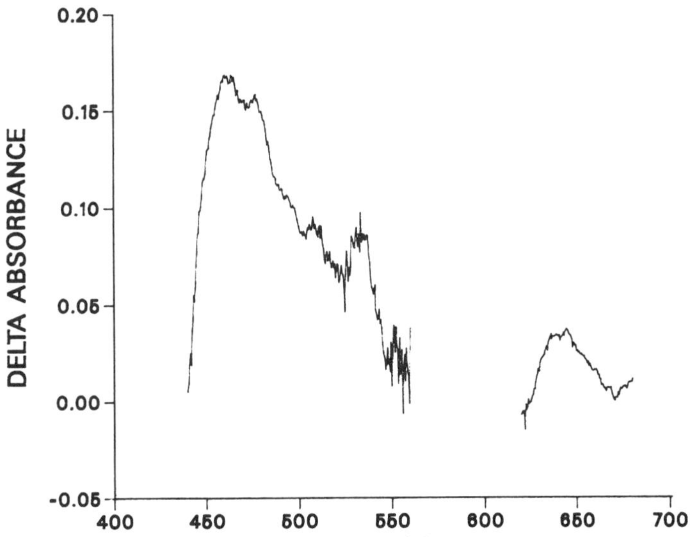

Figure 3. Transient absorption spectrum of a 2 x 10^{-4} M solution 1 in butyronitrile at 100 ps following a 0.3 mJ, 0.5 ps, 600 nm laser flash. Filters that reject stray excitation light cut out the 580-620 nm wavelength region, while the sharp cutoff at 440 nm is due to the intense absorption of the porphyrin Soret band at 419 nm.

similar to the published spectrum of $ZnTPP^+$.(15) Specifically, the molar extinction coefficient for $ZnTPP^+$ is 3.3 x 10^4 at 460 nm and 10^4 at 640 nm.(15) The distinct band in the 600 - 700 nm region of the spectrum is highly characteristic of $ZnTPP^+$ formation. Both $^1ZnTPP^*$(16) and $^3ZnTPP^*$(17) possess extinction coefficients that are about 10 times smaller than that of $ZnTPP^+$ in the 600 - 700 nm region. Thus, the $ZnTPP^+$ spectrum strongly dominates the spectrum in Figure 3. Similar transient spectra develop and decay for 2 and 3 with kinetics that are shown in Figure 4.

The decay of the transient absorption of 1 at 640 nm is shown in Figure 4A. It is clear that the radical ion pair produced following excitation of 1 decays cleanly to ground state. In Figure 4B the rise of the absorbance at 640 nm due to formation of the radical ion pair of 2 as well as its decay to ground state are both much slower than those of 1 in Figure 4A. Since the rate constant for formation of the $ZnTPP^+$ band at 640 nm as shown in Figure 4C is smaller than that of 1 and 2, direct intersystem crossing of $^1ZnTPP^*$ to $^3ZnTPP^*$ is very competitive with the formation of the radical ion pair. Nanosecond flash photolysis measurements show that the absorbance at 640 nm due to $ZnTPP^+$ decays in 10 ns.

The dependence of electron transfer rate constants on distance has been described by Equation 3:

$$k_{et} = \nu e^{-\alpha r} \tag{3}$$

where the pre-exponential factor, ν includes the dependence of the electron transfer reaction on the electron exchange matrix element, the Franck-Condon factors, and the free energy of reaction, α is a constant that depends on the overlap of the wavefunctions for the donor and acceptor, and r is the donor-acceptor distance. Thus, a plot of ln k_{et} vs r should have a slope equal to $-\alpha$ and an intercept of ln ν. The rate constants determined for both photoinduced charge separation and subsequent radical ion pair recombination in 1-3 are plotted as a function of the edge-to-edge distance between the π systems of the donor and acceptor in Figure 5. The slope of the least squares fits to the data yield $\alpha = 2.3$ and $\nu = 8 \times 10^{12}$ sec^{-1} for the charge separation reaction and $\alpha = 2.1$ and $\nu = 4 \times 10^{11}$ sec^{-1} for charge recombination reaction.

Since the values of ν depend on several factors noted above, in the absence of additional data such as the temperature dependence of the electron transfer rate constants for 1-3 it is difficult to analyze the apparent difference between ν for the charge separation reaction and that of the radical ion pair recombination reaction. However, the difference between these two values of ν is not unreasonable given that the charge separation involves oxidation of an excited state of the donor, while radical ion pair recombination involves two ground state radicals. Small changes in the nuclear coordinates of the donor and acceptor for these two reactions should be sufficient to produce the observed difference in ν. The electronic coupling factor between $^1ZnTPP^*$ and AQ should be different than that between $ZnTPP^+$ and AQ^-.

The data clearly show that the rates of both of these electron transfer reactions diminish by about a factor of 10 for each saturated carbon atom between the edge of the π system of the donor and that of the acceptor. This decrease is more pronounced than that observed in small aromatic donor-acceptor molecules.(18) In these molecules the energy of the lowest excited singlet state of the electron donor is about 3.5 eV and the energy of

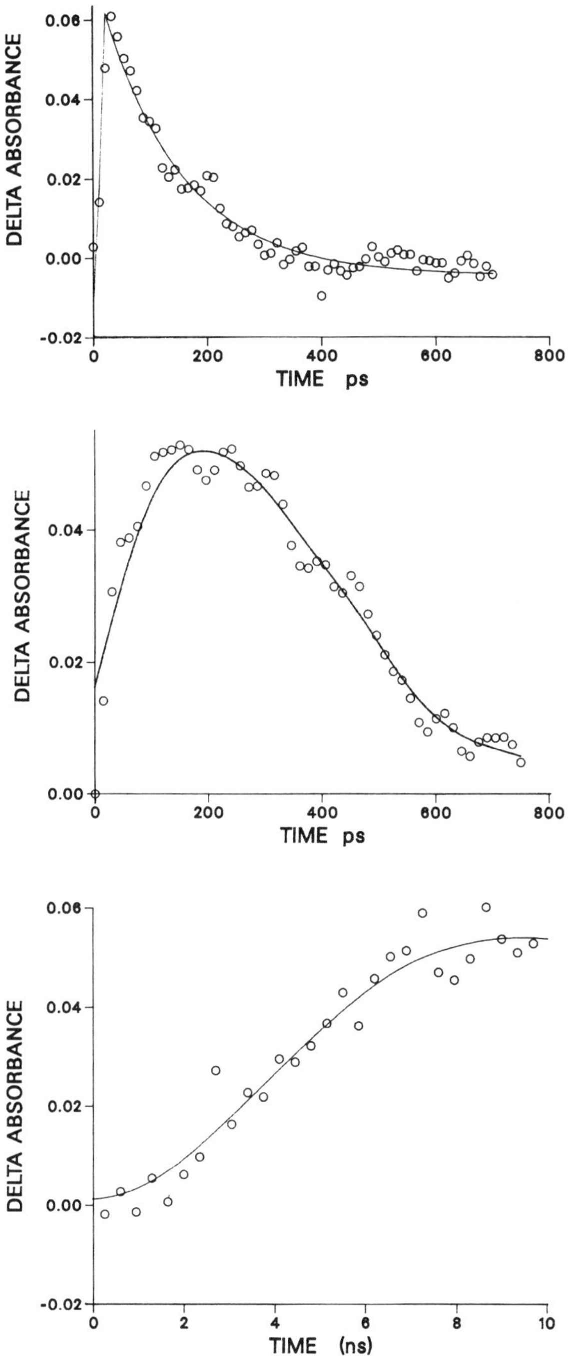

Figure 4. Transient absorption changes at 640 nm for 1-3 in butyronitrile following a 0.3 mJ, 0.5 ps, 600 nm laser flash: A) 1, B) 2, C) 3.

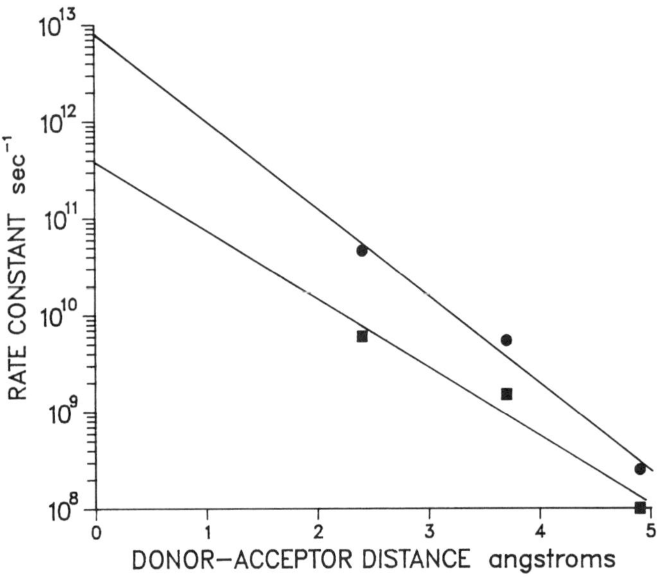

Figure 5. Plot of log k_{et} for charge separation, ●, and log k_{et} for radical ion pair recombination, ■, vs edge-to-edge distance between the π systems of the ZnTPP donor and anthraquinone acceptor in compounds 1-3 in butyronitrile at 21°.

the radical ion pair state is about 3 eV. These energies are substantially higher than the 2.07 eV S_1 energy and the 1.64 eV radical pair energy for 1-3. Thus, mixing of the high energy states involved in the electron transfer reactions of small aromatic molecules with the high energy states of the spacer should be stronger than mixing of the corresponding low energy states of 1-3 with those of the spacer. This mixing results in participation of the intervening spacer bonds in the electron transfer reaction. The weak participation of the spacer bonds of 1-3 in the electron transfer reaction results in α values for the data presented in Figure 5 that are somewhat higher than most α values previously reported experimentally.(19)

The mechanism of electron transfer may involve either direct overlap of the π orbitals of the donor with those of the acceptor or the admixture of spacer orbitals with these orbitals.(20) The fact that there appears to exist an absolute dependence of the electron transfer rate on the donor-acceptor energy levels relative to those of the intervening spacer suggests that the spacer orbitals participate in the electron transfer reaction. In most rigid spacer molecules thus far studied the σ orbitals of the spacer are in the correct geometry to promote electronic interaction between atoms that are several bonds removed from one another.(21) This interaction is closely related to the phenomenon of spin density propagation along a w-plan hydrocarbon backbone which results in hyperfine coupling of the unpaired electron spin at a radical center to a nuclear spin several bonds away from the radical center.(21) Both electron-nuclear spin coupling and electron transfer have similar dependencies on the orientation of the spacer orbitals.(22)

Conclusion

Our data show that a molecular spacer consisting of a rigid saturated hydrocarbon attenuates the rate of both photoinduced charge separation and subsequent radical ion pair recombination in porphyrin-quinone donor-acceptor molecules by about an order of magnitude for each intervening saturated carbon atom between the donor and the acceptor. This strong distance dependence is most likely due to the fact that the radical ion pair states in our porphyrin-quinone model systems are sufficiently low in energy that they mix weakly with the σ orbitals of the saturated hydrocarbon spacer. Since the energies of the radical ion pair states in photosynthetic electron transfer proteins are similar to those of the model systems studied here, our results suggest that saturated bonds of the protein backbone that lie between electron donors and acceptors in these proteins may attenuate the rate of electron transfer reactions between these donors and acceptors to a degree similar to that observed in our model systems.

Acknowledgment

The authors wish to thank Mr. W. A. Svec for his assistance in the preparation of the compounds described in this work. This work was supported by the Division of Chemical Sciences, Office of Basic Energy Sciences, U.S. Department of Energy under Contract W-31-109-Eng-38.

Literature Cited

1. Deisenhofer, J.; Epp, O.; Miki, K.; Huber, R.; Michel, H. J. Molec. Biol. 1984, 180, 385-398.

2. (a) Tabushi, I.; Koga, N.; Yanagita, M. Tetrahedron Lett. 1979, 257-260. (b) Dalton, J.; Milgrom, L.R. J. Chem. Soc. Chem. Comm. 1979, 609-610. (c) Kong, J.L.Y.; Loach, P.A. J. Heterocycl. Chem. 1980, 17, 737. (d) Ho, T.-F.; McIntosh, A.R.; Bolton, J.R. Nature, 1980, 286, 254-256. (e) Ganesh, K.N.; Sanders, J.K.M. J. Chem. Soc. Chem. Comm. 1980, 1129-1131. (f) Harriman, A.; Hosie, R.J. J. Photochemistry 1981, 15, 163-167. (g) Netzel, T.L.; Bergkamp, M.A.; Chang, C.K.; Dalton, J. J. Photochemistry 1981, 451-460. (h) Nishitani, S.; Kurata, N.; Sakata, Y.; Misumi, S.; Migita, M.; Okada, T.; Mataga, N. Tetrahedron Lett. 1981, 2099-2102. (i) Migita, M.; Okada, T.; Mataga, N; Nishitani, S.; Kurata, N.; Sakata, Y.; Misumi, S. Chem Phys. Lett. 1981, 84, 263-266. (j) Kong, J.L.Y.; Spears, K.G.; and Loach, P.A. Photochem. Photobiol. 1982, 35, 545-553. (k) Ganesh, K.N.; Sanders, J.K.M. J. Chem. Soc. Perkin Trans. I 1982, 1611-1615. (l) Ganesh, K.N.; Sanders, J.K.M. J. Chem. Soc. Perkin Trans. I 1982, 1617-1624. (m) Bergkamp, M.A.; Dalton, J.; Netzel, T.L. J. Am. Chem. Soc. 1982, 104, 253-259. (n) Lindsey, J.S.; Mauzerall, D.C. J. Am. Chem. Soc. 1982, 104, 4498-4500. (o) Lindsey, J.S.; Mauzerall, D.C. J. Am. Chem. Soc. 1983, 105, 6528-6529. (p) McIntosh, A.R.; Siemiarczuk, A; Bolton, J.R.; Stillman, M.J.; Ho, T.-F.; Weedon, A.C. J. Am. Chem. Soc. 1983, 105, 7215-7223. (q) Siemiarczuk, A.; McIntosh, A.R.; Ho, T.-F.; Stillman, M.J.; Roach, K.J.; Weedon, A.C.; Bolton, J.R.; Connally, J.S. J. Am. Chem. Soc. 1983, 105, 7224-7230. (r) Nishitani, S.; Kurata, N.; Sakata, Y.; Misumi, S.; Karen, A; Okada, T; Mataga, N. J. Am. Chem. Soc. 1983, 105, 7771-7772. (s) Moore, T.A.; Gust, D; Mathis, P.; Mialocq, J.-C.; Chachaty, C.; Bensasson, R.V.; Land, E.J.; Doizi, D.; Liddell, P.A.; Lehman, W.R.; Nemeth, G.A.; Moore, A.L. Nature 1984, 307, 630-632. (t) Joran, A.; Leland, B.A.; Geller, G.G.; Hopfield, J.J.; Dervan, P.B. J. Am. Chem. Soc. 1984, 106, 6090-6091. (u) Bolton, J.R.; Ho, T.-F.; Liauw, S.; Siemiarczuk, A.; Wan, C.S.K.; Weedon, A.C. J. Chem Soc., Chem. Commun. 1985, 559-560. (v) Leland, B.A.; Joran, A.D.; Felker, P.M.; Hopfield, J.J.; Zewail, A.H.; Dervan, P.B. J. Phys. Chem. 1985, 89, 5571-5573.
3. Wasielewski, M.R.; Niemczyk, M.P. J. Am. Chem. Soc. 1984, 106, 5043-5044.
4. Wasielewski, M.R.; Niemczyk, M.P. J. Am. Chem. Soc. 1985, 107, 1080-1082.
5. The donor-acceptor distances were estimated using Corey-Pauling-Koltun molecular models.
6. Reesor, J.W.B.; Smith, J.G.; Wright, G.F. J. Org. Chem. 1954, 19, 940-956.
7. Rieche, A.; Gross, H.; Hoft, E. Chem. Ber. 1960, 93, 88-94.
8. Adler, A.D.; Longo, F.R.; Finarelli, J.D.; Goldmacher, J.; Assour, J.; Korsakoff, L. J. Org. Chem. 1967, 32, 476.
9. Newman, H. Synthesis 1972, 692-693.
10. Wasielewski, M.R.; Smith, R.L.; Kostka, A.G. J. Am. Chem. Soc. 1980, 102, 6923-6928.
11. Seybold, P.G.; Goutermann, M. J. Molec. Spectroscopy 1969, 31, 1.
12. Fleming, G.R. Adv. Chem. Phys. 1982, 49, 1-45.
13. Provencher, S.W. J. Chem. Phys. 1976, 64, 2772.
14. Weller, A. Z. Phys. Chem. N.F. 1982, 133, 93.
15. Fajer, J.; Borg, D.C.; Forman, A.; Dolphin, D.; and Felton, R.H. J. Am. Chem. Soc. 1970, 92, 3451-3459.
16. M.R. Wasielewski, unpublished results.
17. Pekkarinen, L; Linschitz, H. J. Am. Chem. Soc. 1959, 82, 2407.

18. Hush, N.S.; Paddon-Row, M.N.; Cotsaris, E.; Oevering, H.; Verhoeven, J.W.; Heppener, M. Chem. Phys. Lett. 1985, 117, 8-11.
19. Miller, J.R.; Beitz, J.V.; Huddleston, R.K. J. Am. Chem. Soc. 1984, 106, 5057-5068.
20. Hush, N.S. Coord. Chem. Rev. 1985, 64, 135-157.
21. Paddon-Row, M.N. Accounts Chem. Res. 1982, 15, 245.
22. Siders, P.; Cave, R.J.; Marcus, R.A. J. Chem. Phys. 1984, 81, 5613-5624.

RECEIVED April 8, 1986

PICOSECOND SPECTROSCOPY

12

Picosecond Excited-State Relaxation of Some Iron Porphyrins and Hemoproteins

Karl D. Straub[1] and Peter M. Rentzepis[2]

[1]John L. McClellan Memorial Veterans' Hospital and University of Arkansas for Medical Sciences, Little Rock, AR 72205

[2]Department of Chemistry, University of California at Irvine, Irvine, CA 92664

The excited state spectra and kinetics of decay of cytochrome c, myoglobin and Fe(II) protoporphyrin IX dimethyl ester (Fe(II)ProtoDME) have been studied by picosecond spectroscopy. The existence of different oxidation states, spin multiplicities, and axial ligands in these compounds affords a way to study the effects of different electronic configurations on the excited state decay. The simplest complex, FeII Proto DME, in benzene (no axial ligands) and spin state, S=2, has an excited state absorption band, 450-500 nm, which decays with a time constant of about 10 psec while the ground state Q band does not reappear until 45 psec. Similar kinetics are seen in Fe(II) cytochrome c and on a somewhat slower time scale with Fe(II) myoglobin. These kinetics indicate a rapid decay of the excited state with relaxation through the singlet manifold (S=0) or the triplet manifold (S=2) by way of π^* - d or d - d intermediate transitions. Excitation of the bipyridine complex of Fe(II) Proto DME results in return to ground state in ~40 psec with formation of the ground-state 5 ligand species. Return to the 6 ligand ground state occurs in 250-300 psec. Fe(III) porphyrins regardless of spin state (S=1/2; S=5/2) have relaxation kinetics which are much faster than the resolution of our experimental conditions.

The large number of electronic configuration of iron porphyrins with oxidative states of 2^+ or 3^+, high and low spin forms, and charge transfer states with different axial ligands offer the possibility of a number of non-radiative decay pathways (1). In

0097-6156/86/0321-0168$06.00/0

particular, the effect of spin state on intersystem crossing rate and the availability of charge transfer states of the iron porphyrins makes it possible that there will be different relaxation mechanisms for various iron porphyrin complexes. Electronic excited states in iron porphyrins are thought to relax rapidly. In general, these compounds have been classified as "non-radiative" (2) since their luminescence is not detectable by ordinary means. However, by using line width measurements and background fluorescence from resonance Raman spectroscopy, Adar *et al* (3) have been able to determine that reduced cytochrome b_5 has a weak luminescence (quantum yield 10^{-5} - 10^{-6}) while oxidized cytochrome b_5 and myoglobin should have no luminescence in the visible range. Direct measurement of relaxation of the excited state of iron porphyrins and hemoproteins confirms that the Fe(III) complexes have very fast relaxations (<6 psec) (4). Axial ligand ejection from Fe(II) porphyrins and hemoproteins also occurs on a picosecond scale. Dissociation of CO, O_2, and NO from hemoglobin and myoglobin takes place within 15 psec of photoexcitation (5-7). The ejection of axial ligands such as piperidine has also been shown to take place within the 35 psec excitation pulse for Fe(II) porphyrins (8).

While it is apparent that there are several pathways for relaxation of the excited state of iron porphyrins, the exact steps involved in these different pathways are still to a great extent unknown. It is the purpose of this study to explore the role of spin state and axial ligands in the decay process of the excited states of iron porphyrins.

Materials and Methods

Picosecond spectroscopy was performed using an apparatus described by Reynolds and Rentzepis (9). The duration of the excitation pulse was 25 psec at 530 nm. Pulse intensity was about 1.5 m joules. Dependence of relaxation time on power from 0.3 up to 2.0 m joules revealed no effects of pulse intensity on decay times. The concentration of the iron porphyrins was adjusted to be between 0.2 and 0.3 OD for 2 mm cell at 530 nm. All relaxation times were calculated from the first order kinetic curves of excited state decay or ground state reappearance. This procedure eliminates error in delay times between the excitation and different wavelength probe pulses ("chirp") since constant delay times are subtracted out of the kinetic curves. There may, however, be some error introduced in the shorter decay times because of the excitation pulse and the probe pulse may overlap at the earliest points of the kinetic curve calculations.

Myoglobin (equine skeletal muscle), cytochrome c (type IV from horse heart), and protoporphyrin IX dimethyl ester, DME, were purchased from Sigma Chemical Co. All other solid chemicals were reagent grade. Pyridine was highest grade spectroscopic material. Benzene was redistilled over metallic sodium under N_2. All ferrous compounds were prepared in a glove bag with 95 percent N_2 5 percent H_2 circulated over a palladium catalyst to remove traces of O_2. Cuvettes containing ferrous porphyrins were sealed with a glass-encased teflon screw-type plunger-seal used in vacuum hydrolysis. The cap was then sealed under parafin wax by inverting the top of the assembly in a beaker filled with melted parafin. This type of double-seal gave very good protection against O_2. Fe(III) protoporphyrin IX was prepared after the method of Falk (10). The Fe(III) complex was converted to Fe(II) using the method of Brault and Rougee (11) in the N_2-H_2 atmosphere glove bag. The oxidized cytochrome c and myoglobin were converted to the Fe(II) form by using a 10-fold excess of Na S_2O_4 and immediately desalting on a G-25 column in 0.05 M phosphate buffer at pH 7 in the N_2-H_2 atmosphere glove bag.

Results

Fe(III) Hemoproteins:

The Fe(III) cytochrome c with CN^-, N_3^-, H_2O or F^- as ligand species all relaxed with a very fast relaxation (<6 psec). Even during the duration of the pulse very little change in the absorbance spectrum could be seen. Similar results were obtained with Fe(III) myoglobin having F^- or CN^- as ligands. For the complexes of Fe(III) cytochrome c the energy levels of the low-lying charge transfer states change almost 3000 cm^{-1} (12) with the cytochrome c iron in the low spin state (S=1/2) only. On the other hand, myoglobin in the oxidized state undergoes transition from low spin, S=1/2, in the CN^- complex to high spin S=5/2 in the F^- complex (13). Thus neither spin state nor shift in the low lying charge transfer bands affected the very rapid, < 6 psec, decay of the excited state of ferric cytochrome c or oxidized myoglobin.

Fe(II) Protoporphyrin IX DME in benzene:

The relaxation of Fe(II) protoporphyrin dimethyl ester in benzene is seen in Figure 1. It can be seen that in this complex which has no axial ligands and a high spin state, S=2 (14), there is an excited state absorption band which extends from ~ 520 nm into the blue. The peak is not seen because of experimental cut-off in the spectrum at 450 nm. This difference spectrum also indicates an absorption band extending to the blue from 620 nm and interupted by the strong ground state bleaching at 570 nm. The peak of the excited state absorption band seems to be centered at about 585-590 nm.

If the relaxation rate is plotted (Fig. 2) at the blue absorption band at 460 nm the lifetime is approximately 21 psec. However, the lifetime of the bleaching at 570 nm is 46 psec. These relaxation times differ enough to make it highly probable that two separate excited states are being seen. Thus an excited state absorption decays (21 psec) before reappearance of the ground state (46 psec).

Fe(II) Cytochrome c:

Reduced cytochrome c which has two axial ligands and a low spin state, S=0, has a picosecond relaxation spectrum very similar to the Fe(II) protoporphyrin DME in benzene with a blue absorption

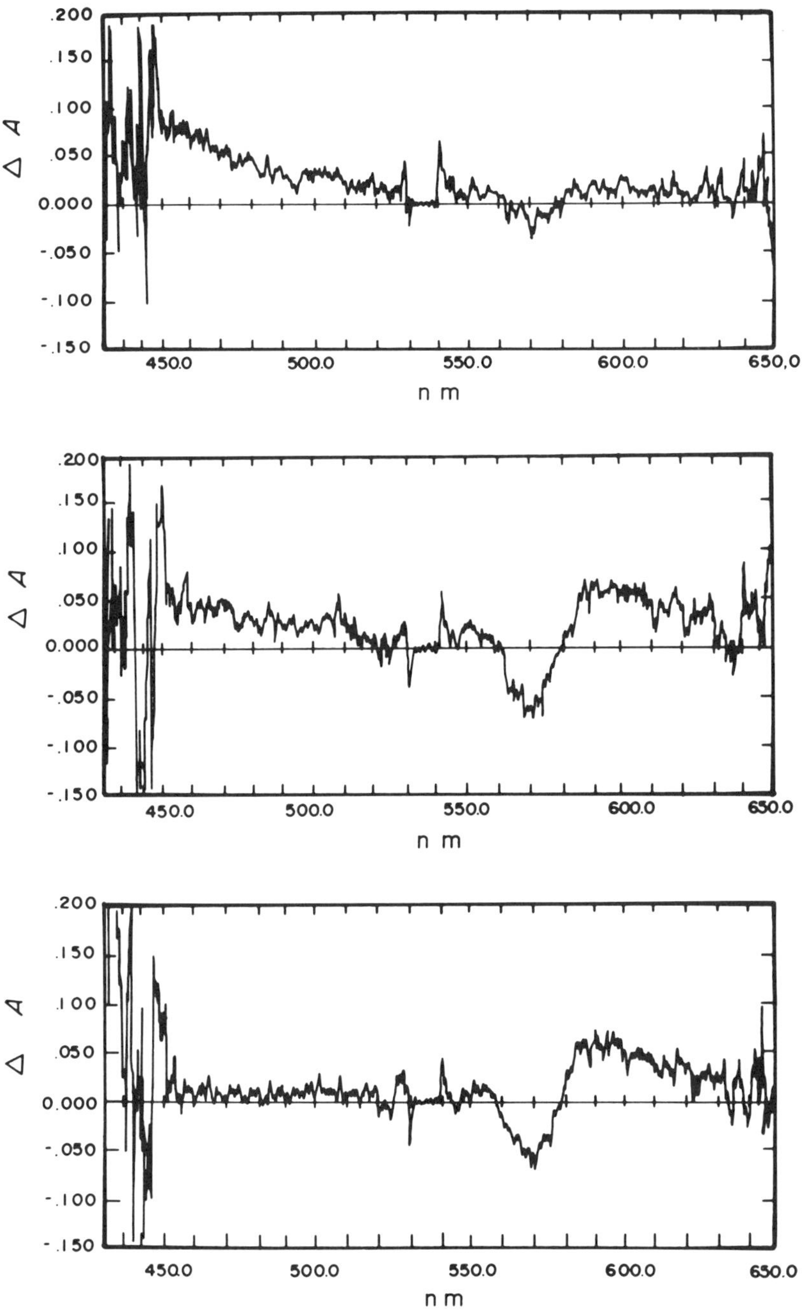

Figure 1: Picosecond spectra of Fe(II) protoporphyrin DME in benzene. Top - 0 psec (center of 25 psec excitation pulse); Middle - 8 psec; Bottom - 25 psec.

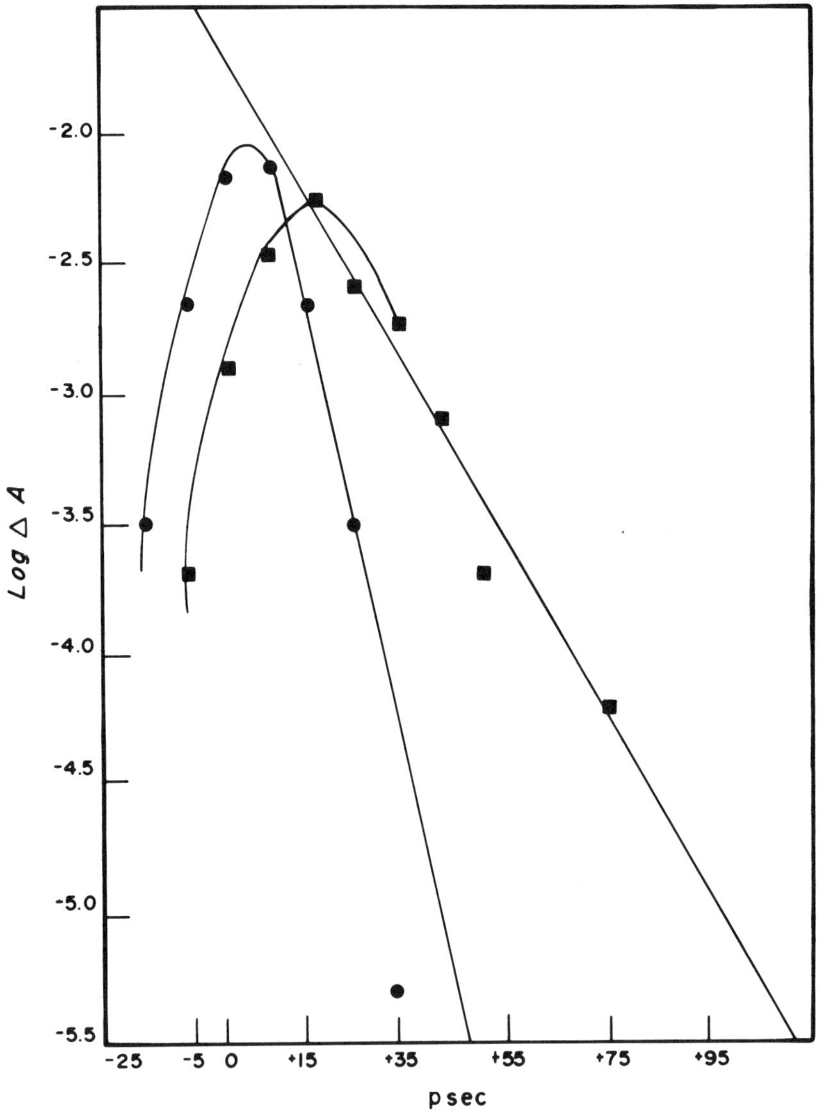

Figure 2: Log absorbance vs. time plots for Fe(II) protoporphyrin DME in benzene. 530 nm excitation; circles - 460 nm absorption; squares - 570 nm bleaching.

band extending from 495 nm to above 450 nm and visible band extending from 600 nm to the blue. The apparent peak at 555 nm is interupted by strong ground state bleaching of the Q band. Plots of the excited state absorption at 450 nm give a relaxation time of 20 psec while that at 570 nm has a relaxation time of 33 psec (Fig. 3). The relaxation of cytochrome c is complicated by a persistent low intensity absorption at both 450 nm and 570 nm which lasts longer than 100 psec. However, in the early evolution of the decay process Fe(II) cytochrome c and Fe(II) protoporphyrin DME in benzene have absorptions of an excited state which decays before the reappearance of the ground state.

Fe(II) Myoglobin:

Deoxymyoglobin has an excited state spectrum without any strong ground state bleaching. Thus, the 450 absorption band and a broad band at 600-650 nm can be seen but the fused Q band in the ground state must be counterbalanced by a like absorption in the excited state so that no bleaching is seen in the 500-570 nm range. The decay from the excited state occurs with a 34 psec time constant in the 460 nm excited state absorption band but is delayed to 60 psec in the broad 600 nm absorption band as shown in Fig. 4.

Fe(II) Protoporphyrin IX DME in Pyridine:

Fe(II) protoporphyrin IX DME in pyridine has strong absorption changes on excitation at 530 nm. An absorption band at 450 nm appears but broad-banded bleaching occurs with one band at about 580 nm and another at 520 nm and then a strong ground state Q band bleaching at 558 nm. The strong absorption band observed from 565 to 620 nm is similar in Fe(II) Protoporphyrin DME in benzene and reduced cytochrome c. The picosecond evolution of the excited state difference spectrum is shown in Fig. 5. The decay of the 450 nm peak and the decay of bleaching at 555 nm are biphasic. The initial decay rate is 41 psec for the 450 nm peak at 48 psec for the 550 nm bleaching (Fig. 6). The longer lived state has a lifetime of 304 psec at 450 nm and 230 psec at 550 nm. The longer time constant corresponds well with the time required for ligand recombination of 285 $\pm$ 25 psec for Fe(II) protoporphyrin DME in piperidine reported by Dixon et al (8). However, these authors observed a single fast decay of the 450 nm peak which was less than 35 psec.

Discussion

In general, the rapid decay of the excited state of metalloporphyrins is thought to be brought about by rapid intersystem crossing because of spin-spin and spin-orbit effects and (d,d), (d,π*) and (π,d) states which can form a ladder down to the ground state thus lowering the energy gap and increasing the relaxation rate (15-20). For the Fe(III) cytochrome c and myoglobin there is good reason to believe that its iron provides such intermediate energy levels for rapid decay of the excited state, regardless of the spin state (1).

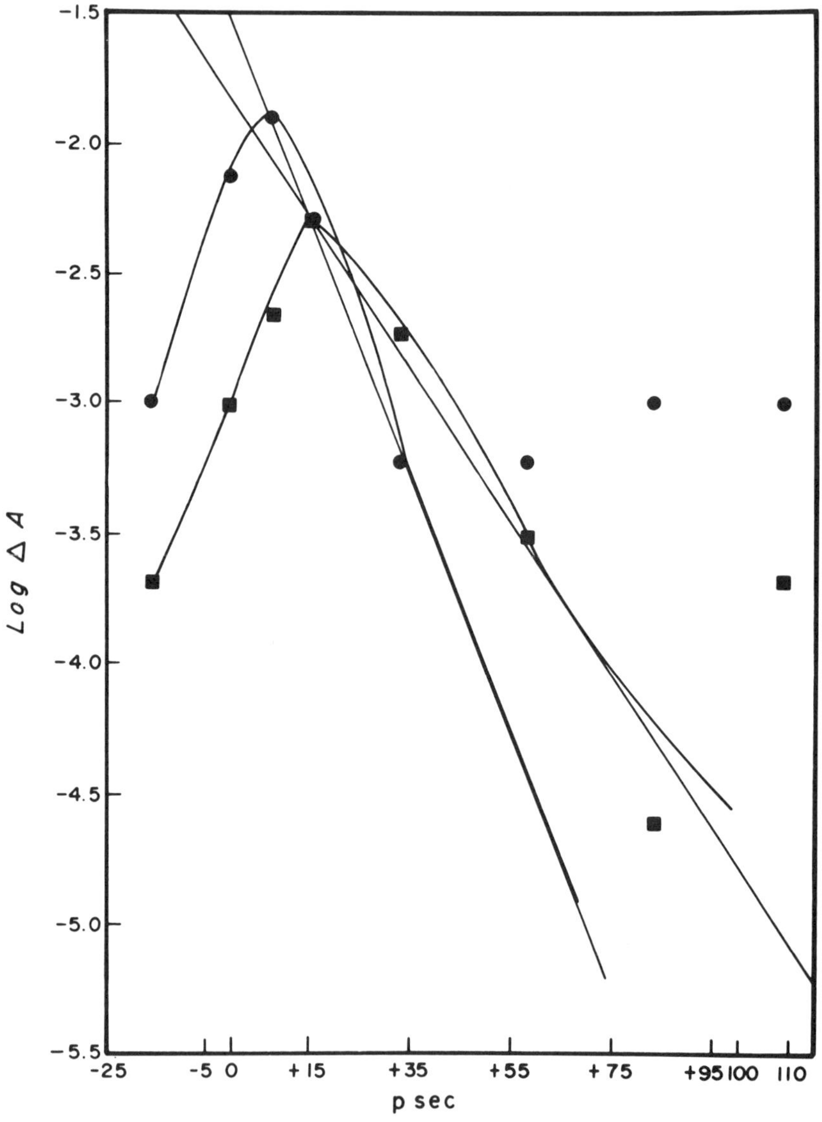

Figure 3: Log absorbance vs. time plots for Fe(II) cytochrome c. 530 nm excitation; circles - 450 nm absorption; squares - 570 nm bleaching.

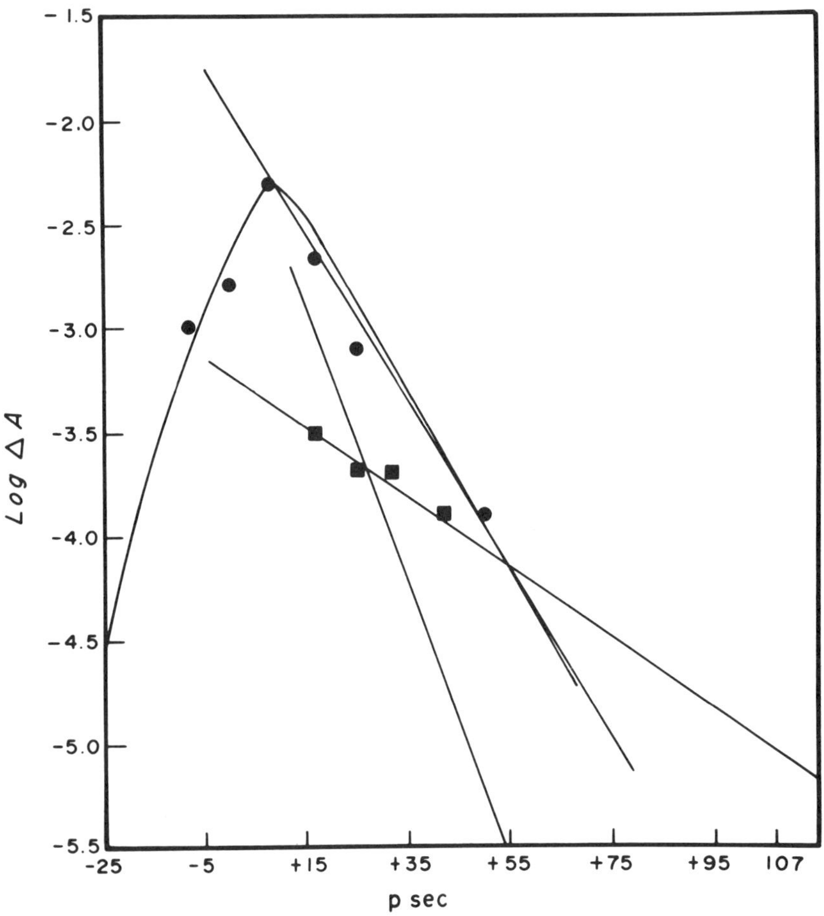

Figure 4: Log absorbance vs. ime plots for Fe(II) myoglobin (deoxy). 530 nm excitation; circles - 460 nm absorption; squares - 600 nm absorption.

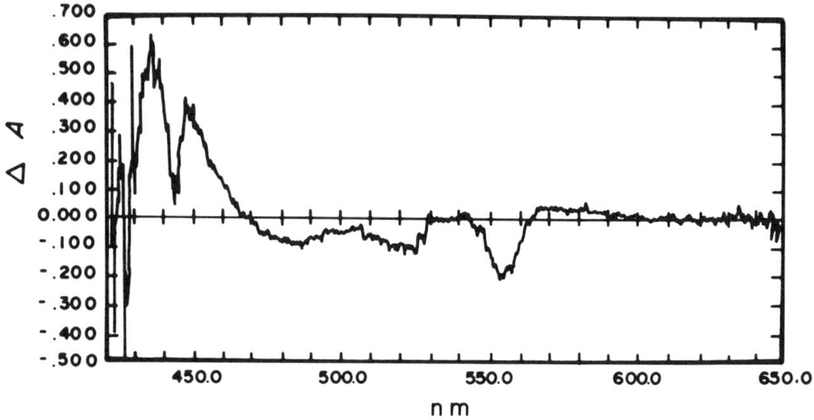

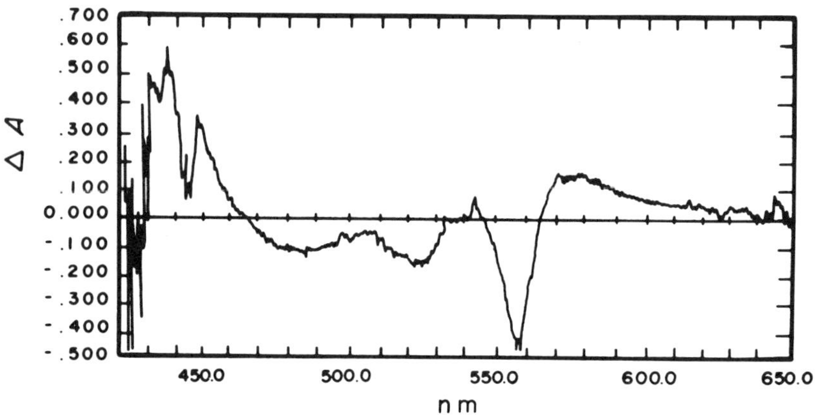

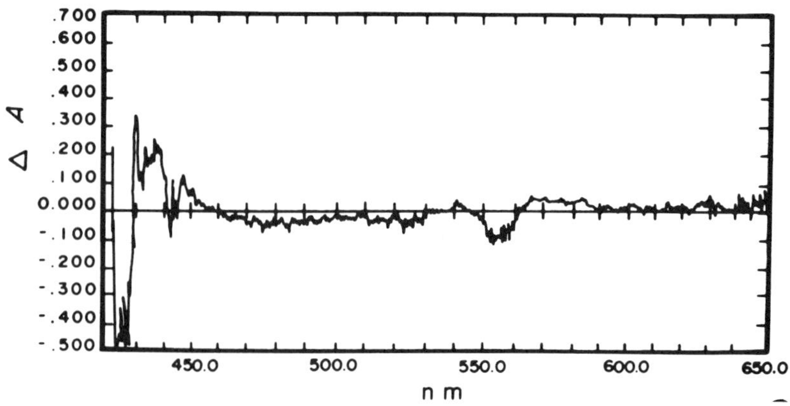

Figure 5: Picosecond spectra of Fe(II) protoporphyrin DME in pyridine. Top - 0 psec (center of 25 psec excitation pulse); Middle - 25 psec; Bottom - 200 psec.

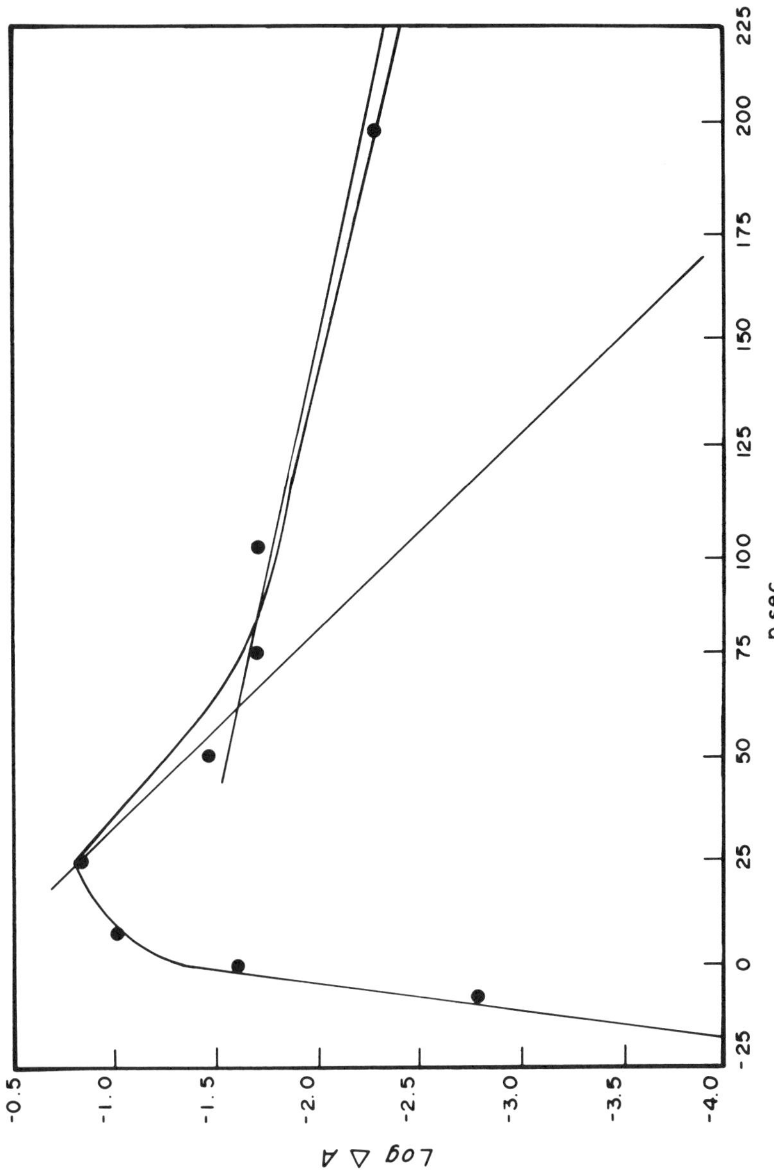

Figure 6: Log absorbance of Fe(II) protoporphyrin DME in pyridine. 530 nm excitation; 555 nm bleaching.

However, the rapid decay of Fe(II) protoporphyrin IX DME in benzene where no axial ligands are present indicates that energy levels between the $\pi \rightarrow \pi^*$ transitions are available even without axial ligands. Based on the temperature independent quadrapole splitting in the Mossbauer spectra Dolphin et al (21) have concluded that there are no low-lying excited states in the square planar Fe(II) octaethylporphyrins. Calculations by Zerner and Gouterman (22) and Zerner et al (23) show that the d orbitals are shifted into such intermediate levels only if the Fe^{++} atom is displaced from the plane of the porphyrin ring. Such a shift in d orbital energy levels would also result in lifting the orbital degeneracy thus allowing the high spin state to exist. This 4-ligand state is known to be high spin, S=2, (14) which confirms that the degeneracy of the d levels is lifted. The placement of the Fe^{++} atom outside the plane of the ring may also reduce the symmetry from D_{4h} to C_{4v}.

Evolution of the Excited State of Iron Porphyrins: In all the Fe(II) porphyrins studied the blue absorption increase between 450-500 nm is seen. Its initial lifetime is between 20 and 40 psec and is independent of spin-state, S=0 or S=2, and the number of axial ligands 4, 5, or 6. Absorption in this region is seen for $\pi \rightarrow \pi^*$ singlet and triplet absorption for other metalloporphyrins (16,18,24). Certainly the assignment of this absorption could be complicated by the formation of π cation and anion radicals which absorb in the region (25,26) and there is already evidence for ground state delocalization of the low spin electron of Fe(III) in myoglobin CN- complex onto the porphyrin ring (27). Nonetheless, we propose that the most likely origin of this band is the excited state singlet or triplet.

In most open-shell metalloporphyrins the first excited singlet state is rapidly converted to the triplet state by intersystem crossing due to strong pertabations of the unpaired spins and/or magnetic effects of unquenched orbital angular momentum. Thus, paramagnetic metalloporphyrins rarely have detectable fluorescence (28,29) and picosecond studies of Cu(II), (18,20), Ag(II) (18), and Co(II) porphyrins (30), confirm that the intersystem crossing $^2S_1 \rightarrow {}^2T_1$ occurs faster than the excitation pulses (6-35 psec). In addition to the large effect of unpaired electrons on the intersystem crossing, picosecond studies have also been able to demonstrate the added effect of orbital angular monetion in increasing the intersystem crossing in isoelectronic diamagnetic open-shelled species such as Pd(II) and Pt(II) (19). For the iron porphyrins excited state relaxation is rapid even when the ground state is diamagnetic and the unquenched orbital angular momentum is less than that of the diamagnetic Pd(II) complexes which have intersystem crossing rates slow enough (20 psec) to be measured. If the intersystem crossing rate with diamagnetic iron porphyrins is slow then the observed correspondence in relaxation times between S=0 and S=2 iron(II) porphyrins may not necessarily indicate that the relaxation pathways are the same.

The assignment of relaxation pathway for the S=0 state might be $S_1 \gg S_d \gg S_0$ with no triplet state involved. The π^*-d CT transitions would be chiefly responsible for quenching the lifetime of S_1. The resonance Raman linewidth measurements of Adar et al (3) indicate that cytochrome b_5 and by analogy other similar cytochromes such as cytochrome c have first excited state lifetimes which are on order of magnitude longer than high-spin deoxyhemoglobin. It is possible, therefore, that for the S=0 iron-porphyrins the relaxation occurs through the singlet manifold. We have postulated the same mechanism for non-radiative decay to account for our observations of picosecond spectra of dimagnetic Ni(II) protoporphyrin DME.

The bleaching in the ground state Q bands can be associated with the disappearance of the ground state and the decrease in the bleaching is then associated with return of the ground state. It should be noted, however, that d-d transitions may occur on the way to the ground state without being seen in the visible spectrum (31). It is difficult to unambigiously assign the relaxation of the 520 - 570 nm bleaching to reformation of the ground state, but we believe that this is the most likely event. Thus, the decay pathway of the high spin (S=2) Fe(II) porphyrins would be $\pi \gg \pi^*$ (S_0 - S_1) which undergoes intersystem crossing to the triplet or multiplet state within the pulsewidth of the excitation followed by a decay to the metal porphyrin charge transfer states $\pi^* \gg$ d in ~ 20 psec and subsequently generation of the ground state in 40 psec. The slower decay of both states in myoglobin may be due to shifts in the orbital energies in myoglobin which increase the energy gaps in the d orbitals. Alternatively the 5th ligand of myoglobin could be ejected during this time. Loss of the 5th ligand would generate the 4 ligand state seen with Fe(II) protoporphyrin DME in benzene. The Q bands should be seen during the time the ligand is released (40-60 psec) but no Q band absorption is seen over that time unless it is hidden by the filter at 530 nm.

The relaxation kinetics of Fe^{++}(II) protoporphyrin DME in pyridine are consistant with a ligand dissociation from 6 ligand to 5 ligand state during the first 40 psec and with formation of the 5 ligand ground state which undergoes recombination with pyridine to regenerate the 6 ligand complex in 250 - 300 psec as has been shown to be the case for Fe^{++} porphyrin with other basic nitrogenous ligands by Dixon et al (8). However, the biphasic absorbance at 450 nm has a long-lived lifetime of the same order as the postulated 5 ligand ground state. Either the 450 absorbance is part of the red shifted Soret band of the 5 ligand ground state or there is a long-lived excited state. We prefer the former explanation so that the 450 absorbance represents two different species - first a $\pi \gg \pi^*$ transition which lasts 40 psec in accord with the other Fe(II) porphyrins studied and secondly the longer decay representing the disappearance of a red-shifted ground state Soret band of the 5 ligand species as the 6 ligand ground state reappears.

Dissociation of axial ligands has been followed by picosecond spectroscopy for a number of metalloporphyrins. For the well-known photodissociation of O_2 and CO from hemoglobin and myoglobin the photoproducts appear very early < 10 psec. Dissociation of basic axial ligands such as pyridine and piperadine occurs within the lifetime of the excited state for Ni(II), Co(III) as well as for Fe(II) porphyrins. Whether the ejected species is "hot" with energy from the electronic deactivation of the porphyrin is not known, but the dissociation process does not appear to be dependent upon the wavelength of the excitation pulse (30,32).

The ease of formation of porphyrin radicals and the existence of ground state charge transfer into the axial ligands (33) makes it necessary to keep in mind the possibility of ligand ejection accompanied by oxidation-reduction during the excited states of metalloporphyrins.

From our observations, we feel it is unlikely that photo-induced redox reactions are occurring in the present studies. However, at wavelength < 300 nm photodissociation of oxyhemoglobin results in the formation of the O_2^- radical anion (34).

Literature Cited

1. Adar, F. In "The Porphyrins"; Dolphin, D., Ed.; Academic: New York, 1978, Vol. III, Chapt. 2, p. 167.
2. Gouterman, M. In "The Porphyrins"; Dolphin, D., Ed.; Academic: New York, 1978, Vol. III, Chapt. 1, p. 1.
3. Adar, F., Gouterman, M., Aranowitz, S. J. Phys. Chem., 1976, 80, 2184-91.
4. Huppert, D., Straub, K.D., Rentzepis, P.M. Proc. Natl. Acad. Sci. USA, 1977, 74, 4139-43.
5. Noe, L.J., Eisert, W.G., Rentzepis, P.M. Proc. Natl. Acad. Sci. USA, 1978, 75, 573-77.
6. Reynolds, A.H., Rand, S.D., Rentzepis, P.M. Proc. Natl. Acad. Sci. USA, 1981, 78, 2292-6.
7. Cornelius, P.A., Hochstrasser, R.M., Steele, A.W. J. Mol. Biol., 1983, 163, 119-28.
8. Dixon, D.W., Kirmaier, C., Holten, D. J. Am. Chem. Soc., 1985, 107, 808-13.
9. Reynolds, A.H., Rentzepis, P.M. Biophys. J., 1982, 38, 15-18.
10. Falk, J.E. "Porphyrins and Metalloporphyrins"; Elsevier; Amsterdam, 1964, p. 244.
11. Brault, D., Rougee, M. Nature New Biol., 1973, 241, 19-20.
12. Schejter, A., Eaton, W.A. Biochemistry, 1984, 23, 1081-84.
13. Beetlestone, J., George, P. Biochemistry, 1964, 3, 707-14.
14. Brault, D., Rougee, M. Biochemistry, 1974, 13, 4598-4602.
15. Antipas, A., Dolphin, D., Gouterman, M., Johnson, E.C. J. Am. Chem. Soc., 1978, 100, 7705-09.
16. Tsvirko, M.P., Stelmakh, G.F., Pyatosin, V.E. Chem. Phys. Letts., 1980, 73, 80-83.
17. Antipas, A., Gouterman, M. J. Am. Chem. Soc., 1983, 105, 4896-4901.

18. Kobayashi, T., Huppert, D., Straub, K.D., Rentzepis, P.M. J. Chem. Phys., 1979, 70, 1720-26.
19. Kobayashi, T., Straub, K.D., Rentzepis, P.M. Photochem. Photobiol., 1979, 29, 925-31.
20. Kim, D., Holten, D., Gouterman, M. J. Am. Chem. Soc., 1984, 106, 2793-98.
21. Dolphin, D., Sams, J.R., Tsin, T.B., Wong, K.L. J. Am. Chem. Soc., 1976, 98, 6970-75.
22. Zerner, M., Gouterman, M. Theor. Chim. Acta, 1960, 4, 44-63.
23. Zerner, M., Gouterman, M., Kobayashi, H. Theor. Chim. Acta, 1966, 6, 363-400.
24. Magde, D., Windsor, M.W., Holten, D., Gouterman, M. Chem. Phys. Letts., 1974, 29, 183-88.
25. Dolphin, D., Muljiani, Z., Rousseau, K. Ann. NY Acad. Sci., 1973, 206, 177-98.
26. Felton, R.H. In "The Porphyrins"; Dolphin, D., Ed.; Academic; New York, 1978; Vol. 5, p. 53.
27. Shulman, R.G., Glarum, S.H., Karplus, M. J. Mol. Biol., 1971, 57, 93-115.
28. Gouterman, M., Mathies, R.A., Smith, B.E., Caughey, W.S. J. Chem. Phys., 1970, 52, 3795-3802.
29. Eastwood, D., Gouterman, M. J. Mol. Spectroscopy, 1970, 35, 359-75.
30. Tait, D.C., Holten, D., Gouterman, M. Chem. Phys. Letts., 1983, 100, 268-72.
31. Ake, R.L., Gouterman, M. Theor. Chim Acta, 1969, 15, 20-42.
32. Kim, D., Holten, D. Chem. Phys. Letts., 1983, 98, 584-89.
33. Bajdor, K., Kincaid, J.R., Nakamoto, K. J. Am. Chem. Soc., 1984, 106, 7741-47.
34. Demma, L.S., Salhany, J.M. J. Biol. Chem., 1977, 252, 1226-30.

RECEIVED July 3, 1986

13

Intermediates Found in the Picosecond Photodissociation of the CO and O_2 Forms of Synthetic Heme Complexes

K. Caldwell[1], L. J. Noe[1], J. D. Ciccone[2], and T. G. Traylor[2]

[1]Department of Chemistry, The University of Wyoming, Laramie, WY 82071
[2]Department of Chemistry, University of California at San Diego, La Jolla, CA 92037

We have investigated the picosecond photodissociation of the O_2 and CO forms of a number of synthetic and natural heme complexes that differ in the geometry and strain associated with the imidazole-iron (porphyrin) interaction using the standard Nd:Glass laser pump-probe absorption method. Our results indicate that it takes some picoseconds for these complexes to dissociate and further suggest that a 'pseudo' four coordinate complex may be present as a photointermediate in the synthetic compounds with the strained imidazole geometry.

In this report we would like to discuss our results on the picosecond photodissociation experiments of the CO and O_2 forms of a number of synthetic reversible oxygen carriers (1,2) and compared them to earlier picosecond absorption work on the same derivatives of the natural forms of hemoglobin. The latter work has provided us with a better understanding of the details of the photodissociation in terms of the sequential evolution of four photointermediates which were experimentally isolated, characterized, and kinetically analyzed (3,4).

Because of the differences between the early picosecond evolution of the intermediates found using absorption spectroscopy for Hb, Mb, and the α and β chains of Hb (5), we became interested in examining the effects of tertiary constraint of the proximal imidazole on the course of the dissociation, effects that could more easily be examined by studying appropriate synthetic analogues. Additionally, our interest in elucidating the role that the imidazole plays in modulating the dissociation process was prompted by the results of recent ps resonance Raman work (6-10). For example, Findsen et al. (6,10) present results that show the Fe in MbCO has relaxed out of place by 20 to 30 ps based on analysis of Fe-His resonantly enhanced modes using Raman shifted Soret excitation, whereas in HbCO other RR work reveals that such relaxation does not occur to the fullest extent possible before about 20 ns (7-9). These studies clearly show the importance of the tertiary structural constraints,

0097-6156/86/0321-0182$06.50/0

particularly ones involving the Fe-His linkage, in the out of plane relaxation process.

So far two important conclusions have emerged from our comparative studies on several dozen natural and synthetic compounds: That the effects of strain on the heme are important in terms of variations found in the dynamics and mechanism of early events of the picosecond dissociation and that it takes some picoseconds for these complexes to dissociate, although the process begins in the subpicosecond regime. The synthetic compounds used in this study mimic the 'R' and 'T' states of hemoglobin depending on whether the imidazole-heme geometry is strain free or not (11-15). As our results will show, this particular structural variation dramatically effects the course of the dissociation. We have assigned an intermediate found early in the dissociation of the strained CO-complexes to a pseudo four-coordinate complex. This particular intermediate is absent in the unstrained CO-complexes and as well as in all the O_2-complexes, strained or not. Our experimental procedure used to characterize these photointermediates as they evolve employs the standard pump-probe method by following absorption changes in the $\pi\pi^*$ Soret band after initiating the dissociation with a 531 nm pulse directed to the Q-band. A mode-locked Nd^{3+}-phosphate glass laser is used for the experiments. Our analysis of the results is somewhat different than usual in that we obtain modeled spectra for comparison to the experimental spectra over a wide range of pump-probe delays using a numerical integration procedure (3,4). In this procedure we convolute the pump and probe pulses with a kinetic model for the photodissociation, that is consistent with the sequential evolution of the intermediates observed experimentally, as these pulses pass through the sample.

As the picosecond dynamics of the CO and O_2 ligand photodissociation in heme-complexes is a topic of current theoretical and experimental interest, certain other work having a bearing on the results in this report should be mentioned. The question of what orbital promotion(s) in the Fe(II)(d^6) six-coordinate manifold lead to photodissociation of the CO or O_2 complexes is more easily answered for the CO than the O_2 parent molecules. The extended Huckel calculations of Zerner et al. (16) have been used for many years to interpret spectroscopic results, particularly with reference to the heme-CO complexes. Their calculation, for example, suggests that population of the $d_{xz,yz}$ (or d_π) $\rightarrow$ d_{z^2} transition (octahedral symmetry) will lead to photodissociation of the CO-complex. This Huckel calculation was updated by Eaton et al. (17) with particular emphasis on the d-d metal and d-ligand change transfer transitions for the HbO_2, HbCO and deoxy complexes. These theoretical results combined with their experimental CD and MCD spectra allowed spectral assignments to be made for certain of d-d and d-ligand transitions. More recently, Waleh and Loew (18-20) using INDO-SCF-MO-LCAO-CI calculations concluded that both the $d_\pi \rightarrow d_{z^2}$ and the $d_{x^2-y^2} \rightarrow d_{z^2}$ (pyrrole N-atoms bisect the xy axes) are the best photodissociative candidates since these transitions decrease in energy as a function of increasing Fe-CO distance. They prefer the $d_\pi \rightarrow d_{z^2}$ transition over the other possibility because it has higher oscillator strength and better bond weakening characteristics. These authors also suggest that the singlet, $d_{\pi} \longrightarrow d_{z^2}$ transition is the best candidate for the photodissociating state, rather than intersystem crossing to lower triplet states, or through $\pi\pi^*$ porphyrin transitions as Hoffman and Gibson suggest (21).

There has also been a number of experimental investigations detailing the picosecond photodissociation of the O_2 and CO forms of the natural complexes using absorption techniques. Representative citations and discussion may be found elsewhere (3,4). Recent femtosecond absorption experiments on HbCO, MbCO and CO-protoheme reported by Martin et al. (22) provide evidence that each of these species forms a deoxy photoproduct with a time constant of 350 fs after excitation with 250 fs pulses at 307 nm. Support of this particular experimental result is provided by a molecular dynamics trajectory simulation study which suggests that it is possible for the iron to relax out of the plane of the porphyrin by 350 fs (23). Our results and conclusions differ somewhat from these investigations, although our results suggest that the first stable photoproduct appears in the subpicosecond regime.

Experimental

The double-beam phosphate-glass Nd^{3+} laser apparatus used in these experiments has been described elsewhere (3,4). Experiments were of the pump-probe type done at a rate of one per minute using a 531 nm (8 ps FWHM, TEM_{oo}, $\leq$ 1 mj) pulse to initiate the photodissociation by pumping the Q-band of the parent six-coordinate CO or O_2 complex. A continuum light pulse, generated from a D_2O cell and appropriately filtered to examine the Soret region, provided the absorbance versus 42 nm assay of sample-photoproduct concentrations at specific probe-pump maxima pulse delays ranging from -20 to 50 psec. Negative delay settings refer to experiments where the maximum of the probe pulse reaches the sample cell (1 mm optical path) before the maximum of the pump pulse; a configuration in which the tail of the probe pulse interrogates excited sample on the leading edge of the pump pulse. These negative delay experiments, conducted over a suitable range, are critical to the detection and kinetic analysis of photointermediates that appear early in the dissociation, ones that have lifetimes considerably shorter than the 8 ps FWHM pulse width.

The equilibrium ground-state spectra of the parent six-coordinate and deoxy five-(or four-)coordinate species were obtained using a Tungsten lamp placed before the I_o/I beam splitter. Earlier work in our laboratory (3,4) has established that the experimental spectra of transient protoproducts can be resolved as they sequentially evolve during the dissociation process. At relatively long delays > 20 psec after excitation, the spectrum of the stable photoproduct HB^* is obtained by adding ground state to the ΔA difference curve: $\Delta A(HB^*\text{-}HBCO) + A(HBCO) = A(HB^*)$. At earlier delays the procedure of isolating a particular photoproduct is the same except that spectra of other intermediates produced must be taken into account in a manner consistent with the quantum yield of dissociation and conservation principles. Thus the band shape, the experimental parameters λ_{max}, FWHM, and an estimate of ε_{max}, as well as an estimate of photoproduct lifetime is obtained by a work-up of the experimental data.

The experiments on CO forms of the synthetic compounds were done with sample concentrations adjusted to ~ 50 μM to give an absorbance of 1 over a 1 mm optical path at the Soret maximum. For the oxygen experiments, the samples were prepared as the CO complexes and covered with a balloon containing ~ 3:1 mixture of CO to O_2. An additional photographic strobe lamp having a flash duration of ~ 25 μs was integrated into the system and adjusted to pump the sample 0.45 ms before the picosecond pump pulse arrived, allowing sufficient time for the production of the heme-dioxygen complex by means of the following reactions: $HBCO \xrightarrow{h\nu} HB + CO$ and $HB + O_2 \longrightarrow HBO_2$. The 0.45 ms period allows sufficient time for the formation of HBO_2 with only a minor fraction of this complex returning to the more stable CO form (11).

The compounds were received from T. G. Traylor's laboratory in the Fe(III) chloride form. Approximately .4 mg of this crystal was dissolved in absolute methanol, degassed, and brought back to atmospheric pressure with argon. Aqueous solutions of cetyltrimethylammonium bromide (CTAB) were prepared, filtered and bubbled with carbon monoxide gas to saturate the system. Sodium dithionite was added for the reduction of Fe(III) to Fe(II) and the methanol solution of the heme was added to the CTAB to create an aqueous solution of ~ 10^{-5} - 10^{-4} molar heme-CO. CTAB produces a micellar system with a hydrophilic exterior and an hydrophobic interior pocket in which the heme complex is contained. Because of the possibility of reactions in which the heme complex binds water, experiments were also done on compounds prepared with benzene or toluene in place of CTAB. All experiments were done at room temperature, ~ 20°C. At the CTAB concentrations used, precipitation of the micelles occurs at temperatures below 20°C.

Results

To examine the effects of imidazole strain on the picosecond dissociation of reversible oxygen carriers, the CO and O_2 forms of four synthetic hemes shown in Fig. 1 were studied: 1, 1-ST, 1-ET and 1-AD. The first three compounds were studied in aqueous solution and the last in toluene. Compound 1 is a chelated protoheme containing vinyl side chains. The imidazole is attached to the porphyrin through an amide linkage that minimizes the interaction or strain between the imidazole and the porphyrin-Fe(II)heme. This molecule serves as a model for R-state hemoglobin (11,12). 1-ST is a chelated protoheme in which face strain has been introduced to the complex by means of a methylimidazole. The compound 1-ET is one of two unchelated hemes studied. It is prepared in the presence of an external base, 1,2-dimethylimidazole. Due to the possibility of binding two base molecules axially to the heme, this bulky disubstituted base is used. It introduces strain into the heme-imidazole

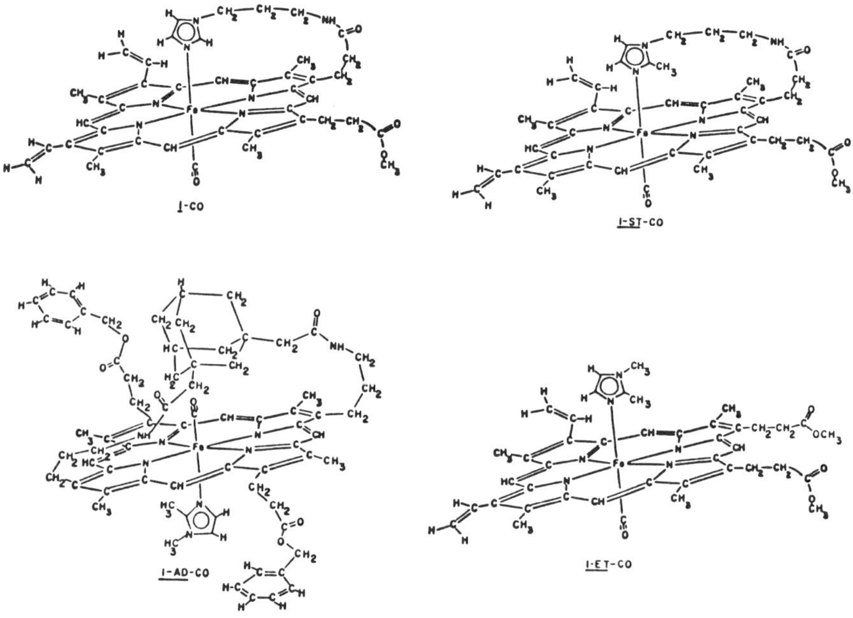

Figure 1. Structures of some of the synthetic compounds studied. The iron-imidazole bond in 1-CO is unstrained while it is strained in 1-ET-CO, 1-ST-CO and 1-AD-CO.

bond providing a model for T-state hemoglobin as well as preventing the binding of a second base to the complex. The second unchelated protoheme studied, 1-AD, is one in which a bulky nonaromatic adamantane cyclophane group has been introduced distally to provide protection for CO or O_2 binding site. Due to the possible interactions with water, this compound was prepared in anhydrous toluene and 1,2-dimethylimidazole was again used for the proximally bound external base.

Experiments on 1-CO using benzene in place of CTAB were also done to examine the effects solvent and environment on the photodissociation. None were found. The photointermediates arrived at the same time, had the same peak wavelengths, extinction coefficients and band shape. In so far as the dynamics observed in these experiments are independent of CO pressure and since there is no detectable geminate CO recombination, it is reasonable to expect effects on the photodissociation due to solvation to be minimal as diffusion has not yet occurred on the time scale studied.

To study the possibility of electronic effects on the photodissociation localized to the porphyrin ring, we examined 2-CO, a compound similar to 1-CO except for the substitution of ethyl groups in place of vinyl groups on the outside of the porphyrin ring. Earlier bimolecular kinetic studies (11,15) had shown that this change in the electron distribution of the porphyrin had no effect on the reaction of CO with the heme. Our results also indicate that this perturbation has no apparent effect on the picosecond photodissociation.

Examination of the transient ΔA spectra reveals that the synthetic complexes photodissociate according to one of two pathways: (I) A—>AS—>B—>C—>D or (II) A—>AS—>X—>B—>C—>D. AS, X, B, C and D represent intermediates that have been experimentally observed and isolated in various time frames up to 50 ps. The assignment of these will be discussed shortly in terms of octahedral notation. The six-coordinate parent ground state ^1A, is designated A.

Pathway I was observed for all the O_2-complexes studied, strained or unstrained, as well as for the unstrained CO-complexes. This particular pathway is the same one observed in the photodissociation of the natural heme complexes (3,4) (HbCO, MbCO, HbO_2 and MbO_2) with the exception that there is no detectable geminate recombination to the limit of our experiment, 50 ps. Pathway II, observed for the strained-CO complexes, reveals the presence of a fifth intermediate X found early in the dissociation that is either absent or undetectable in the natural or synthetic heme complexes following pathway I. The kinetics associated with the evolution of these intermediates will be discussed shortly. First, it is appropriate to examine in some detail the experimental ΔA difference spectra of two representative complexes, 1-CO and 1-ET-CO. A discussion of 1-ST-CO and 1-ET-O_2 is also included for comparative purposes.

1-CO. The experimental ΔA spectra for 1-CO are shown in Fig. 2 over delays ranging from -12 ps to 50 ps. The 512 point spectra shown in this figure as well as in Fig. 5 for 1-ET-CO have been smoothed using a 31 point procedure. This allows us to determine the spectral parameters λ_{max}, FWHM, and ε_{max} of the photointermediates more accurately and facilitates the comparison of the experimental spectra to the kinetically modeled ΔA spectra to establish rate constants and intermediate lifetimes. For 1-CO the spectroscopic parameters of the ground state and intermediates in terms of λ_{max} (nm), FWHM (nm) and ε_{max} (upper estimates in M^{-1} cm^{-1}) are respectively: A(ground) (12) -419, 11, 203K; AS-419, 15, 205K; B-430, 32, 75K; C-442, 24, 160K; D-435, 20, 175K. The first event observed at the -12 ps delay is a positive absorbance change at the low wavelength end of the spectrum. This increase in absorbance is an indication of the presence of the first intermediate AS. AS has an absorption maximum at the same wavelength as the six coordinate ground state species, 419 nm, with a slightly higher extinction coefficient and a broader band width. AS is not long lived and by -8 ps the induced absorbance present at the low wavelength end of the spectrum has been replaced by the disappearance or bleaching of the ground state spectrum. The -4 ps transient spectrum reveals a form of the complex having a maximum positioned at a longer wavelength has grown in. After removal of the ground state spectrum from the -4 ps difference spectrum, this second intermediate, B, is seen as a weak band with a λ_{max} at about 430 nm. B is replaced with a photoproduct that has a maximum at the far red end of the spectrum by 8 ps. Isolation of this intermediate C from the 8 ps difference spectrum reveals a species that has a λ_{max} centered at 442 nm with an extinction coefficient of approximately three-quarters that of the ground state. This band at the red end of the spectrum gradually blue shifts and narrows clearly revealing the presence of the final intermediate of the photodissociation process, D, which has an extinction coefficient that is similar to that for C with an absorption maximum approximately 7 nm lower, at 435 nm. D is highly developed by 50 ps, and with only slight differences, it is identical to the ground state deoxy 1 species. This in fact is the case for all complexes studied; the five-coordinate deoxy ground state of the parent complex is fully developed in less than 100 ps, a result very much different from hemoglobin where it takes greater than 20 ns for the iron to relax out of the plane of the porphyrin. The spectra of the 1-CO intermediates are shown in Fig. 3.

1-ST-CO. Experiments done at the earliest delays of -12 and -8 ps show that this compound is dissociating similarly to 1-CO. At first, there is an increased absorbance at the wavelength of the ground state species. However, as the photodissociation proceeds, a noticeable difference becomes apparent in the transient spectra between this compound and the unstrained species. Although the second intermediate B can be seen to be appearing by 0 ps, the disappearance of the ground state species does not progress as it did in the previous compounds. At a time of 0 ps the disappearance of the ground state complex is only about 50% complete. This indicates that either there is a delay in the disappearance of the ground state species that could, for example, be caused by a bottleneck in the path of the dissociation, or that there is another intermediate present in this photodissociation of the strained compound that is not present in the unstrained compounds. Removing the ground state A and the intermediate B from the -4 ps difference spectra reveals a

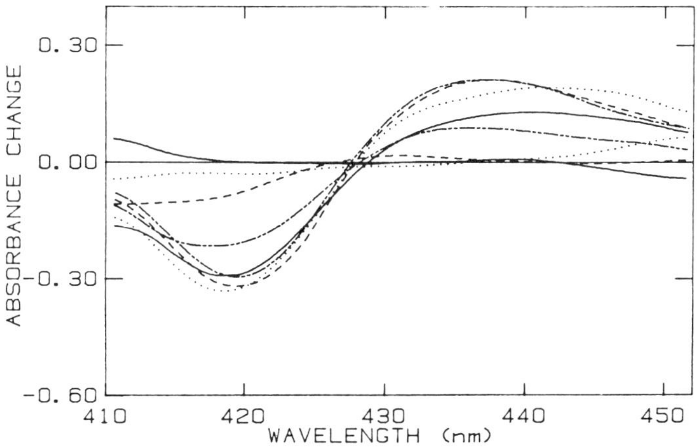

Figure 2. Experimental spectra for 1-CO. Delays range from -12 to 50 ps; -12, +4 (solid); -8, +8 (dotted); -4, +20 (dashed); 0, +50 (broken). Experimental data has been smoothed (see text).

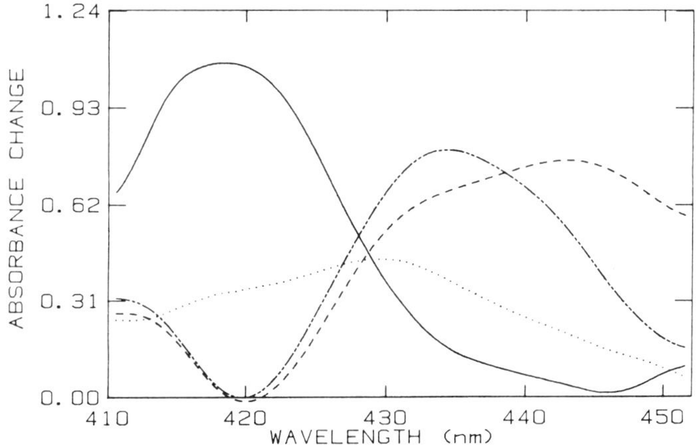

Figure 3. Spectra of the 1-CO intermediates: AS (solid), B (dotted), C (dashed) and D (broken).

photointermediate in the same wavelength range as the ground state species. This intermediate has a spectrum that is characterized by a double maximum and an extinction coefficient approximately one-half of that of the ground state compound. The new intermediate appears in the reaction pathway after the formation of the species AS, but before B. The pathway for dissociation then continues as it does in the unstrained compounds. X reacts to produce B, then C is then observed, with the shifting and narrowing of the band characteristic of the final reaction product D. The intermediates, B, C and D, are, within the accuracy of the experiment, sequentially equivalent to those produced in the photodissociation of 1-CO, with differing dynamics and spectral parameters.

1-ET-CO. One of the advantages of examining this complex lies in the fact that the imidazole is incorporated as an external base that is not tethered to the porphyrin, allowing the ground state spectrum of the four-coordinate species to be easily obtained. Fig. 4 reveals that the spectrum of the four-coordinate ground state is characterized by a double-peaked band in approximately the same wavelength range as the six-coordinate ground state.

What can again be seen from work-up of the transient absorption spectra of this strained complex is the early appearance of a new photoproduct X in the dissociative pathway that is not present in the unstrained CO-complexes. The difference spectrum for the 0 ps delay shows only ~ 30% ground state bleaching. Removing the remaining ground state and the intermediate B from this difference spectrum reveals the presence of the X characterized by a double peaked band. The smoothed transient absorption spectra and the spectra of the intermediates are shown in Figs. 5 and 6. The spectroscopic parameters for 1-ET-CO, λ_{max}, FWHM, and ε (upper estimates) respectively are: A(ground state) (14) -416, 28, 200K; X-380, 40, 120K; X-429, 25, 70K; B-430, 35, 25K; C-441, 32, 150K; D-426, 78, 165K. Because of background noise problems, we were unable to collect suitable data for this compound before -8 ps. As a result AS has not been seen, though we believe it to be present in earlier time frames, as it is present for the other compounds studied.

1-ET-O_2. This complex is representative of the strained oxy-heme compounds studied, and it gives clear evidence that these compounds photodissociate according to pathway I. Transient absorption data for this compound show no evidence of photoproduct X.

Kinetic Modeling

Kinetic modeling of the transient absorption spectra is necessary in order to obtain estimates of the lifetimes and rate constants associated with the evolution of the intermediates observed during the dissociation. A detailed description of the modeling may be found in our work on the picosecond photodissociation of the natural carboxy- and oxy- heme complexes (3,4). The model is designed to treat the dissociation as a series of first or second order steps indicated in pathways I and II, ones that are consistent with the transient absorption data and with arguments based on bottlenecks

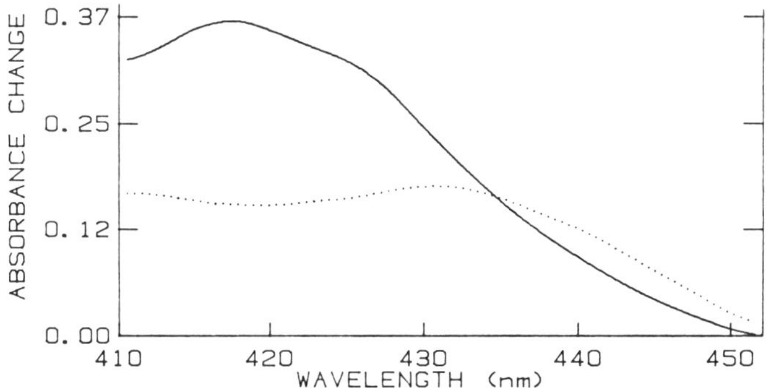

Figure 4. Ground state spectra of the four-coordinate form of 1-ET and 1-ET-CO: 6-coordinate (solid) and 4-coordinate (dotted).

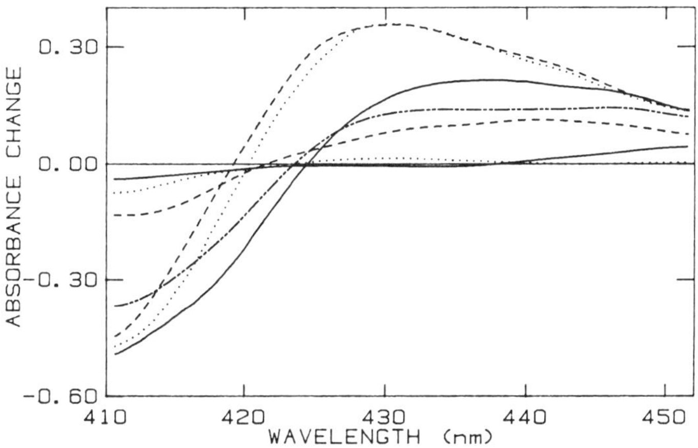

Figure 5. Experimental spectra for 1-ET-CO. Delays range from -8 to 50 ps; -8, +8 (solid); -4, +20 (dotted); +4 (broken); 0, +50 (dashed). Experimental data has been smoothed (see text).

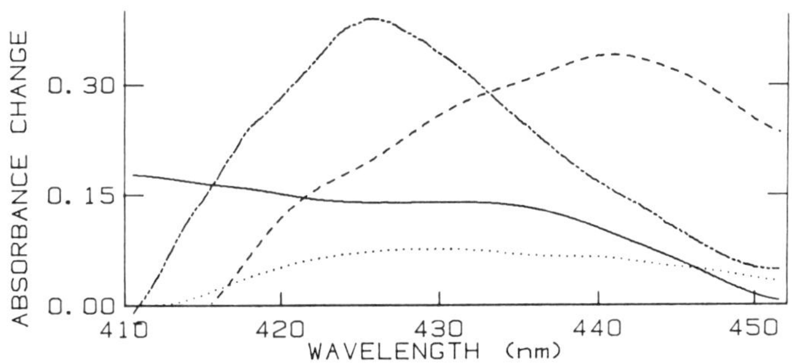

Figure 6. Spectra of the 1-ET-CO intermediates: X (solid), B (dotted), C (dashed), D (broken). Because of the noisy conditions we were unable to collect data for delays < -8 ps. As a result AS was not seen, though we believe it to be present in these earlier time forms.

encountered in the radiationless dissipation of energy in the mainfold of ligand-field states, carboxy, oxy and deoxy, available to the photodissociation. We have used simplified octahedral energy level diagrams derived from extended Huckel calculations (17,24,25) for this purpose. Based on direct and favorable comparisons between the results and analysis of our photodissociation work on the natural heme-complexes and the results described above, we propose two kinetic schemes which adequately interpret our observations:

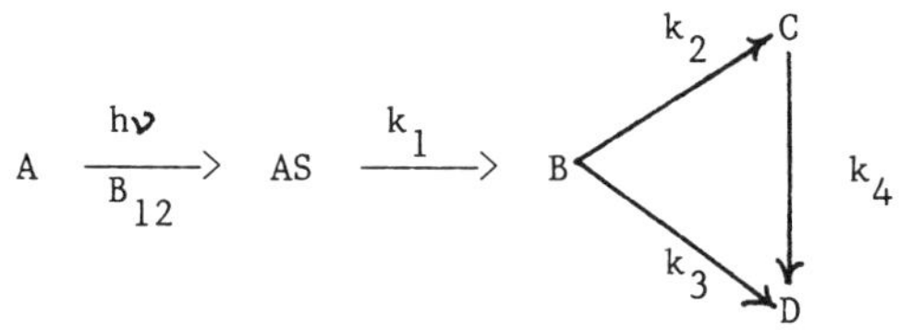

(1)

for the unstrained CO and strained or unstrained O_2 complexes, and

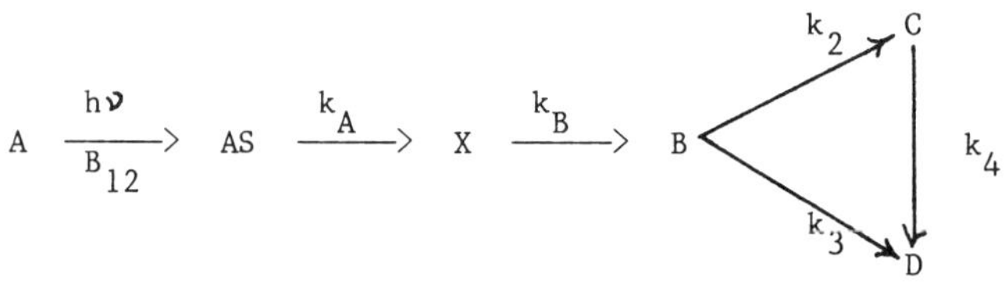

(2)

for the strained CO complexes. Since geminate recombination was not observed for times < 50 ps, the rapid recombination step O_2 (or) $CO + C \xrightarrow{k_5} A$ has been eliminated. Additionally, our experimental limit of 50 ps precluded examination of the slow O_2 (or) $CO + D \xrightarrow{k_6}$ recombination path from consideration (3,4).

The analysis of the early < 10 ps transient ΔA data is complicated by the fact that the pump and probe pulses each have 8 ps FWHM widths. Because of this, the pump pulse will excite different molecules at different times over a range of more than 8 ps. As the probe pulse passes through the sample it will see molecules at different stages of excitation and decay which in effect extends the range of the laser shot to over 16 ps and makes interpretation of the early time frame spectra difficult. This complication is resolved by analyzing the experimental absorbance change, ΔA, (as a function of time delay Δt) using a Runge-Kutte-Nystrom numerical integration procedure that convolutes the pump and probe pulses together with the kinetic models for the photodissociation described above. The experimental parameters used in the modeling procedure include the concentration of the ground state and the λ_{max} and extinction coefficients of the intermediates (see Tables I-IV for example). Initial guesses for the rate constants are also needed. These come from estimates of the intermediate lifetimes after workup of the transient ΔA data. For the calculation, all spectra are assumed to be Gaussian in shape. The rate constants are then adjusted to obtain the best computer fit to the experimental difference spectra for the entire set of probe-pump delays.

The calculation requires two sets of fine differential equations to be solved simultaneously for Mechanism I (two sets of six equations for Mechanism II). One set describing the interaction of the pump pulse with the sample and resultant instantaneous concentrations of species A through D, and the second set, the apparent concentrations of these species as the probe pulse passes through the sample. Certain assumptions must be made in order to simplify the calculations. Amongst them, the light pulses are assumed to be an ensemble of photons whose cross section is Gaussian in shape; the excitation path is perpendicular to the sample cell and the volume swept out by the pump pulse, though in reality slightly conical in shape, is assumed to be cylindrical (the path of the interrogation pulse is considered to be cylindrical and coaxial with the pump pulse while in reality there is a 7° angle to the pump pulse and further it is focused in the sample cell, resulting in an hour-glass shaped volume); and finally the assumption is made that optical pumping caused by the probe pulse is negligible. The rate at which the ground state is depleted is expressed as a second order rate equation, in terms of the laser pulse and the ground state concentration, $-\ dA(t)\ /\ dt = B_{12}\ \rho(t)\ A(t)$. $\rho(t)$ is the photon density. B_{12} is the Einstein coefficient of absorption for the pump wavelength calculated as $B_{12} = \varepsilon_{531}\ c\ \ln(10)\ /\ N_0$ where c is the velocity of light and N_0 is Avogadro's number. Other details of the modeling may be found in reference 3.

There are a number of limitations in the modeling procedure that affect the accuracy of the curve fit between the modeled and experimental difference spectra. As a result, the rates and lifetimes derived from the fit should be viewed as a means of achieving a better understanding of how tertiary structure effects the dynamics or pathway of the photodissociation from a comparison of such data for complexes or classes of complexes, rather than as absolute rate data. One of the limitations affecting the quality of the fit is the assumption of a Gaussian shape to describe the spectral bands of the intermediates and six-coordinate ground state. Another limitation is that the fit of the computer generated spectra in some cases was found to be best at time delays slightly different than those in the actual difference spectra. For example difference spectra for time delays of 0, +4 and +8 in some instances were found to be best fit by computer modeled spectra for time delays of +2, +6 and +10 ps. Such errors can be expected due to the experimental inaccuracy of the Duguay shutter timing procedure. The relative time changes between the delay times, in this case 4 ps, are considered to be accurate. A third problem encountered is due to the group velocity dispersion or chirp in the pulse probe in the red end of the wavelength range studied. The variations in velocity for various media in the laser apparatus were obtained and the chirp was calculated to be 0.6 ps from the generation of the white light pulse in the continuum cell to the time the probe pulse interacts with the sample. For an intermediate such as AS which absorbs at the same wavelength as the ground state no delay caused by chirp should be observed. For the intermediate C, however, whose λ_{max} is red shifted 20-30 nm from the ground state, the chirp would result in an additional delay in the arrival of the intermediate of approximately .5 ps. This type of delay was corrected for in the modeling procedure.

Under these conditions and limitations, the computer modeled spectra for the compounds 1-CO and 1-ET-CO discussed earlier are shown in Figs. 7 and 8. For clarity, each figure contains curves generated for five of the highest delay settings, although curves for delays up to -20 ps were generated to achieve a good fit. The rate constants and lifetimes that provided the best fit to the experimental ΔA curves are listed in Tables I-III. For comparative purposes we have included lifetime data for some of the natural hemes studied in Table IV. As the jitter in the pulse duration is nearly 1 ps, the rate constants k_1 and k_A are clearly approximations and are used only to obtain a good fit to the experimental data.

Table I. Rate Constants for Complexes Dissociating According to Mechanism I $(ps)^{-1}$

Compound	k_1	k_2	k_3	k_4
1-CO	4.00	0.5	0.05	0.07
1-O_2	4.00	0.05	0.01	0.02
1-ET-O_2	4.00	0.05	0.02	0.02
1-AD-O_2	4.00	0.05	0.02	0.03
1-ST-O_2	4.00	0.5	0.02	0.03
2-CO	4.00	0.5	0.05	0.07

Table II. Rate Constants for Complexes Dissociating According to Mechanism II Showing the Pseudo Four Coordinate Intermediate $X(ps)^{-1}$

Compound	k_A	k_B	k_2	k_3	k_4
1-ST-CO	4.00	0.2	0.02	0.03	0.03
1-ET-CO	4.00	0.2	0.08	0.01	0.03
1-AD-CO	4.00	0.1	0.04	0.02	0.05

Discussion

With reference to the octahedral energy level diagrams for oxy, carboxy and deoxy hemoglobin derived from extended Huckel calculations (17,24,25) and our earlier work on the natural heme complexes, the results of this study can be discussed.

Intermediates for the CO Complexes. Absorption of a 531 nm photon in the Q-band would produce a vibronically excited singlet state

Table III. Lifetimes of Intermediates (ps)

Compound	τ_{AS}	τ_X	τ_B	τ_C^b
1-CO	0.25	----	2	15
2-CO	0.25	----	2	15
1-ST-CO	0.25	5	20	33
1-ET-CO	0.25	5	11	33
1-AD-CO	0.25	10	17	20
1-O_2	0.25	----	16	50
1-ST-O_2	0.25	----	2	33
1-ET-O_2	0.25	----	14	50
1-AD-O_2	0.25	----	2	33

Table IV. Lifetimes of Intermediates Found in the Photodissociation of Some Natural Heme Complexes (ps) (3,4)

Compound	τ_{AS} $(k_1)^{-1}$	τ_B $(k_2+k_3)^{-1}$	τ_C $(k_4+k_5)^{-1}$
HbCO	0.4	1.2	2.3
HbO_2	0.3	5.6	1.1
HbαCO[3]	0.3	5.6	33.3
HbαO_2[3]	0.3	2.9	28.6
MbCO	0.3	4.8	4.8
MbO_2	0.3	2.2	2.4

through an allowed $\pi\pi^*$ transition localized in the plane of the porphyrin. Rapid vibronic relaxation would follow the absorption process, to the vibrationless origin of the ^1Q state, and then internal conversion to the $^1T_1$ state would take place. This relaxation will occur on a sub-picosecond time scale, so that the first bottleneck expected to be seen would be the lowest excited state $^1T_1$. This assumes that $^1T_1$ is not the photoactive state that promotes the photodissociation as proposed by Waleh and Loew (18-20). The first intermediate seen in very early time frames, AS, is manifest by an induced absorption in the area of the peak wavelength of the ground state $^1A_1$. This transient is seen in all the compounds, synthetic and natural, in which very early delay times were obtained. At this point it should be emphasized that the presence of intermediates in

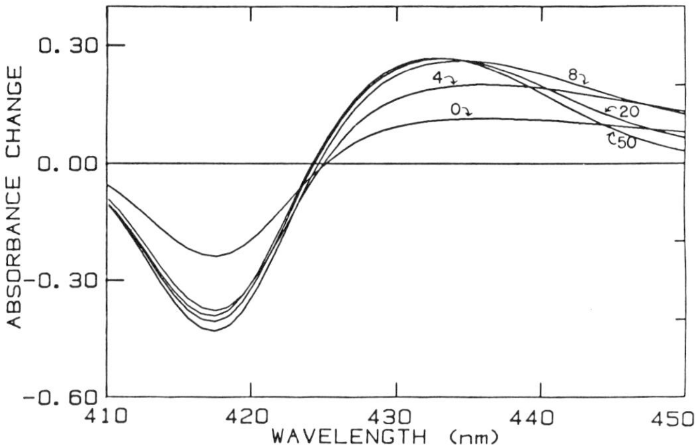

Figure 7. Modeled spectra for 1-CO at the delays indicated.

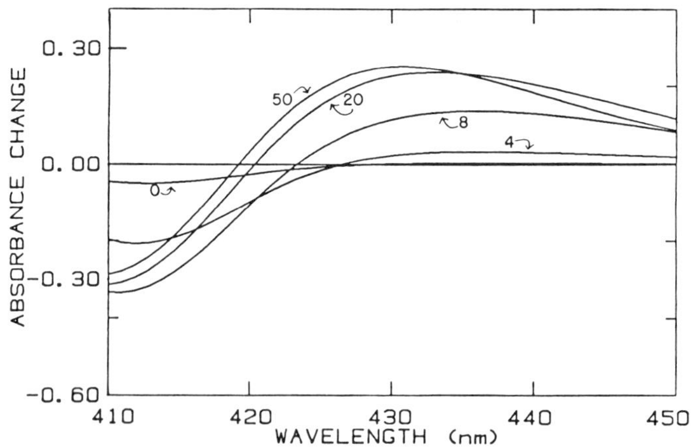

Figure 8. Modeled spectra for 1-ET-CO for the delays indicated.

our experiments are manifest through transient $\pi\pi^*$ absorption changes. It is possible that this induced absorption could be an artifact. Since we observe no induced absorption at any delay setting under conditions of solvent only, our inclination is to discount this possibility. In any event, the presence or absence of AS is not critical to subsequent kinetic analysis of the photodissociation as its lifetime is short, < 1 ps. However our evidence of AS is in agreement with the Martin et al. (22) observation indicating the presence of photoproduct as early as 350 fs after excitation. We are inclined to assign AS to an intermediate originating from the lowest excited singlet state of the 6-coordinate complex, $^1T_1$. However there is the possibility that it could be accounted for by an $S_1 \longrightarrow S_N$ type transition originating from ^1Q.

Once relaxation to the lowest excited singlet has occurred the next intermediate expected would be due to the bottleneck in the six-coordinate manifold caused by intersystem crossing into the $^3T_1$ and $^5T_2$ states that lie below $^1T_1$. $^3T_1$, and $^5T_2$ are very close in energy and it is likely that the radiationless transition from the singlet $^1T_1$ state would result in the population of both of these states with interconversion between the two. We suggest that photoproduct B originates from a transient species composed of a mixture of these six-coordinate states characterized by a low extinction coefficient and a maximum wavelength found about midway between $^1A_1$ and the λ_{max} of the next intermediate C. Intersystem crossing from $^1T_1$ to the $^3T_1$ in the six-coordinate manifold does not necessarily lead to a configuration that is dissociative, since the d_{z^2} orbital is not occupied: $d_\pi^4 d_{x^2-y^2}^1 d_{xy}^1 d_{z^2}^0$. However, conversion into the $^5T_2$ state may, in the six coordinate manifold, result in an unstable dissociative configuration, as the parent quintet state originates via $d_{\pi,x^2-y^2} \longrightarrow d_{z^2}, d_{xy}$ orbital promotions giving rise to the occupancy $d_\pi^3 d_{x^2-y^2}^1 d_{xy}^1 d_{z^2}^1$. As Waleh and Loew (18-20) have shown, both the $d_\pi \longrightarrow d_{z^2}$ and $d_{x^2-y^2} \longrightarrow d_{z^2}$ decrease in energy as a function of increased Fe-CO bond distance promoting the dissociation process. Unfortunately, the combined effect of the imidazole - iron - CO (or O_2) stretch on these transitions has not yet been reported. We may assume though, that the $^5T_2$ is photoactive and population of it could result in the tetragonal displacement of both axial ligands. These arguments would be applicable to both the CO and O_2 ligands and the imidazole. Thus we consider the intermediate B to be an unresolved mixture of the $^5T_2$ and $^3T_1$ states with the possibility of interconversion between them. When the potential surfaces of these states cross over into corresponding ones in the five coordinate manifold that are highly dissociative (because of the effect of the axial displacement on the d_{z^2}, d_π and $d_{x^2-y^2}$ orbitals) photodissociation of the axial ligand occurs producing one triplet and one quintet according to the Wigner Rules:

HBCO $(^3T_1) \longrightarrow$ HB $(^3T_1)$ + CO $(^1\Sigma)$ and HBCO $(^3T_2) \longrightarrow$ HB $(^5T_2)$ + CO $(^1\Sigma)$. Of the two states produced $^3T_1$, with the antibonding d_{z^2} orbital populated because of the inversion of the d_{xy} and the d_{z^2} orbitals, is higher in energy than the quintet and would be expected to intersystem cross to the quintet in time. The evolution of the intermediates shows the production of a species at the red end of the spectrum which progressively narrows and whose maximum wavelength shifts to a lower λ. Intermediate C has been seen in the photodissociation of both the CO and O_2 forms of hemoglobin (3,4). This shifting of the wavelength maximum can be attributed to the production and subsequent reaction of an intermediate with a relatively high λ_{max}. We assign C to $^3T_1$.

C can then undergo further intersystem crossing to the $^5T_2$ state, D. The final intermediate produced in the photodissociation of the synthetic hemes, D, has been shown to be, within experimental error, equivalent to the ground deoxy state. D correlates with the production of the quintet state, both directly from B and through further reaction of the deoxy state C. Therefore we assign D to $^5T_2$ and suggest that the equivalency of D to the ground deoxy state supports the fact that the iron has fully relaxed out of the heme plane. Thus, for the unstrained hemes, the photodissociative pathway is entirely consistent with that for hemoglobin (3,4), namely

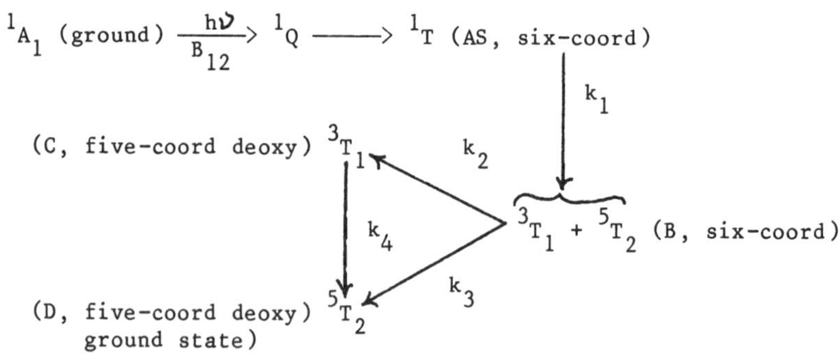

where the rate constants and lifetimes are listed in Tables I and III.

For the CO forms of the strained compounds studied, the photodissociative pathway is noticeably different in the sequential appearance of an intermediate not seen in the unstrained synthetic complexes or in the natural heme complexes. After relaxation to the lowest excited singlet $^1T_1$, the strained compounds show evidence of an intermediate X before intersystem crossing to B. In all strained compounds tested, this photoproduct is characterized by a spectrum having a double maximum in the region of the six-coordinate ground state species. Because of similarities in its spectral characteristics to the corresponding four-coordinate (axially unligated) band found in the ground state species, we are inclined to suggest that this intermediate is indicative of a pseudo four-coordinate species. The similarities in spectral band shape are apparent in Fig. 6 and 8. Since AS and B are spectrally very similar to the corresponding intermediates in both the unstrained and the natural hemes, and since these intermediates are both considered to be six-coordinate species, it is highly unlikely that X is a true four-coordinate complex. Instead it is more probable that it is a six-coordinate species in which the Fe-CO and Fe-Imidazole bonds are stretched and weakened to the stage where the complex behaves more like it is four-coordinate than six-coordinate. Our experiment follows the evolution of these intermediates by monitoring the $\pi\pi^*$ porphyrin Soret absorption changes as a function of time. Since such changes in the Soret wavelength maximum, intensity and band shape are known to be extremely sensitive to the state of coordination of the heme complex, this could account for the small differences between the spectrum of the intermediate X and the ground state of the four-coordinate species.

It is difficult, if not impossible, to justify a theoretical mechanism for the formation of this pseudo four-coordinate species, X, without first a great deal more theoretical work, which might involve extensions of existing investigations or new approaches that would examine the effect of geometry on the dissociation of both axial ligands, and second, new time resolved spectroscopic investigations of the photodissociation by monitoring, directly, various critical d-d transitions or d-ligand charge transfer transitions or from direct structural studies employing resonance Raman technique(s). However, we might envision the formation of X by invoking the Waleh-Loew (18-20) $^1T_1$ ($d_\pi \longrightarrow d_{z2}$) suggestion that has already been discussed. The essential difference here is that there is considerable strain on the Fe-imidazole bond resulting in corresponding and favorable adjustments to the disposition of the $d_\pi \longrightarrow d_{x^2-y^2}$ transition energies, making the $^1T_1$ route competitively dominant over that of the bottleneck mechanism. Thus it can be supposed that rapid conversion from ^1Q into one of the $^1T_1$ states below it would result in a pseudo four-coordinate complex where the axial bonds to the iron have weakened considerably. It is likely that this pseudo geometry would be followed by movement of the iron relative to the plane of the porphyrin to release this strain and return the complex to the six-coordinate species B. Consequently the lifetime of B would necessarily be lengthened, as verified in Table III. The dissociation then proceeds according to the mechanism discussed earlier.

Intermediates for the O_2 Complexes. The results obtained on the oxygen forms of the compounds studied revealed a similar set of intermediates, AS, B, C and D, though the spectral parameters and rates associated with formation of the intermediates are in many cases noticeably different from those of the CO forms. As C and D are postdissociative intermediates, it would be expected that the spectral characteristics of these intermediates should be fairly similar for either the CO or the O_2 forms of the parent compounds, and they are. For example, the interconversion rates k_4 for C—>D are very similar. Assigning the intermediate B to the same states proposed for the CO complexes cannot be done, as the mechanism for photodissociation is undoubtedly different. The process is complicated by the fact that there are low lying singlet and triplet charge-transfer states present in the six-coordinate oxygen species that open up several competitive routes leading to photodissociation. Possibilities for these competitive paths have been discussed by Waleh and Loew (18-20) who conclude that the higher energy channel, which is not accessible by the Q-band pump, is $d_\pi \rightarrow d_{z^2}$ in nature. In the octahedral scheme, this possibility competes with the $d_{\pi,x^2-y^2} \rightarrow d_{z^2,xy}$ promotion associated with the quintet state. They suggest that competitive decays to the lower charge transfer states from this transition and to the ground state from these low lying CT states account for the low quantum yield of O_2 photodissociation. It is also worth noting that these low lying CT states are not readily accessible from the $\pi \rightarrow \pi^*$ porphyrin singlet due to the small value of the Franck-Condon spatial overlap between the two states.

Taking these facts into consideration, AS can be tentatively assigned to $S_1 \rightarrow S_N$ type transition arising from the lowest excited singlet ^1Q. The quintet $^5T_2$ and the triplet $^3T_1$ states in the six-coordinate manifold are much farther apart in the oxygen complexes than they are in the CO form, with the $^5T_2$ state lying very close in energy to the singlet ^1Q state. Because of this, the intermediate B, considered a mixture of the two states, would be expected in the oxygen compounds to consist mainly of the triplet $^3T_1$ state mixed with the CT states. Dissociation of the ligand will again produce the triplet $^3T_1$ and the quintet $^5T_2$ states of the deoxy heme. The $^3T_1$ deoxy heme will then decay into the ground deoxy state, $^5T_2$. A comparison of the final intermediate D with the intermediate D for the carbon monoxide experiments show that, within experimental error, these states are identical, as would be expected if the species D represented the ground deoxy state of the heme.

Studies on the oxygen forms of the strained compounds show no evidence for the intermediate X. One possibility is that the reaction X—>B is extremely fast as might be expected for the essentially geminate recombination of $^3O_2$ back to the heme due to favorable spin allowed pathways:

$$HBO_2\ (^5T_2) \rightarrow HB\ (^3T_1)\ \ldots.\ O_2\ (^3\Sigma_g^-)$$

and $$HBO^2\ (^3T_1) \longrightarrow HB\ (^5T_2)\ \ldots\ O_2\ (^3\Sigma^-_g)\ .$$

Another possibility is that the direct dissociation of the oxygen is more sensitive to strain than in the carbon monoxide complexes and is simply more efficient and competitively dominant over the pseudo four-coordinate pathway.

Geminate Recombination. Both the experimental and the modeling results discussed earlier indicate that the short term O_2 + C —> A geminate recombination pathway is inoperative for any of the synthetic compounds studied. If there is any recombination at all, the estimate for the return is less than 5% by 50 ps. The CO forms of these compounds show no evidence of short term geminate recombination as well. In the modeling studies the short term recombinant pathway was nevertheless included for completeness since return to the ground state with lifetimes of less than 20 ps could effectively compete for C. Not surprisingly, the results contrast sharply with the short term geminate recombination pathway found in hemoglobin and myoglobin (3,4). The absence of fast geminate recombination in the synthetic compounds is most likely due to the absence of the protein surrounding the heme pocket in terms of the role it plays in trapping the oxygen near the heme.

Kinetic Modeling. For all the synthetic heme complexes studied, the lifetime of the first intermediate AS was on the order of 250 fs in agreement with the femtosecond study of Martin et al. (22). As the jitter in the pulse duration is on the order of 1 ps, the AS lifetime is very approximate and chosen to obtain the best fit to the experimental spectra.

The lifetimes for the intermediate B, the species considered to be a mixture of the lowest excited triplet and quintet states of the six coordinate species, vary widely between the compounds. This lifetime $\tau_B = 1/(k_2 + k_3)$ ranges from $\sim$ 2 ps to as high as 20 ps for the various complexes studied, with an average lifetime of 9.5 ps for this intermediate. One fact that should be noted is the lifetime of B in the carbon monoxide forms of the compounds studied. For the two unstrained chelated hemes 1-CO and 2-CO the lifetime for this intermediate is very short, on the order of only 2 ps, indicating a fast dissociation of the axial ligand CO from a photoactive state. However in the remaining CO complexes the situation is much different. The three other carbon monoxide complexes studied are strained compounds that photodissociate through X, a photoproduct that we have assigned as a pseudo four-coordinate intermediate. In these complexes the lifetime for B is found to be 5-10 times longer than in the unstrained species. X has been postulated to a species in which the axial bonds of the iron to both the CO and imidazole are considerably weakened. This geometry would likely be followed by a movement of the iron to release strain on the complex and returning it to the six coordinate species. As mentioned earlier this process should result in a lengthening of the lifetime of B, which is indeed what has been found in these cases.

The lifetime of intermediate C for these complexes should be found to be independent of the ligand present on the parent species.

Since C has been assigned to a triplet deoxy state in which the axial ligand has been dissociated. As can be seen in Table III, in most cases lifetimes τ_C are found to compare favorably between the carbon monoxide and oxygen forms of the synthetic complexes. In all cases the rate constants used had an accuracy of only one significant figure, resulting in an accuracy no better for the lifetimes of these states. One noticeable discrepancy in τ_C occurs between the chelated protoheme 1-CO and the oxygen form of this compound, 1-O_2.

Conclusions

Using standard pump-probe picosecond methods, we have examined the picosecond photodissociation of the O_2 and CO forms of a number of natural (3-5) and synthetic heme complexes by following $\pi\pi^*$ Soret absorption changes localized to the plane of the porphyrin. Keeping in mind that the absorption characteristics of this band are highly sensitive to the coordination and geometry of the heme, our collective results for the natural and synthetic complexes clearly reveal the importance of heme geometry and strain on the dynamics and mechanisms of the dissociation. We have interpreted our results on the synthetic and natural heme complexes as occurring through a series of sequential steps that are evident in the transient photoproducts which we have isolated and characterized spectroscopically and kinetically. Of the four intermediates observed in the unstrained and natural heme complexes, we have assigned the first two as originating from excited predissociative states found in the six-coordinate manifold and the last two as postdissociative states found in the five-coordinate manifold. This interpretation of the photodissociation of the CO complexes is consistent with a series of bottlenecks encountered during the deactivation and dissociation processes. Though the bottleneck possibility was first discussed by Green and Hochstrasser (24), this work, along with our earlier work (3,4), is the first to assign the AS and B intermediates, as well as provide evidence for a pseudo four-coordinate intermediate found early in the dissociation of certain synthetic complexes having strained imidazole-Fe linkages.

The differences that we see in the picosecond photodissociation between the natural and synthetic complexes may be linked to important structural differences in the heme pocket or constraints on the tertiary geometry of the heme imposed by the protein. In particular we would like to comment on the proximal imidazole. Recent theoretical (26), x-ray structural (27), and resonance Raman work (6-10) all suggest that the affinity of the sixth axial-ligand for the heme is critically dependent on the tertiary heme structure that originates from the linkage of the imidazole to the heme. In the natural complexes, this imidazole (His F8) and the F helix to which it is bound undergo major structural movements; these changes for the α and β subunits include a 1-Å translation of the F helix across the porphyrin face and a movement of the imidazole from a position that is asymmetric with respect to the porphyrin nitrogens to one that is more symmetric. In this regard, Tables III and IV show some trends. Preliminary results (5) indicate that intermediate C in the α-Hb complex has a much longer lifetime than the other natural complexes studied, possibly because of quaternary constraint(s) present in the other natural complexes studied are absent or at least relaxed in the α-complex. Similarly we find that all the synthetic complexes are characterized by relatively long C lifetimes where such constraint is absent. It appears that imidazole strain is most effective in lengthening the lifetime of B in the CO forms of the synthetic complexes studied.

The results presented in this work coupled with our earlier work (3,4) suggest that the proposal offered by Waleh and Loew with regard to dissociation is not necessarily correct for the unstrained synthetic and natural compounds studied. Their results suggest that population of the $^1T_1$ state ($d^2_{x^2-y^2}\, d^3_\pi\, d^1_{z^2}$) found just below the ^1Q state should lead to prompt dissociation though the singlet $d_\pi \longrightarrow d_{z^2}$ pathway. If this pathway dominates, then pumping the ^1Q band would be followed by rapid radiationless singlet interconversion to $^1T_1$ leading to a fast, possibly subpicosecond, dissociation. Possibly with sufficient strain in the Fe-imidazole bond, favorable adjustments to the $d_\pi \longrightarrow d_{z^2}$ and $d_{x^2-y^2}$ transitional energies occur allowing the $^1T_1$ route to competitively dominate over the bottle neck mechanism as suggested by our evidence that supports a pseudo four-coordinate complex. Otherwise, our results on the synthetic and natural complexes show that the photodissociation initiated by pumping the Q band, as judged by the lifetime of B, takes some picoseconds to occur and depends on the heme geometry. To the extent that the photodissociation has its beginnings in the subpicosecond regime, we are in agreement with the femtosecond report of Martin et al. (22). However, in their study on CO complexes of Hb and Mb a deoxy-like species appeared with a time constant of 350 fs which had spectral characteristics that did not change for 100 ps. At the moment, we cannot explain the difference between our results and this femtosecond observation in regard to the evaluation of band shape in the transient difference spectra. Earlier picosecond work showed evidence and discussed the development of the deoxy photoproduct band shape changes that we have identified with intermediates C and D (24,25). As the ligand-field level system is complex, the possibility exists that at the 307 nm pump wavelength of the femtosecond experiment a fast photodissociative channel is available that is not otherwise accessible with the Nd^{3+} 353 or 531 pump wavelengths.

Acknowledgments

The authors would like to thank Andy Hutchinson and Chris Guest for their valuable assistance in the preparation and discussion of various aspects of this work. We would also like to thank the National Science Foundation (PMC 7816199) and to the University of Wyoming for support of this work.

Literature Cited

1. Caldwell, K.; Noe, L. J.; Traylor, T. G. "Photodissociation of Carbon Monoxide Forms of Synthetic Heme Complexes Using Picosecond Absorption Spectroscopy", unpublished results, National ACS Meeting, Miami Beach, April 1985.
2. Noe, L. J.; Caldwell, K.; Hutchinson, J. A. "Intermediates Found in the Picosecond Photodissociation of the CO and O_2 Derivatives of the Natural and Synthetic Heme Complexes", unpublished results, 40th Northwest ACS Regional Meeting, Sun Valley, June 1985.
3. Hutchinson, J. A.; Noe, L. J. J. Quant. Elect. 1984, QE-20, 1353.
4. Hutchinson, J. A.; Traylor, T. G.; Noe, L. J. J. Am. Chem. Soc. 1982, 104, 3222.
5. Guest, C.; Hutchinson, J. A.; Noe, L. J., unpublished results.
6. Findsen, E. W.; Scott, T. W.; Chance, M. R.; Friedman, J. M. J. Am. Chem. Soc. 1985, 107, 1335.
7. Scott, T. W.; Friedman, J. M. J. Am. Chem. Soc. 1984, 106, 5677.
8. Stein, P.; Turner, J.; Spiro, T. G. J. Phys. Chem. 1982, 86, 168.
9. Turner, J.; Stong, J. D.; Spiro, T. G.; Nagumo, M.; Nicol, M. F.; El-Sayed, M. A. Proc. Natl. Acad. Sci. USA 1981, 78, 1313.
10. Findsen, E. W.; Friedman, J. M.; Ondrias, M. R.; Simon, S. R. Science 1985, 229, 661.
11. Traylor, T. G. Acc. Chem. Res. 1981, 14, 102.
12. Traylor, T. G.; Chang, C. K.; Geikel, J.; Berzinis, A. P.; Mincey, T.; Cannon, J. J. Am. Chem. Soc. 1979, 101, 6717.
13. Ciconne, J.; Traylor, T. G., private communication.
14. Deardruff, L.; Traylor, T. G., private communication.
15. Traylor, T. G.; White, D. K.; Cambell, D. H.; Berzinis, A. P. J. Am. Chem. Soc. 1981, 103, 4932.
16. Zerner, M.; Gouterman, M.; Kobagashi, H. Theor. Chim. Acta 1966, 6, 363.
17. Eaton, W. A.; Hanson, L. K.; Stephens, P. J.; Sutherland, J. C.; Dunn, J. B. R. J. Am. Chem. Soc. 1978, 100, 4991.
18. Waleh, A.; Loew, G. A. J. Am. Chem. Soc. 1982, 104, 2346.
19. Waleh, A.; Loew, G. A. Ibid. 1982, 104, 2352.
20. Waleh, A.; Loew, G. A. Ibid. 1982, 104, 3513.
21. Hoffman, B. M.; Gibson, Q. H. Proc. Natl. Acad. Sci. USA 1978, 75, 21.
22. Martin. J. L.; Mighs, A.; Poyart, C.; Lecarpentier, Y.; Astier, R.; Antonetti, A. Proc. Natl. Acad. Sci. USA 1983, 80, 173.
23. Henry, E. R.; Levitt, M.; Eaton, W. A. Proc. Natl. Acad. Sci. USA 1985, 82 2034.
24. Green, B. I.; Hochstrasser, R. M.; Weisman, R. B.; Eaton, W. A. Proc. Natl. Acad. Sci. USA 1978, 75, 5255.
25. Chernoff, D. A.; Hochstrasser, R. M.; Steel, A. W. Hemoglobin and Oxygen Binding Ho, ed. 1982, Elsevier, North Holland, 345-350.
26. Gelin, B. R.; Karplus, M. Proc. Natl. Acad. Sci. U.S.A. 1977, 74, 801.
27. Baldwin, J.; Chothia, C. J. Mol. Biol. 1979, 129, 175.

RECEIVED July 18, 1986

14

Picosecond Measurements of Electron Transfer in Bacterial Photosynthetic Reaction Centers

Dewey Holten[1], Christine Kirmaier[1], and William W. Parson[2]

[1]Department of Chemistry, Washington University, St. Louis, MO 63130
[2]Department of Biochemistry, University of Washington, Seattle, WA 98195

We have examined the temperature dependence of the rate of electron transfer to ubiquinone (Q) from the bacteriopheophytin (BPh) that serves as an initial electron acceptor (I) in reaction centers from *Rhodopseudomonas sphaeroides*. The rate *increases* by a factor of about two as the temperature is reduced from 295 to 100 K and then remains constant down to 5 K. Theories for electron transfer can account for the unusual form of the temperature dependence if the reaction is activationless. However, a quantitative fit is obtained only if the frequencies of some vibrational modes (having an average energy of ~ 300 cm^{-1}) are allowed to change when an electron moves from I^- (BPh^-) to Q. Alternatively, the non-Arrhenius temperature dependence can be accounted for if BPh and Q move closer together by about 1 Å as the temperature is reduced. It appears that no more than half the reorganization energy is due to molecular changes in the electron carriers; the rest is apparently associated with other types of motions, some of which may involve the protein.

We also have found an unexpected detection-wavelength dependence of the observed kinetics at both room and low temperature. This observation can be rationalized under the assumption that the kinetics at some wavelengths contain contributions from molecular readjustments (reorganizations) involving the pigments and/or the protein in response to electron transfer.

Reaction centers from photosynthetic organisms are specialized pigment-protein complexes in which photon energy is converted into chemical energy (1). This is accomplished by a series of rapid electron transfer reactions that produce a spacially-separated oxidized donor and a reduced electron acceptor (2). Reaction centers from the purple photosynthetic bacterium *Rhodopseudomonas sphaeroides* contain four molecules of bacteriochlorophyll (BChl), two of bacteriopheophytin (BPh), one tightly-bound or primary ubiquinone (Q), a

0097-6156/86/0321-0205$06.00/0

nonheme iron atom and three polypeptides (3). Photoexcitation prepares the excited singlet state (P*) of the primary electron donor (P), a complex involving two of the BChls (4-6). An electron arrives in about 4 ps (7-9) on an early electron acceptor I producing the radical-pair state P^+I^- (10-19). The electron is transferred from I^- to Q with a time constant of about 200 ps at room temperature (10,11,14,17,20-22). The transient-state spectra and kinetics associated with the reaction $P^+I^- \rightarrow P^+Q^-$ are a major focus of this article.

It has been suggested that P^+BChl^- (where BChl is one of the two monomeric or "accessory" BChls that are not part of P) is a transient state prior to P^+I^- (14,16,19), although the evidence supporting this view has been criticized (23,24). Recent subpicosecond studies find no evidence for P^+BChl^- (8,9). These new results do not preclude some involvement of a monomeric BChl in the early photochemistry, only that P^+BChl^- apparently is not a kinetically resolved transient state. Perhaps P* itself contains some charge-transfer character between its component BChls, or between P and one or both of the monomeric BChls (8,9,25-27). One of the two monomeric BChls apparently can be removed by treatment of the reaction center with sodium borohydride (28) and subsequent chromatography, with no impairment of the primary electron transfer reactions (29). Thus, at present it appears that P^+I^- is the first resolved radical-pair state, and it forms with a time constant of about 4 ps in *Rps. sphaeroides*.

The intermediary acceptor I is thought to be a complex involving a BPh and one of the monomeric BChls (18,30-34). This view is based on the observation that the reduction of I causes absorption changes in regions of the spectrum attributable to BPh and also in regions normally ascribed to BChl. It has been proposed from nanosecond studies on reaction centers having Q reduced that P^+I^- is a thermal mixture of P^+BPh^- and P^+BChl^-, with about 60% of the added electron density on the BPh at room temperature (33). Our recent kinetic measurements (22) and low-temperature picosecond photodichroism studies (35) suggest that P^+I^- is P^+BPh^- in *Rps. sphaeroides*. However, the reduction of BPh appears to affect the BChls strongly. Large absorption changes at near-infrared wavelengths normally ascribed to the BChls accompany the reduction of I, but the decay of the absorption changes is somewhat faster than the kinetics associated with the electron transfer from I^- (BPh^-) to Q. We have tentatively ascribed these extra kinetics to nuclear relaxations (readjustments) involving the BChls and/or P (or P^+) in response to the formation of P^+I^- (P^+BPh^-). The second BPh molecule, which is not an electron carrier, also appears to be affected by charge separation, and a slower relaxation appears to be associated with it. This tentative interpretation of the unexpected detection-wavelength dependence of the kinetics implies that motions of the pigments and the protein accompanying electron transfer may affect the transient-state spectra and kinetics (22,35).

In order to learn more about the importance of vibrational motions in the primary photochemistry, we have measured the temperature dependence of the rate of electron transfer from I^- (BPh^-) to Q over the range 295 to 5 K in polyvinyl alcohol films (22). The rate increases by a factor of about two as the temperature is reduced from 295 K to ~ 100 K, and then remains constant down to 5 K; the time constant below 100 K is ~ 100 ps. These results agree in some respects and disagree in others with earlier, less extensive studies (17,21), as discussed elsewhere (22). Here we discuss briefly the temperature dependence in terms of theories for electron transfer in order to see what new information can be obtained regarding the mechanism of charge separation.

Experimental Section

The picosecond apparatus and experimental details have been described (18,22). Samples were excited with 30-ps 600-nm excitation flashes (< 100 μJ; 1.5-2 mm diameter) at 5 Hz. The maximum standard deviation in ΔA was typically ± 0.005. *Rps. sphaeroides* reaction centers (36) and polyvinyl alcohol (PVA) films (17) were prepared as described previously.

Results

Transient State Spectra. Figure 1 shows the near-infrared difference spectra for the formation of states P^+I^- at 30 ps (A) and P^+Q^- at 1.6 ns (B) in *Rps. sphaeroides* reaction centers imbedded in a PVA film at 295 K (dashed) and at 5 K (solid). Figure 2 shows the corresponding difference spectra in the visible region on an expanded ΔA scale for a film having approximately three times the optical density. The difference spectra are much better resolved at 5 K than at 295 K. Detailed discussion of the absorption changes and comparisons with spectra taken under other conditions have been given (22,35). For the present discussion we note the major features that change during the time in which the electron moves from I^- (BPh^-) to Q. In the near infrared (Figure 1) between 30 ps and 1.6 ns there are *net* absorption increases near 795 and 765 nm. The change near 795-nm is probably associated with one of the two monomeric BChls or P (or P^+). The change in absorption near 765 nm probably reflects decay of bleaching in the Q_Y ground-state absorption band of the "long-wavelength" BPh. (The small trough near 750 nm can be associated with the "short-wavelength" BPh.)

In the visible region two major bands change significantly as P^+I^- (Figure 2A) gives way to P^+Q^- (Figure 2B); the broad "anion" band (34,37) centered near 665 nm decays and the bleaching near 545 nm recovers. The absorption changes in the BPh Q_X region can be seen more clearly in Figure 3, which compares the 5-K transient-state spectra (A) and their difference (B) with the 5-K ground-state spectrum. Clearly only the long-wavelength (545-nm) BPh is bleached significantly when I is reduced (22,35), in agreement with other work (13,20,30,31,38). The 530-nm Q_X band of the short-wavelength BPh is affected by the oxidation of P and/or reduction of Q (Figures 2B and 3A), also in agreement with earlier work (39).

Temperature and Detection-Wavelength Dependence of the Kinetics. Figures 1-3 show that there are four major regions in which kinetics can be measured during the time in which the electron moves from I^- (BPh^-) to Q: in the 545- and 765-nm regions of BPh ground-state absorption, in the 665-nm anion band, and near 795-nm, a region normally associated with the BChls. The measurements at all wavelengths and temperatures are described well by first-order kinetics.

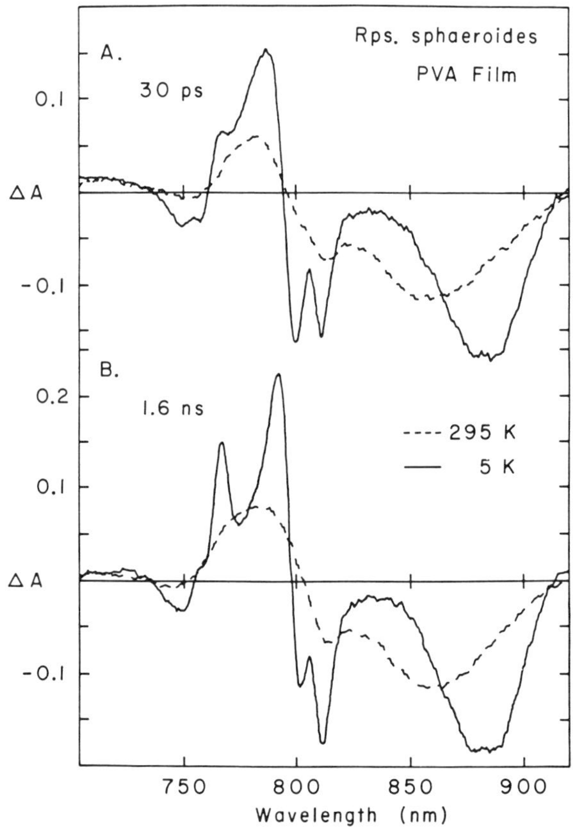

Figure 1. Near-infrared absorption changes for *Rps. sphaeroides* reaction centers imbedded in a PVA film at 30 ps (A) and 1.6 ns (B) with respect to the center of a 30-ps 600-nm excitation flash, at 295 K (dashed) and 5 K (solid). Reproduced with permission from Ref. 22. Copyright 1985, Elsevier Science Publishers.

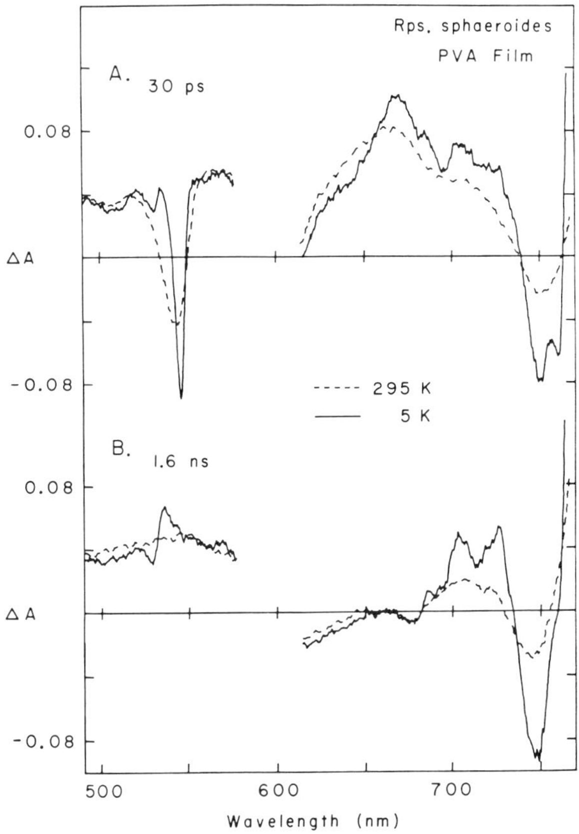

Figure 2. Absorption changes in the visible region acquired as in Figure 1, except that the PVA film had approximately three times the optical density. Reproduced with permission from Ref. 22. Copyright 1985, Elsevier Science Publishers.

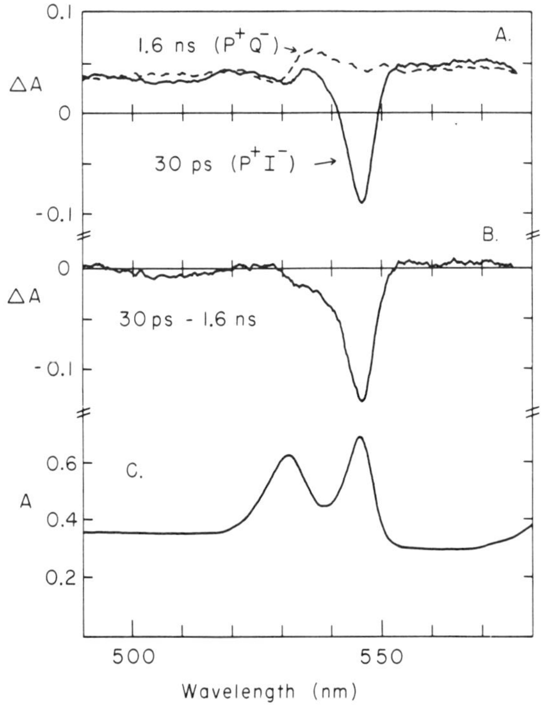

Figure 3. (A) Absorption changes in the BPh Q_X region at 30 ps and 1.6 ns at 5 K. (B) Difference between the 30-ps and 1.6-ns spectra. (C) Ground-state spectrum of the same PVA film in this region at 5 K. Reproduced with permission from Ref. 35. Copyright 1985, Elsevier Science Publishers.

However, the time constants measured in the four regions are not the same at a given temperature (Table 1).

Table 1. Average time constants (in picoseconds) as a function of temperature and detection wavelength in *Rps. sphaeroides* reaction centers.

	BPh Q_X (545 nm)	BPh^- (665 nm)	(BPh Q_Y) 758 nm	765 nm	($BChl/P$) 795 nm
PVA Films:					
5 K	103 ± 10	92 ± 7	140 ± 20	104 ± 11	73 ± 12
76 K	98 ± 7	98 ± 8	134 ± 18	108 ± 12	69 ± 10
295 K		229 ± 25			154 ± 15
Buffer					
285 K	253 ± 10	207 ± 12		249 ± 12	150 ± 17

Analysis of these results (22,35) indicates that the rate of the process $P^+I^- \rightarrow P^+Q^-$ is best measured via the decay of the broad 665 nm anion band (Figure 2). This band is relatively insensitive to temperature and appears to have only one underlying component; the time constants measured across the 665-nm band (650-680 nm) are the same within experimental error at a given temperature. The band is well separated from ground-state absorptions, which are sensitive to temperature, electrochromic, excitonic and other effects.

Figure 4 shows the temperature dependence of the decay kinetics measured in PVA films in the 665-nm band (triangles). Some measurements at lower temperatures in the 545-nm band are also shown (circles). The electron transfer rate constant increases by a factor of ~ 2 as the temperature is lowered from 295 K to ~ 100 K, at which point the rate becomes independent of temperature within experimental error. The solid curve through the experimental data in Figure 4 is a theoretical fit that will be discussed below.

The time constants measured near 665 and 795 nm at 295 K in PVA film are in good agreement with those measured at 285 K in flowed buffer (Table 1). Even though the kinetics are different at different wavelengths, the temperature dependence in each region appears to be similar; the time constant decreases by about a factor of two between 295-285 and 76 K, and is the same at 76 and 5 K.

Discussion

Temperature Dependence of the Kinetics. In an earlier article (22) we gave a detailed analysis of the temperature dependence of the rate of electron transfer from I^- (BPh^-) to Q (Figure 4). Here we summarize some of the important considerations and discuss further the possible insights that the temperature and detection-wavelength dependence of the kinetics may give into the molecular mechanism of electron transfer and the overall charge separation process.

Theories for photosynthetic electron transfer normally begin with the premise that the electron transfer is nonadiabatic (40-48). The

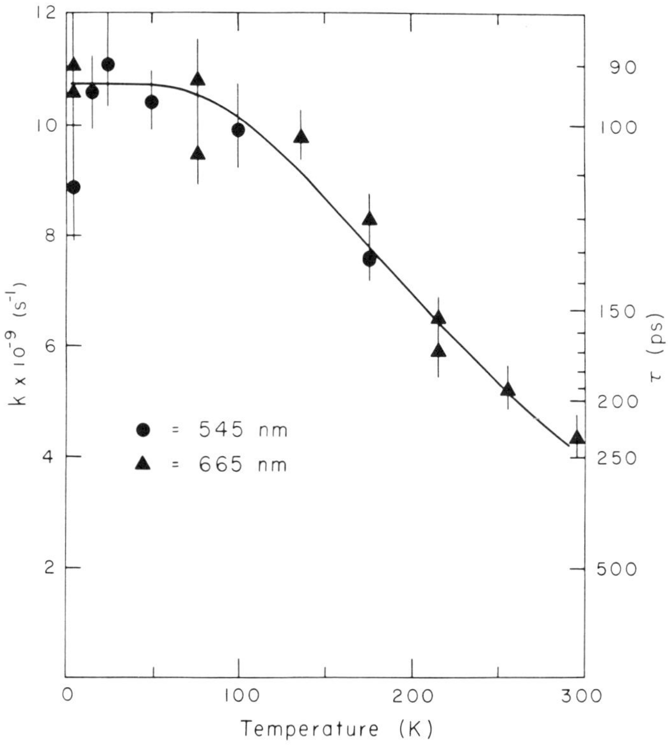

Figure 4. Temperature dependence of the rate of electron transfer from BPh^- to Q in *Rps. sphaeroides* PVA films measured at 545 nm (circles) and 665 nm (triangles). The solid curve is a fit using the theory of Kakitani and Kakitani (44) with $\langle\nu_e\rangle/c$ = 250 cm^{-1}, $\langle\nu_q/c\rangle$ = 1550 cm^{-1}, S = 0.6, ΔG(295 K) = 4100 cm^{-1}, ΔE_1 = 2400 cm^{-1}, and V = 30 cm^{-1} (23). Reproduced with permission from Ref. 22. Copyright 1985, Elsevier Science Publishers.

rate of electron transfer is expressed in terms of the Fermi golden-rule: $W = 4\pi^2 h^{-1} V^2 F$, where h is Planck's constant, V is the electronic (tunneling) matrix element between the donor and acceptor, and F is the thermally-weighted Franck-Condon (vibrational overlap) factor.

We have considered two possible interpretations for the temperature dependence of the rate of electron transfer from I^- to Q shown in Figure 4. First, the pigments might move closer together (22,49) or into an increasingly more favorable geometry as the temperature is reduced. The electronic factor (V) is thought to decrease exponentially with distance: $V = V_0 \cdot \exp(-\beta R)$, with $\beta \sim 1$ $Å^{-1}$ (41,50). The increase in rate by a factor of ~ 2 would mean that the distance between I^- and Q decreases by about 1 Å as the temperature is lowered from 295 to 100 K. This interpretation requires that the conformation remain constant below 100 K, where the rate is independent of temperature. (A possible argument against this interpretation is that the rate of this electron transfer reaction in Rps. viridis reaction centers appears to be independent of temperature (51).)

An alternative interpretation is that V is independent of temperature, and that the variation of the electron transfer rate with temperature is contained in the thermally-weighted Franck-Condon factor (40-45). On the basis of this interpretation the non-Arrhenius temperature dependence can be explained if the electron transfer reaction is activationless. This means that the nuclear potential surface for the product state (P^+Q^-) crosses near the minimum of the potential surface for the reactant state (P^+I^-). In other words, the (free) energy change in the reaction equals the total (pigment + protein) reorganization energy. A non-Arrhenius temperature dependence can be explained since at low-temperature ($kT \ll h\nu$) only the zero-point vibrational level(s) of the reactant state will be populated. The rate will be maximal for a given electronic factor V, since the vibrational overlap (Franck-Condon factor) will be optimum. The rate will decrease as the temperature is raised, since higher vibrational levels of the reactants will be thermally populated. These higher vibrational states have poorer overlap with the product vibrational states.

The most simple view is that the reorganization energy is a measure only of the configurational changes (changes in bond lengths and bond angles) accompanying electron transfer. Using this approach Jortner proposed that high-frequency quantum (pigment) modes having an average value of $\langle\nu_q\rangle/c \sim 400$ cm^{-1} make the dominant contribution to the Franck-Condon factor (40,41). Sarai included explicit contributions from the low-frequency (protein) modes (42,43). The observed critical temperature of ~ 100 K for the $P^+I^- \rightarrow P^+Q^-$ reaction (Figure 4) fixes $\langle\nu_q\rangle/c \sim 250$-$300$ cm^{-1}. Both Jortner's and Sarai's theories predict an increase in rate of at most a factor of ~ 1.15 as the temperature is lowered from 295 K to 100 K, as compared to the measured factor of ~ 2 (2.3 in Figure 4).

It should be noted here that the high-temperature ($kT > h\nu$) forms of the expressions in these theories (40-43) and in others (48) predict that the rate will be proportional to $1/\sqrt{T}$ for an activationless reaction (the exponential term containing the activation energy divided by kT will be unity). If taken at face value, these high-temperature (classical) expressions give an increase in rate of $\sqrt{3}$ =

1.7 between 300 and 100 K. However, they cannot account for the critical temperature and the temperature independence below 100 K, which are associated with nuclear tunnelling, a quantum phenomenon. Also with $\langle\nu\rangle/c \sim 250\text{-}300\ cm^{-1}$ derived from the critical temperature and $kT \sim 70\text{-}200\ cm^{-1}$ between 100 and 295 K, the high-temperature expressions introduce error since the relation $kT > h\nu$ is not met. The full expressions give only a 15% increase in rate between the two temperatures, as mentioned above.

One way to obtain a larger negative temperature dependence while having a critical temperature of ~ 100 K is to include a change in entropy in the electron transfer reaction. Kakitani and Kakitani (44) recently developed a theory in which vibrational frequency changes as well as displacements are allowed. This means that the reactants and products can have different densities of vibrational states and that their potential energy surfaces do not have the same shape. The electron transfer reaction thus can involve a change in entropy. These authors derive an expression identical to Jortner's, except that the energy change ΔE is replaced by the free energy change ΔG, and the entropy change makes ΔG a function of temperature. The vibrational modes were artificially separated into two classes, those that undergo displacements and those that experience changes in frequency. Kakitani and Kakitani (44) and Warshel (51) argue that modes having $\langle\nu_q\rangle/c = 1000\text{-}1600\ cm^{-1}$, involving in-plane skeletal vibrations of the atoms in the conjugated π-system, are likely to be the most strongly coupled to electron transfer involving porphyrins and define the reaction coordinate. These high-frequency modes are not excited significantly above their zero-point levels, even at 300 K ($kT \sim 200\ cm^{-1}$). Therefore, the temperature dependence of the electron transfer rate is not determined by the thermal population of these modes, as it is in theories of Jortner and Sarai. Instead, the temperature dependence is determined by the thermal population of the modes that experience changes in frequency. Since the protein undoubtedly has a very large number of low-frequency modes, the entropic contribution to the free energy could be substantial even though the average frequency change may be relatively small. The critical temperature of ~ 100 K means that these modes must have an average energy $\langle\nu_e\rangle/c \sim 250\text{-}300\ cm^{-1}$.

The solid curve in Figure 4 is a computer fit using the theory of Kakitani and Kakitani. Details of the analysis are discussed elsewhere (22). The increase in rate with decreasing temperature means that the average vibrational frequency $\langle\nu_e\rangle$ decreases as the electron spacing means that the density of vibrational states is larger in the products than in the reactants, corresponding to an entropy increase as a result of the electron transfer. The theoretical fit gives an electronic factor $V = 15\text{-}30\ cm^{-1}$. Using the simple formula given above that V decreases exponentially with distance, and using $V_0 = 10^5\ cm^{-1}$ (41), we calculate R = 8-9 Å. This is close to the edge-to-edge distance between BPh and Q found in the crystal structure of the *Rps. viridis* reaction center (52). However, the calculated result may not be highly accurate because the simple relationship assumes a specific electron-tunneling barrier height and contains no orientation factor (40).

The total reorganization energy is equal to the free energy change for an activationless electron transfer reaction (40-48). In Marcus theory (48,53) this condition is expressed as $\Delta G = -\lambda$, where

λ is the reorganization energy. In the theories of Jortner (40,41) and Kakitani and Kakitani (44) the reorganization energy is equal to $S \cdot h\langle\nu_q\rangle$, where S is the electron-vibration coupling strength and $h\langle\nu_q\rangle$ is the energy of the vibrational mode that is displaced along the reaction coordinate. In discussing electron transfer from I^- (BPh^-) to Q above we used $h\langle\nu_q\rangle$ = 1550 cm^{-1} (Figure 4), although a range of values could be used to fit the data (22). Kakitani and Kakitani (44) chose an average value of 1000 cm^{-1} and Warshel used 1400 cm^{-1} (46). The coupling strength for porphyrins has been estimated to be S = 0.5-1 (42,43,46,54), consistent with the relatively small changes in bond lengths and angles observed upon oxidation or reduction of porphyrins (55-57). The value of S should be somewhat larger for quinone (58,59); Kakitani and Kakitani estimated it to be 1, compared to 0.5 for porphyrin (44). Thus S ~ 1.5 appears to be a reasonable estimate for electron transfer from BPh to Q. Together with $h\langle\nu_q\rangle$ = 1550 cm^{-1} we estimate that the reorganization energy associated with the electron carriers is ~ 0.3 eV. Thus, for the $P^+I^- \rightarrow P^+Q^-$ reaction, for which the free energy change is 0.5-0.7 eV (60-62), it appears that about half of the reorganization energy may be associated with other types of motions, some of which may involve the protein. However, a more definitive analysis must await a more detailed understanding of the actual molecular changes that accompany electron transfer in the reaction center.

Detection Wavelength Dependence of the Kinetics. The decay time of 100 ps measured at low temperature in the 665-nm anion band appears to be the same as the kinetics of the absorption changes in the 545- and 765-nm ground-state bands of BPh (Table 1). These results are consistent with the view that the BPh that receives an electron has its Q_X and Q_Y bands at 545 and 765 nm respectively (resolved at low temperature), and forms the anion radical (BPh^-) that gives rise to the 665-nm band in P^+I^- (P^+BPh^-). The slower (~ 135 ps) decay kinetics measured at 758 nm could reflect an additional relaxation involving the Q_Y band of the second BPh. Such a longer-lived component apparently does not contribute (within our resolution) to the kinetics measured at 545 nm at 5 K or 76 K; the Q_X bands of the two BPhs are well separated at low temperature (Figure 3). At higher temperatures there appears to be substantial overlap of the absorption bands (and absorption changes) in both the Q_X and Q_Y regions of the BPhs. The measurements on the flowed reaction centers at 285 K gave an average time constant of ~250 ps at both 765 and 545 nm (Table 1), slightly longer than the average 207-ps time constant across the 665-nm feature that we ascribe to the actual rate of electron transfer from I^- (BPh^-) to Q. The slower component of the kinetics in the 750-770 and 530-550 nm regions of BPh ground-state absorption could, again, reflect a relaxation involving the shorter-wavelength BPh. This relaxation could involve readjustment of the protein to accomodate the new charge distribution on the pigments following electron transfer. Kleinfeld et al (63) have proposed that the formation of P^+Q^- causes a conformation change in the reaction center.

At all temperatures, the decay times measured in the 795-nm region (Fig. 3) are significantly shorter than the time constants measured in the 545-nm, 765-nm, and 665-nm regions (Table 1). The absorption changes near 795-nm do not appear to be due to decay of the anion radical $BChl^-$. If this were the case, then one would expect to measure some component of the 795-nm kinetics at 665 nm. In addition, photodichroism measurements at low temperature suggest that the 665-nm absorption band of P^+I^- is due predominantly to BPh^- (35). One possibility to explain these results is that the absorption change at 795 nm is due mainly to a change in the interaction of BChl with P, with the BPh that undergoes reduction, or with the protein. The kinetics measured near 795-nm thus could reflect movement of the BChl(s) or readjustment of the protein following formation of P^+I^- or even following absorption of a photon. A second, but related, possibility is that the kinetics may be due to a relaxation involving P (or P^+). This could involve movement of the protein (35) and/or readjustment of the two rings of P with respect to one another (64). A (neutral) momomer-like absorption band of P is thought to lie in this region (39) and this band may continue to relax after P^+I^- (P^+BPh^-) forms.

Recent studies show that *Chloroflexus aurantiacus* reaction centers exhibit a detection-wavelength dependence of the kinetics similar to that observed here for *Rps. sphaeroides* (Table 1); the time constants in each of four similar wavelength regions are about 115 ps longer in *Chloroflexus* (65). If our tentative interpretation of the detection-wavelength dependence of the kinetics is correct, then it suggests that the early electron transfer reactions may not occur from thermally-equilibrated states of the pigments and the protein. This would greatly complicate assignment of the time evolution of absorption changes simply as pigment oxidations and reductions. Further work is underway to help understand more clearly how nuclear motions of the pigments and/or the protein contribute to the primary charge separation process.

Acknowledgments

This work was supported by grants PCM-8302477 (DH) and PCM-8312371 (WWP) from the National Science Foundation.

Literature Cited

1. Okamura, M.Y., Feher, G. and Nelson, N., in Govindjee (ed.) (1982) "Photosynthesis: Energy Conversion by Plants and Bacteria"; Academic Press, NY pp. 195-272.
2. Parson, W.W. and Ke, B., in Govindjee (ed.) (1982) "Photosynthesis: Energy Conversion by Plants and Bacteria", Academic Press, NY, pp. 331-385.
3. Straley, S.C., Parson, W.W., Mauzerall, D.C. and Clayton, R.K. (1973) Biochim. Biophys. Acta 305, 597-609.
4. Norris, J.R., Uphaus, R.A., Crespi, H.L. and Katz, J.J. (1971) Proc. Natl. Acad. Sci. USA 71, 4897-4900.
5. Norris, J.R., Druyan, M.E. and Katz, J.J (1973) J. Am. Chem. Soc. 95, 1680-1682.
6. Feher, G., Hoff, A.J. Isaacson, R.A. and Ackerson, L.C. (1975) Ann. N.Y. Acad. Sci. 244, 239-259.
7. Holten, D., Hoganson, C., Windsor, M.W., Schenck, C.C., Parson, W.W., Migus, A., Fork, R.L. and Shank, C.V. (1980) Biochim. Biophys. Acta 592, 461-477.
8. Woodbury, N.W., Becker, M., Middendorf and Parson, W.W. (1976) Biochem. 24, 7516-7521.

9. Martin, J.L., Breton, J., Woff, A.J., Migus, A. and Antonetti, A. (1986) Proc. Natl. Acad. Sci. (in press).
10. Rockley, M.G., Windsor, M.W., Cogdell, R.J. and Parson, W.W. (1975) Proc. Natl. Acad. Sci. USA 72, 2251-2255.
11. Kaufmann, K.J., Dutton, P.L., Netzel, T.A., Leigh, J.S. and Rentzepis, P.M. (1975) Science 188, 1301-1304.
12. Dutton, P.L., Kaufmann, K.J., Chance B. and Rentzepis, P.M. (1975) FEBS Lett. 60, 275-280.
13. Kaufmann, K.J., Petty, K.M., Dutton, P.L. and Rentzepis, P.M (1976) Biochem. Biophys. Res. Commun. 70, 839-845.
14. Shuvalov, V.A., Klevanik, A.V., Sharkov, A.V., Matveetz, Y.A. and Kryukov, P.G. (1978) FEBS Lett. 91, 135-139.
15. Moscowitz, E. and Malley, M.M. (1978) Photochem. Photobiol. 27, 55-59.
16. Akhamanov, S.A., Borisov, A.Y., Danielius, R.V., Gadonas, R.A., Kazlowski, V.S., Piskarkas, A.S., and Shuvalov, V.A. (1980) FEBS Lett. 114, 149-152.
17. Schenck, C.C., Parson, W.W., Holten, D., Windsor, M.W. and Sarai, A. (1981) Biophys. J. 36, 479-489.
18. Kirmaier, C., Holten, D. and Parson, W.W. (1983) Biochim. Biophys. Acta 725, 190-202.
19. Shuvalov, V.A. and Klevanik, A.V. (1983) FEBS Lett. 160, 51-55.
20. Pellin, M.J., Wraight, C.A. and Kaufmann, K.J. (1978) Biophys. J. 24, 362-369.
21. Peters, K.S., Avouris, P. and Rentzepis, P.M. (1978) Biophys. J. 23, 207-217.
22. Kirmaier, C., Holten, D. and Parson (1985) W.W. Biochim. Biophys. Acta 810, 33-48.
23. Borisov, A.Y., Danielius, R.V., Kudzmauskas, S.P., Piskarankas, A.S., Razjivin, A.P., Sirutkaitis, and Valkunes, L.L. (1983) Photobiochem. Photobiophys. 6, 33-38.
24. Kirmaier, C., Holten, D. and Parson, W.W. (1985) FEBS Lett. 185, 76-82.
25. Boxer, S.G., Lockhart, D.J. and Middendorf, T.R. (1986) Chem. Phys. Lett. (in press).
26. Meech, S.R., Hoff, A.J. and Wiersma, D.A. (1985) Chem. Phys. Lett. 121, 287-292.
27. Schenck, C.C., Parson, W.W., Holten, D. and Windsor, M.W. (1981) Biochim. Biophys. Acta 635, 383-392.
28. Ditson, S.L., Davis, R.C. and Pearlstein, R.M. (1984) Biochim. Biophys. Acat 766, 623-629.
29. Maroti, P., Kirmaier, C., Wraight, C., Holten, D. and Pearlstein, R.M. (1985) Biochim. Biophys. Acta, 810, 132-139.
30. Tiede, D.M., Prince, R.C. and Dutton, P.L. (1976) Biochim. Biophys. Acta 449, 447-469.
31. Shuvalov, V.A. and Klimov, V.V. (1976) Biochim. Biophys. Acta 440, 587-599.
32. Okamura, M.Y., Isaacson, R.A. and Feher, G. (1979) Biochim. Biophys. Acta 546, 397-417.
33. Shuvalov, V.A. and Parson, W.W. (1980) Proc. Natl. Acad. Sci. USA 78, 957-961.
34. Fajer, J., Brune, D.C., Davis, M.S., Forman, A. and Spaulding, L.D. (1975) Proc. Natl. Acad. Sci. USA 72, 4956-4960.
35. Kirmaier, C., Holten, D. and Parson, W.W. (1985) Biochim. Biophys. Acta 810, 49-61.

36. Schenck, C.C., Blankenship, R.E. and Parson, W.W. (1982) Biochim. Biophys. Acta 680, 44-59.
37. Davis, M.S., Forman, A. and Fajer, J. (1979) Proc. Natl. Acad. Sci. USA 76, 4170-4174.
38. Robert, B. Lutz, M. and Tiede, D.M. (1985) FEBS Lett. 83, 326-330.
39. Vermeglio, A. and Clayton, R.K. (1977) Biochim. Biophys. Acta 461, 159-165.
40. Jortner, J. (1976) J. Chem. Phys. 64, 4860-4867.
41. Jortner, J. (1980) J. Am. Chem. Soc. 102, 6676-6686.
42. Sarai, A. (1979) Chem. Phys. Lett. 63, 360-366.
43. Sarai, A. (1980) Biochim. Biophys. Acta 589, 71-83.
44. Kakitani, T. and Kakitani, H. (1981) Biochim. Biophys. Acta 635, 498-514.
45. Hopfield, J.J. (1974) Proc. Natl. Acad. Sci. USA 71, 3640-3644.
46. Warshel, A. (1980) Proc. Natl. Acad. Sci. USA 77, 3105-3109.
47. Blankenship, R.E and Parson, W.W. (1979) in "Photosynthesis in Relation to Model Systems" (Barber, J., ed.) Elsevier Scientific, Amsterdam 71-114.
48. Marcus, R.A. and Sutin, N. (1985) Biochim. Biophys. Acta 812, 295-322.
49. Hales, B.J. (1976) Biophys. J. 16, 471-480.
50. Beitz, J.V. and Miller, J.R. (1979) J. Chem. Phys. 71, 4579-4595.
51. Kirmaier, C., Holten, D. and Parson, W.W. unpublished results.
52. Deisenhofer, J., Epp, O., Miki, K., Huber, R. and Michel, H. (1984) J. Mol. Biol. 180, 385-398.
53. Marcus, R.A. (1964) Ann. Rev. Phys. Chem. 15, 155-196.
54. Felton, R.H. (1978) in "The Porphyrins" (Dolphin, D., ed.) Academic Press, New York, Vol. V, Chapter 3.
55. Spaulding, L.D., Eller, P.G., Bertrand, J.A. and Felton, R.H. (1974) J. Am. Chem. Soc. 96, 982-987.
56. Takeno, T. and Dickerson, R.E. (1981) J. Mol. Biol. 153, 95-115.
57. Churg, A.K., Weiss, R.M., Warshel, A. and Takano, T. (1983) J. Chem. Phys. 87, 1683-1694.
58. Efrima, S. and Bixon, M. (1974) Chem. Phys. Lett. 25, 34-37.
59. Gouterman, M. and Holten, D. (1977) Photochem. Photobiol. 25, 85-92.
60. Woodbury, N.T. and Parson, W.W. (1984) Biochim. Biophys. Acta 767, 345-361.
61. Arata, H. and Parson, W.W. (1981) Biochim. Biophys. Acta 638, 201-209.
62. Woodbury, N.W., Gunner, M., Prince, R.C., Dutton, P.L. and Parson, W. W. (1986) Biochim. Biophys. Acta (submitted).
63. Kleinfeld, D., Okamura, M.Y. and Feher, G. (1984) Biochem. 23, 5780-5786.
64. Vermeglio, A. and Paillotin, G. (1982) Biochim. Biophys. Acta 681, 32-40.
65. Kirmaier, C., Blankenship, R.E. and Holten, D. (1986) Biochim. Biophys. Acta (in press).

RECEIVED July 15, 1986

15

Two-Photon Absorption and Radiationless Transitions of Porphyrins

Seiji Tobita[1], Yoshizumi Kajii[2], and Ikuzo Tanaka[2]

[1]Gunma Technical College, Toribacho, Maebashi, Gunma, Japan
[2]Department of Chemistry, Tokyo Institute of Technology, Ohokayama, Meguroku, Tokyo, Japan

$S_2 \rightarrow S_0$ fluorescence and radiationless transitions from the S_1 state of porphyrins have been studied in order to reveal photodynamics of porphyrins. The S_2 state fluorescence of zinc(II)-tetraphenylporphin is caused even by the excitation to the S_1 state. Two-photon absorption and optical-optical double resonance studies show that a stepwise two-photon absorption through the S_1 state is a main process populating the S_2 state. The absorption cross sections σ_{1n} of the $S_n \leftarrow S_1$ transition are determined from the kinetic analyses based on the two-photon absorption measurement. In contrast, tetraphenylporphin scarcely show the S_2 state fluorescence. In order to elucidate the effect of the imino hydrogens on the radiationless transitions, deuterium isotope effect on the relaxation processes from the lowest excited singlet state of metal free porphyrins has been investigated.

Photodynamics of metalloporphyrins have been extensively investigated on account of its importance in the understanding of photosynthesis and other processes of biological importance (1). Particular attension has been paid to the reason why the excited metalloporphyrins possess unique characteristics from the viewpoint of redox (2-4), energy transfer (5), and other photodynamical processes (6,7). In comparison with the considerable knowledge accumulated on the photochemical properties of the lowest excited states, little has been known on the $S_2 \rightarrow S_0$ fluorescence and $S_1 \rightsquigarrow S_0$ internal conversion processes which can also be regarded as unusual characters of metalloporphyrins.

The S_2 state fluorescence of metalloporphyrins was first noticed by Bajema et al. (8) and later the photophysical parameters concerned with the S_2 state were determined on several metalloporphyrins (9,10). $S_2 \rightarrow S_0$ emission from large molecules in condensed phases has also been recognized in many other organic compounds, e.g., azulene (11), thiocarbonyl compounds (12), and several polyenes (13). However, in some metalloporphyrins one can observe the S_2 state fluorescence even after the excitation to the S_1 state (14-17), that is, the blue

0097-6156/86/0321-0219$06.00/0

emission can be obtained from the visible excitation light. Possible mechanisms to account for this anomalous behavior is investigated by measuring the laser induced fluorescence of zinc(II)-tetraphenylporphin (18,19). The results show that the S_2 state fluorescence is mainly caused by stepwise two-photon absorption through the S_1 state and there exists a minor contribution of triplet-triplet annihilation. The measurement of the S_2 state fluorescence is applied to the determination of absorption cross sections for the $S_n \leftarrow S_1$ transition of ZnTPP.

Porphyrin free bases are characterized by central imino hydrogens which have been proved to play an important role not only in electronic structures but also in the radiationless transitions from the excited states. Several groups have substantiated that deuteration of imino hydrogens decreases the rate constant of $T_1 \rightsquigarrow S_0$ intersystem crossing (20-23). This was attributed to the N-H stretching mode of the center of the porphin ring which is presumed to act as an accepting mode in the $T_1 \rightsquigarrow S_0$ intersystem crossing. Recently, it was shown that in deuterated tetraphenylporphin the fluorescence quantum yield and lifetime, and the triplet yield increase in a similar manner relative to the undeuterated one (24). Such a deuterium isotope effects on the S_1 state suggested the participation of $S_1 \rightsquigarrow S_0$ internal conversion in the energy dissipating processes. The effect of isotopic substitution on the photophysical parameters of S_1 state is investigated in order to elucidate the mechanism of radiationless transitions in metal free porphyrins.

Experimental

Tetraphenylporphin (H_2TPP) was prepared from pyrrole and benzaldehyde (25). The tetraphenylchlorin contamination was oxidized by use of 2,3-dichloro-5,6-dicyano-p-benzoquinone (26). Octaethylporphin (H_2-OEP)was prepared by the method of Paine et al.(27) from 3,4-diethyl-2-ethoxy-carbonyl-5-methyl-pyrrole (28). Zinc(II)-tetraphenylporphin (ZnTPP) was prepared by refluxing H_2TPP and zinc acetate in dimethylformamide (29). ZnTPP was dissolved at a concentration of 10^{-5}-10^{-6} M in EPA (5 parts diethyl ether, 5 parts isopentane, and 2 parts ethanol by volume). H_2TPP and H_2OEP were dissolved at a concentration of 10^{-5}-10^{-6} M in toluene. The deuterated TPP (D_2TPP) and OEP (D_2OEP) on the center imino hydrogens were obtained by adding a few drops of C_2H_5OD to a solution of TPP and OEP in dry toluene (30). OEP deuterated at the meso positions was obtained by keeping OEP in concentrated D_2SO_4 for 24 h (31). All samples were thoroughly degassed by freeze-pump-thaw cycles on a high vacuum line.

Absorption and emission spectra were taken with a JASCO UVIDEC-510 spectrophotometer and a JASCO FP-550A spectrofluorometer, respectively. The fluorescence quantum yields were determined relative to 9,10-diphenylanthracene or rhodamine 640 using the optically dilute method (32). The transient absorption was measured by a nanosecond laser flash photolysis system using a nitrogen pumped dye laser as an excitation source (33). Transient signals were accumulated by an Iwatsu DM-901 digital memory combined with a personal computer. Two-photon measurements upon laser illumination were carried out using a nitrogen laser (Molectron UV-22) and a nitrogen pumped dye laser (Molectron DL-14). The optical-optical double resonance (OODR) measurements were carried out using two-color laser beams (λ_1 = 540 nm; λ_2 = 640 nm) from two sets of dye heads which were pumped simultaneously with the same nitrogen laser. The fluorescence lifetimes of metal free porphyrins were measured by using a single-photon counting method. A dye laser (Spectra-Physics model 375) pumped by a mode-locked Ar-ion laser (Spectra-Physics model 171) was used for the excitation source. A single pulse of the dye laser was extracted by cavity dumper (Spectra-Physics model 344). The pulse width was ca. 20 ps and the repetition rate was 0.8 MHz.

S_2 Emission of ZnTPP by Two-Photon Absorption Mechanism

ZnTPP and several metalloporphyrins show typical $S_2 \rightarrow S_0$ emission by ultraviolet irradiation with stationary light (8-10). Similar violations of Kasha's rule were favored for molecules having relatively large energy differences between their S_1 and S_2 states (e.g., 14000 cm^{-1} in azulene (34) and 6900 cm^{-1} in ZnTPP). In the case of metalloporphyrins, another origin is found in the large absorption cross section for the $S_2 \leftarrow S_0$ transition (ca. 2.3 x 10^{-15} cm^2 $molecule^{-1}$ in ZnTPP) because the natural radiative rate from the S_2 state could be so large to compete with the fast internal conversion. The S_1 and S_2 fluorescence quantum yields determined were ϕ_1 = 0.07 and ϕ_2 = 0.001, respectively. The fluorescence lifetime τ_1 from the S_1 state was obtained to be 1.8 ns by using a second harmonic (527 nm, pulse width : ca. 4 ps) of Nd^{3+} phosphate glass picosecond laser, while the lifetime of the S_2 state fluorescence was too short to be determined (35).

$S_2 \rightarrow S_0$ fluorescence in condensed media has so far been found in several types of molecules. However, metalloporphyrins are contrasted with these compounds by another arresting feature such that the S_2 fluorescence can be observed even upon photoexcitation to the S_1 state. Stelmakh and Tsvirko have first noticed the anomalous $S_2 \rightarrow S_0$ fluorescence in metalloporphyrins (15,16). Figure 1(a) shows the fluorescence spectra of ZnTPP in EPA taken by the 540 nm excitation of a nitrogen pumped dye laser. The fluorescence band at around 430 nm observed by visible excitation is safely assigned to the S_2 state fluorescence. The laser power dependence of the fluorescence intensity is quadratic at low power density of excitation (<5 x 10^{15} photons cm^{-2} $pulse^{-1}$) but shows typical saturation effect with increasing the laser intensity. It should be emphasized here that the S_2 fluorescence of ZnTPP can be observed without focusing of the laser beam.

Possible mechanisms of excitation to the upper electronically excited states can be classified as follows:

[A] triplet-triplet annihilation

$$S_0 + h\nu \rightarrow S_1 \rightsquigarrow T_1 \; , \; T_1 + T_1 \rightarrow S_2 + S_0$$

[B] stepwise two-photon absorption through the S_1 state

$$S_0 + h\nu \rightarrow S_1 \; , \; S_1 + h\nu \rightarrow S_n \rightsquigarrow S_2$$

[C] $T_n \rightsquigarrow S_2$ inverse intersystem crossing following stepwise two-photon absorption through the T_1 state

$$S_0 + h\nu \rightarrow S_1 \rightsquigarrow T_1 \; , \; T_1 + h\nu \rightarrow T_n \rightsquigarrow S_2$$

If the S_2 fluorescence is mainly produced by the triplet-triplet annihilation (mechanism [A]), the decay profile must depend upon the decay rate constant (k_T) of the first excited triplet state, and delayed fluorescence should be observed in the wavelength region of S_2 state fluorescence (36). The observed decay profile of $S_2 \rightarrow S_0$ fluorescence consisted of two components. The main part had a shorter lifetime than the time width (ca. 6 ns) of the excitation laser pulse and weak delayed fluorescence was detected simultaneously with the fast fluorescence. Figure 1(b) shows the delayed $S_1 \rightarrow S_0$ and $S_2 \rightarrow S_0$ fluorescence spectra taken at 1 μs after the laser pulse excitation of 540 nm. The decay profile of the delayed fluorescence was corresponding to the second-order component of $T_n \leftarrow T_1$ absorption of ZnTPP under the same experimental conditions. Although the trace of delayed $S_2 \rightarrow S_0$ fluorescence was observed, the intensity was negligibly weak compared with the prompt $S_2 \rightarrow S_0$ fluorescence. The time-resolved experiments reveal that the S_2 state fluorescence is caused by mainly the stepwise two-photon absorption due either to mechanism [B] or [C], and the triplet-triplet annihilation is slightly responsible for the S_2 state fluorescence.

If the $S_n \leftarrow S_1$ transitions at around 540 nm are strongly allowed, then mechanism [B] will become more important than mechanism [C], because the latter process involves the inefficient $T_n \rightsquigarrow S_2$ intersystem crossing. The quantum yield of $T_n \rightsquigarrow S_2$ intersystem crossing is generally less than 10^{-3}, except for the heavy atom-containing molecules (37). In order to determine which mechanism is dominant under the present experimental condition the time-resolved OODR study was carried out by using two-color laser pulses. Since the S_1 or T_1 state acts as the intermediate state in the course of the stepwise two-photon absorption, the large difference between their lifetimes was utilized in OODR method. The λ_1(540 nm) and λ_2(640 nm) pulses were used for the $S_1 \leftarrow S_0$ and $S_n \leftarrow S_1$ excitations, respectively, and the fluorescence from the S_2 state was monitored. The time delay between the λ_1 and λ_2 pulses was changed optically to the range 3.0 to 20 ns. The dependence of the intensity of S_2 fluorescence on the time delay is shown in Figure 2(a). The relative increments (ΔF_2) of the S_2 fluorescence due to the second laser pulse λ_2 are plotted as a function of the delay time t_d between λ_1 and λ_2 pulses. Experimental values are shown by open circles, which tend to decrease rapidly with increasing delay time in the range 3.0 to 20 ns. The probability producing the S_2 state by the irradiation of λ_1 and λ_2 pulses at a given delay time t_d can be simulated by

$$P(t_d) = \int I_L(t + t_d) f(t) dt \tag{1}$$

where $I_L(t + t_d)$ is the laser light function , and $f(t)$ is the singlet or triplet response function. The shape of $I_L(t)$ can be approximated by the Gaussian. The response functions of the singlet ($f_S(t)$) and triplet ($f_T(t)$) states are calculated by

$$f_S(t) = \int I_L(t + t') e^{-k_1 t'} dt' \tag{2}$$

and

$$f_T(t) = \int I_L(t + t')(1 - e^{-k_1 t'}) dt' \tag{3}$$

respectively, using singlet decay rate constant (k_1). The calculated curves of $I_L(t)$, $f_S(t)$, and $f_T(t)$ are shown in Figure 2(b), and the $P(t_d)$ values simulated based on these functions are plotted in Figure 2(a) for both the S_1 and T_1 states. It can be found from Figure 2(a) that the $P(t_d)$ curve calculated based on the mechanism [C] tends to increase with increasing delay time, while the corresponding values

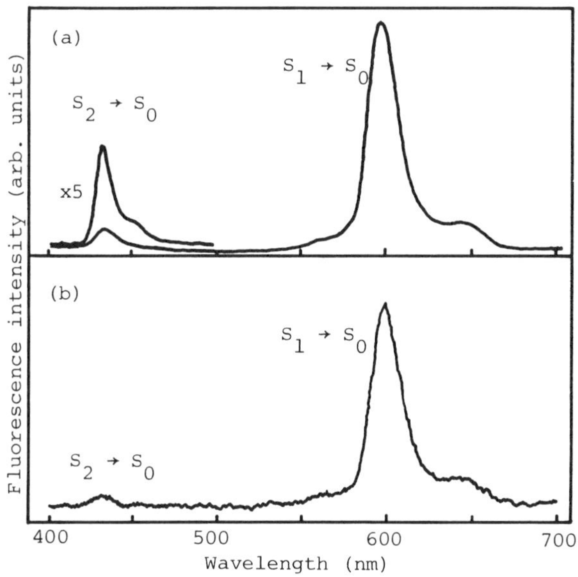

Figure 1. Fluorescence spectra (uncorrected for the spectral response) of ZnTPP in EPA at room temperature taken by the 540 nm excitation of a nitrogen pumped dye laser. (a) normal fluorescence spectra, (b) delayed fluorescence spectra taken at 1 μs after the laser pulse excitation.

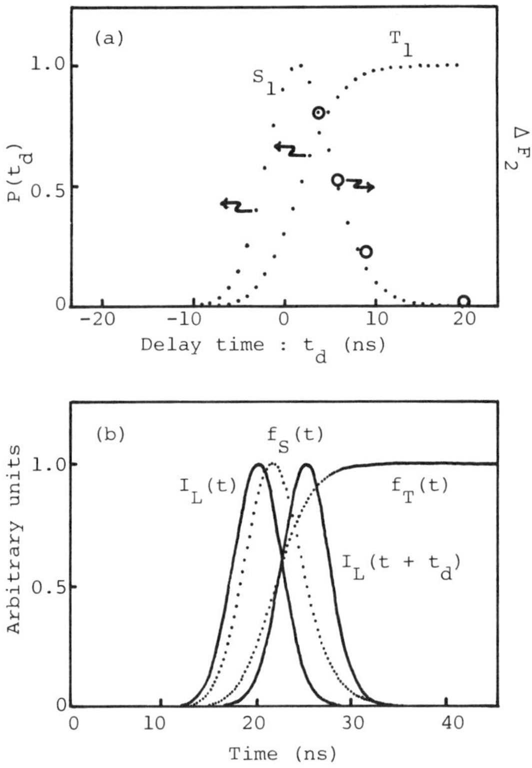

Figure 2. (a) Relative increment (ΔF_2) of the S_2 state fluorescence and simulated formation probability $P(t_d)$ of the S_2 state as a function of the delay time t_d. (b) Laser light functions ($I_L(t)$ and $I_L(t + t_d)$) and convoluted time profiles of excited singlet ($f_S(t)$) and triplet ($f_T(t)$) state concentrations.

for the S_1 state are inclined to fall off rapidly. The experimental results are in accord with the $P(t_d)$ curve for the S_1 state, indicating that in the present experimental conditions the stepwise two-photon absorption $S_0 \rightarrow S_1 \rightarrow S_n$ is a major route of excitation of high-energy S_2 state by low-energy photons.

Stelmakh and Tsvirko have also proposed the mechanism [B] for the production of S_2 state fluorescence in Mg, Zn, and lanthanide-porphyrins complexes (15), and showed the possibility of another mechanism of pumping by triplet-triplet annihilation (16) (mechanism [A]). The latter process was considered to predominate at relatively low values of the excitation densities. However, the contribution of the mecahnism [C] was not determined experimentally. The experimental results in the present work are summarized as follows: (i) The stepwise two-photon absorption by the mechanism [B] operates predominantly at moderate laser power densities. (ii) The mechanism [A] is actually cooperative in support of the result of Stelmakh and Tsvirko though less efficient than the mechansim [B]. (iii) The time-resolved OODR measurement reveals that the mechanism [C] is excluded as a principal channel.

The reason why the mechanism [B] is favored is ascertained in the next section.

Absorption Cross Sections of the $S_n \leftarrow S_1$ Transition of ZnTPP

The result in the previous section implies that ZnTPP possesses extremely high $S_n \leftarrow S_1$ absorption coefficients in the visible region. The direct detection of the $S_n \leftarrow S_1$ absorption in organic compounds has so far been achieved by a nanosecond or picosecond laser flash photolysis method. The general features of transient absorption spectra of metalloporphyrins actually suggest the presence of strong absorption bands in visible or ultraviolet region (38-40). However, as the transient absorption of the S_1 state often overlaps with that of ground state depletion, it is usually difficult to evaluate the absolute absorption cross sections σ_{1n} for the $S_n \leftarrow S_1$ transition by means of the transient absorption method. On the other hand, it is important in the present case to note that the fluorescence is observed not only from the S_1 state but also from the S_2 state with little interference of self-absorption. Thus the σ_{1n} values of ZnTPP can be evaluated by means of two-photon absorption technique.

In the following treatment, only the $S_2 \rightarrow S_0$ fluorescence and $S_2 \rightsquigarrow S_1$ internal conversion were included as energy dissipating processes from the S_2 state. Namely, the other processes such as intersystem crossing and chemical reactions have not been taken into account. If the concentration of S_0, S_1, and S_2 molecules are denoted by $N_0(t)$, $N_1(t)$, and $N_2(t)$, respectively, then the transition rates for $S_1 \leftarrow S_0$ and $S_n \leftarrow S_1$ absorption can be given by $\sigma_{01}I_L(t)N_0(t)$ and $\sigma_{1n}I_L(t)N_1(t)$, respectively, using each absorption cross section. The dependence of S_1 and S_2 fluorescence intensities on the sample concentration was investigated between 10^{-6} - 10^{-4} M, and a sample concentration of less than 10^{-5} M was chosen to minimize the effect of concentration quenching. Two-photon absorption measurements were carried out by the excitation by relatively low power density ($<10^{15}$ photons cm^{-2} $pulse^{-1}$) in order to avoid the saturation effect on the laser power. Under these experimental conditions, the following rate equations for the concentration of S_2 molecule is derived:

$$\frac{dN_2(t)}{dt} = \sigma_{1n} I_L(t) N_1(t) - k_2 N_2(t) \tag{4}$$

From "Equation 4" the time integrated intensity ratio F_2/F_1 between S_2 and S_1 fluorescence can be given by

$$\frac{F_2}{F_1} = \sigma_{1n}\tau_1 \cdot \frac{\phi_2}{\phi_1} \cdot \frac{\int I_L(t) N_1(t) dt}{\int N_1(t) dt} \tag{5}$$

where the integration on the right can be simplified by use of the following constant:

$$\tau^* = \frac{\int N_1(t) dt \cdot \int I_L(t) dt}{\int N_1(t) \cdot I_L(t) dt} \tag{6}$$

Based on the $I_L(t)$ and $N_1(t)$, τ^* was calculated by numerical integration to be 11 ns. Consequently the following relation is obtained:

$$\frac{F_2}{F_1} = \sigma_{1n} \cdot \frac{\tau_1}{\tau^*} \cdot \frac{\phi_2}{\phi_1} \cdot I_L \tag{7}$$

where $I_L = \int I_L(t) dt$ is the integrated laser intensity. The laser power dependence of F_2/F_1 was investigated up to the incident laser power ca. 10^{16} photons cm^{-2} $pulse^{-1}$. A linear relationship was obtained at less than 5 x 10^{15} photons cm^{-2} $pulse^{-1}$. On the basis of the same treatment, the absorption cross sections of $S_n \leftarrow S_1$ transition were determined at several wavelengths in the range of 490 - 560 nm. The σ_{1n} values determined are listed in "Table I" together with the converted values to molar extinction coefficient.

Table I. Absorption Cross Sections (σ_{1n}) and Molar Extinction Coefficients (ε) of $S_n \leftarrow S_1$ Transition of ZnTPP

λ a) nm	ΔE b) cm^{-1}	σ_{1n} 10^{-16} $cm^2 \cdot molec^{-1}$	ε mol^{-1} l cm^{-1}
560	17857	7.57	198000
550	18182	6.88	180000
540	18519	6.84	179000
530	18868	10.2	267000
520	19231	4.82	126000
510	19608	4.74	124000
500	20000	5.66	148000
490	20408	7.04	184000

a) wavelength
b) $S_n \leftarrow S_1$ transition energy

Judging from the large absorption cross section obtained, the $S_n \leftarrow S_1$ transition is expected to be strongly allowed in the Q-band region. The first excited singlet state (Q state) of metalloporphyrins has E_u symmetry. Hence our finding of the strongly allowed $S_n \leftarrow S_1$ transition suggests that a higher excited singlet state of even parity should exist at 32000 - 36000 cm^{-1} above the ground state. The result of MO calculation taking into account doubly excited configurations and higher singly excited configurations actually indicates the presence of a strongly two-photon allowed state ($^1A_{2g}$) in this energy region (19). This result supports the conclusion in the previous section, that is, the extremely large cross sections for the $S_n \leftarrow S_1$ absorption are responsible for the observed S_2 state fluorescence by the two-photon absorption mechanism.

Deuterium Isotope Effect on the $S_1 \rightsquigarrow S_0$ Internal Conversion of Free Base Porphyrins

The absorption spectrum of TPP was not altered by deuteration of center imino hydrogens. Spectral shapes and positions of the two fluorescence spectra agreed closely with each other, while the fluorescence intensity of D_2TPP was 20% stronger than that of H_2TPP. In the case of OEP, a similar tendency was recognized as TPP. Logarithmic plots of the fluorescence decays for the lowest excited singlet state of H_2TPP, H_2OEP, and their deuterium analogues were described by single exponential forms. Fluorescence lifetimes of all compounds were obtained by deconvolution and fluorescence quantum yields were determined relative to rhodamine 640 in ethanol (32). Triplet quantum yields (ϕ_T) were determined relative to benzophenone in benzene by using the transient absorption (33). Quantum yields and lifetimes determined are summarized in "Table II".

Table II. Fluorescence Lifetimes (τ_f), Fluorescence Quantum Yields (ϕ_f), Triplet Quantum Yields (ϕ_T), and Quantum Yields of Internal Conversion (ϕ_{ic})

	τ_f / ns	ϕ_f	ϕ_T	ϕ_{ic}
H_2TPP	12.5	0.10	0.70	0.20
D_2TPP	15.0	0.12	0.84	0.04
H_2OEP	18.6	0.11	0.75	0.14
D_2OEP	19.4	0.12	0.80	0.08
$H_2OEP\text{-}d_4$	20.0	0.12	0.78	0.10
$D_2OEP\text{-}d_4$	21.0	0.13	0.84	0.03

As the absorption spectra did not change after laser light irradiation, no appreciable reaction seems to occur through the S_1 state. Thus, the decay processes from the S_1 state can be assumed to be only fluorescence, intersystem crossing, and internal conversion. The quantum yields of internal conversion (ϕ_{ic}) are, therefore, estimated by the following equation:

$$\phi_{ic} = 1 - [\phi_f + \phi_T] \quad (8)$$

Using the values of quantum yields and fluorescence lifetime, rate constants on the S_1 state were obtained. The results are summarized in "Table III". As seen in "Table III", no isotope effect is observed in the rate constants for the fluorescence and intersystem crossing. On the other hand, in the case of TPP the rate constants of internal conversion show a large isotope effect; that of D_2TPP is is about one order of magnitude smaller than that of H_2TPP. Therefore, the central N-H group is expected to play an important role in internal conversion.

Table III. Rate Constants of Fluorescence (k_f), Intersystem Crossing (k_{isc}), and Internal Conversion (k_{ic})

	k_f $10^6\ s^{-1}$	k_{isc} $10^7\ s^{-1}$	k_{ic} $10^6\ s^{-1}$
H_2TPP	8.0	5.6	16
D_2TPP	8.0	5.6	2.8
H_2OEP	5.9	4.0	7.5
D_2OEP	6.2	4.1	4.1
H_2OEP-d_4 [a)]	6.0	3.9	5.0
D_2OEP-d_4	6.2	4.0	1.4

a)OEP deuterated at meso positions

Two models have been examined to explain the isotope effect: (i) a migration of hydrogen atoms between tautomeric structures which would result in a large difference in the rates of internal conversion, and (ii) an effect of the N-H vibrations which may act as a promoting or an accepting mode for dissipation of the electronic energy of the S_1 state (20). Storm and Teklu (41) first reported the migration of hydrogen atoms in the ground state of H_2TPP, and following them, many groups studied the mechanism of hydrogen migration in free base porphyrins (30,42,43). The rate constant of the migration is very small in the ground state. On the other hand, hydrogen migration in the S_1 state is expected to be faster. However, the rate constant of the tautomerisation in the S_1 state has been determined to be far smaller than that of the fluorescence from measurements of polarized fluorescence (44,45). Therefore, the two hydrogen atoms are considered to be fixed during the fluorescence lifetime (46) and model (i) is ruled out. These facts indicate that the N-H vibrations are a promoting or an accepting mode in the internal conversion. At the present stage, it is difficult to conclude whether the N-H vibrations act as a promoting or as an accepting mode. In the case of $T_1 \rightsquigarrow S_0$ intersystem crossing, the N-H vibrations are regarded as an accepting mode. Hence, it is likely that the N-H vibrations would be an accepting mode in the internal conversion. The stretching vibrational frequencies, ν_{N-H} and ν_{N-D}, are 3318 and 2482 cm^{-1}, respectively (43). Assuming that the N-H vibration is harmonic, the lowest excited singlet state would be isoenergetic with the "(N-H stretching quanta)" = 5 and "(N-D stretching quanta)" = 7 levels of the ground state. If the internal conversion proceeds from the lowest vibrational level of S_1 to the isoenergetic vibrational levels of the N-H (N-D) stretching in the ground state, the Franck-Condon factor for D_2TPP would be much smaller than that of H_2TPP (47).

While in the case of OEP, the rate constant of the internal conversion for D_2OEP was about one-half that of H_2OEP as shown in "Table III". There is not such a marked isotope effect as in the case of TPP but a slight isotope effect can be observed. So, we can conclude that N-H vibrations may play an important role in internal conversion in OEP. The deuterated OEP at meso positions ($OEP-d_4$)

also shows a slight isotope effect in the internal conversion. Therefore, not only N-H vibrations but also C-H vibrations of meso positions are responsible for the internal conversion. In addition to meso C-H groups, OEP has ethyl groups on a porphin ring. vibrational modes of these substituent groups may be also an effective mode in the radiationless transitions. Hence, several vibrational modes of free base porphyrins may be responsible for the weakness of S_2 state fluorescence.

Literature Cited

1. Gouterman, M. In "The Porphyrins"; Dolphin, D., Ed.; Academic: New York, 1978; Vol. 3, p. 1.
2. Nahor, G. S.; Rabani, J. J. Phys. Chem. 1985, 89, 2468.
3. Hoshino, M.; Seki, H.; Shizuka, H. J. Phys. Chem. 1985, 89, 470.
4. Aono, S.; Okura, I.; Yamada, A. J. Phys. Chem. 1985, 89, 1593.
5. Gust, D.; Moore, T. A.; Bensasson, R. V.; Mathis, P.; Land, E. J.; Chachaty, C.; Moore, A. L.; Liddell, P. A.; Nemeth, G. A. J. Am. Chem. Soc. 1985, 107, 3631.
6. Tait, C. D.; Holten, D.; Barley, M. H.; Dolphin, D.; James, B. R. J. Am. Chem. Soc. 1985, 107, 1930.
7. Dixon, D. W.; Kirmaier, C.; Holten, D. J. Am. Chem. Soc. 1985, 107, 808.
8. Bajema, L.; Gouterman, M.; Rose, C. B. J. Mol. Spectrosc. 1971, 39, 421.
9. Martarano, L. A.; Wang, C. P.; Horrocks, W. Jr.; Goncalves, A. M. P. J. Phys. Chem. 1976, 80, 2389.
10. Kurabayashi, Y.; Kikuchi, K.; Kokubun, H.; Kaizu, Y.; Kobayashi, H. J. Phys. Chem. 1984, 88, 1308 ; Ohno, O.; Kaizu, Y.; Kobayashi, H. J. Chem. Phys. 1985, 82, 1779.
11. Beer, M.; Longuet-Higgins, H. C. J. Chem. Phys. 1955, 23, 1390.
12 Plummer, B. F.; Al-Saigh, Z. Y. Chem. Phys. Lett. 1982, 91, 427.
13. Snyder, R.; Arvidson, E.; Foote, C.; Harrigan, L.; Christensen, R. L. J. Am. Chem. Soc. 1985, 107, 4117.
14. Tsvirko, M. P.; Stelmakh, G. F.; Pyatosin, V. E.; Solovev, K. N. ; Kachura, T. F. Chem. Phys. Lett. 1980, 73, 80.
15. Stelmakh, G. F.; Tsvirko, M. P. Opt. Spektrosk. 1980, 48, 185.
16. Stelmakh, G. F.; Tsvirko, M. P. Opt. Spektrosk. 1980, 49, 511.
17. Tsvirko, M. P.; Solovev, K. N.; Stelmakh, G. F.; Phyatosin, V. E.; Kachura, T. F. Opt. Spektrosk. 1981, 50, 555.
18. Tobita,S.; Tanaka, I. Chem. Phys. Lett. 1983, 96, 517.
19. Tobita,S.; Kaizu, Y.; Kobayashi, H.; Tanaka, I. J. Chem. Phys. 1984, 81, 2962.
20. Burgner, R. P.; Goncalves, A. M. P. J. Chem. Phys. 1974, 60, 2942.
21. Burgner, R. P.; Goncalves, A. M. P. Chem. Phys. Lett. 1978, 46, 275.
22. Solovev, K. N.; Knyukshto, V. N.; Tsvirko, M. P.; Gradyushko, A. T. Opt. Spektrosk. 1976, 41, 964.
23. Gradyushko, A. T.; Knyukshto, V. T.; Solovev, K. N.; Shulga, A. M. Opt. Spektrosk. 1978, 44, 458.
24. Kajii, Y.; Obi, K.; Tanaka, I.; Tobita, S. Chem. Phys. Lett. 1984, 111, 347.

25. Adler, A. D.; Longo, F. R.; Finarelli, J. D.; Goldmacher, J.; Assour, J.; Korsakoff, L. J. Org. Chem. 1967, 32, 476.
26. Barnett, G. H.; Hudson, M. F.; Smith, K. M. J. Chem. Soc. Perkin Trans. 1. 1975, 1401.
27. Paine, J. B.; Kirshner, W. B.; Moskowitz, D. J. Org. Chem. 1976, 41, 3857.
28. Wang, C. B.; Cang, C. K. Synthesis 1979, 548.
29. Adler, A. D.; Longo, F. R.; Kampas, F.; Kim, J. J. Inorg. Nucl. Chem. 1970, 32, 2443.
30. Eaton, S. S.; Eaton, G. R. J. Am. Chem. Soc. 1977, 99, 1601.
31. Bonnett, R.; Stephenson, G. F. Proc. Chem. Soc. 1964, 291.
32. Demas, J. N.; Crosby, G. A. J. Phys. Chem. 1971, 75, 991.
33. Tobita, S.; Arakawa, M.; Tanaka, I. J. Phys. Chem. 1984, 88, 2697. ; Tobita, S.; Arakawa, M.; Tanaka, I. J. Phys. Chem. 1985, 89, 5649.
34. Eber, G.; Schneider, S.; Dorr, F. Chem. Phys. Lett. 1977, 52, 59.
35. Even, U.; Jortner, J.; Friedman, J. J. Phys. Chem. 1982, 86, 2273.
36. Birks, J. B.; Moore, G. F.; Munro, I. H. Spectrochim. Acta 1966, 22, 323.
37. Kobayashi, S.; Kikuchi, K.; Kokubun, H. Chem. Phys. 1978, 27, 399.
38. Madge, D.; Windsor, M. W.; Holten, D.; Gouterman, M. Chem. Phys. Lett. 1974, 29, 183.
39. Shepanski, J. F.; Anderson, R. W. Jr. Chem. Phys. Lett. 1981, 78, 165.
40. Bergkamp, M. A.; Dalton, J.; Netzel, Y. L.; J. Am. Chem. Soc. 1982, 104, 253.
41. Storm, C. B.; Teklu, Y. J. Am. Chem. Soc. 1972, 94, 1745.
42. Hennig, J.; Limbach, H. H. J. Am. Chem. Soc. 1984, 106, 292.
43. Limbach, H. H.; Hennig, J.; Stulz, J. J. Chem. Phys. 1983, 78, 5432.
44. Sevchenko, A. N.; Gurinovich, G. P.; Solovev, K. N. Soviet Phys. Doklady 1961, 5, 808.
45. Gouterman, M.; Stryer, L. J. Chem. Phys. 1962, 37, 2260.
46. Niwa, Y.; Kobayashi, H.; Tsuchiya, T. J. Chem. Phys. 1974, 60, 799.
47. Siebrand, W. J. Chem. Phys. 1967, 47, 2411.

RECEIVED April 3, 1986

RESONANCE RAMAN STUDIES

16

Axial Coordination in Nickel Porphyrins and Nickel-Reconstituted Heme Proteins Investigated by Raman-Difference and Transient-Raman Spectroscopy

J. A. Shelnutt[1], K. Alston[2], E. W. Findsen[3], M. R. Ondrias[3], and J. M. Rifkind[4]

[1]Sandia National Laboratories, Albuquerque, NM 87185
[2]Benedict College, Columbia, SC 29204
[3]Department of Chemistry, University of New Mexico, Albuquerque, NM 87131
[4]National Institute on Aging, National Institutes of Health, Baltimore, MD 21224

Nickel-porphyrin complexes and Ni(II)-reconstituted hemoglobin (^{Ni}Hb) and myoglobin (^{Ni}Mb) have been investigated using cw and transient resonance Raman spectroscopy. The state of axial coordination at the metal in these materials can be determined using the characteristic frequencies of the Raman marker lines arising from the porphyrin moiety. The existence of only one axial ligand in the proteins is supported by identification of the axial ligand-Ni stretching vibration by isotopic substitution. The frequency of the Ni-ligand mode is consistent with histidine as the fifth ligand. The effects of the T→R structural change in the protein environment of the Ni porphyrin are observed in the Raman spectrum. T-R structural changes at the metal are associated with changes in the electronic structure of the porphyrin ring. We have also investigated the photoinitiated ligand-release and ligand-uptake processes in excited states of the Ni porphyrins and Ni-reconstituted proteins. The results indicate that the protein matrix effectively controls photoinduced ligation dynamics. Laser excitation of ~10-ns duration induces the release of axial ligands in coordinating solvents, but not for the histidine ligand of Ni-reconstituted hemoglobin and myoglobin. Ligand uptake is observed in ^{Ni}Hb when the 4-coordinate sites are excited.

Nickel porphyrins are of interest because of their occurrance in coal, shale, and petroleum deposits (1) and because of their key role in biological conversion of CO_2 to methane (2-5). The enzyme methylreductase is the nickel-tetrapyrrole-containing enzyme that catalyses the final step, and possibly other steps, in the 6-electron reduction of CO_2 to methane (6). The active site of the methylreductase enzyme contains a nickel-sirochlorin derivative called F_{430} (6-8). Understanding the involvement of F_{430} in methane production is of importance to the natural gas industry for

0097-6156/86/0321-0232$06.00/0

utilizing abundant inorganic resources for production of gaseous fuels.

Recently, we have used resonance Raman spectroscopy to investigate axial ligand processes at the metal in Ni-porphyrin complexes and Ni-reconstituted heme proteins (9). These materials provide useful models for studies of axial ligation processes and for Ni-porphyrin-protein interactions that may be important in the function of the methylreductase enzyme. Also, for the first time, we have used transient Raman spectroscopic techniques to examine the dynamical aspects of axial ligation at Ni porphyrins in coordinating solvents (10) and in the Ni-reconstituted heme proteins (11). The results of transient Raman investigations of the Ni-porphyrins, which show photoinitiated axial ligand release and uptake, have important implications for fuel generating photocycles based on these processes (12).

Materials and Methods

^{Ni}Hb and ^{64}Ni-reconstituted Hb were prepared according to reported methods (13). ^{Ni}Mb was made by the method of Alston and Storm (14) Solutions of the proteins in 0.05 M phosphate buffer at pH 7.5 were used for obtaining spectra. The proteins are stable in air and do not photodecompose. Ni protoporphyrin IX (Ni(ProtoP)), Ni uroporphyrin I (Ni(UroP)), and Ni octaethylporphyrin (Ni(OEP)) were obtained from Porphyrin Products and used without further purification. All solvents were of highest purity obtainable from commercial sources. All materials showed the literature uv-visible absorption spectra. Absorption spectra were obtained on a Perkin-Elmer Model 330 spectrophotometer.

Raman spectra were obtained on selected pairs of samples using a Raman difference spectrometer described previously (15). The 90° scattering geometry was employed. Spectral resolution was 4 cm^{-1}. All spectra were obtained at room temperature. Samples were excited with a krypton ion laser (Coherent 3000K) using the 415.1-, 413.1-, or 406.7-nm laser lines. To avoid heating, the samples were irradiated in a rotating (100 Hz) cell with two compartments. The laser power at the sample was less than 300 mW in a partially defocused beam. The Ni porphyrins are very stable to visible light and even higher powers could be used. No sample decomposition was observed as determined by the lack of changes in successive scans of the Raman spectrum during signal averaging and in the absorption spectrum taken before and after the Raman spectrum.

Nanosecond resonance Raman spectra were obtained using a nitrogen pumped dye laser (Molectron) and a spectrometer system described elsewhere (16). The ~180° backscattering geometry was used. "Low power" laser pulses (~0.3 mJ/pulse) slightly focused with a cylindrical lens (f.l.= 250 mm) gave an energy density of ~ 5 mJ/cm^2. For "high power" spectra, the beam was tightly focused with a spherical lens (f.l.= 100 mm); the photon density is ~ 100 times greater than for the low power spectra. The pulse repetition rate was 10 hz. Static cuvettes held the sample and the temperature was between 20-25 C.

Results and Discussion

State of Axial Ligation in ^{Ni}Hb and ^{Ni}Mb. As we have previously reported, (17-20) axial ligation at the metal for non-hyper transition metal porphyrins (Zn^{2+}, $V{=}O^{2+}$, Ni^{2+}, Cu^{2+}) results in a distinctive pattern of shifts in the Raman electronic structure marker lines. The marker lines are the oxidation-state marker line (21-24) ν_4 and the lines sensitive to porphyrin core size (center-to-nitrogen pyrrole distance) (24,25). The pattern of shifts that indicates axial ligation is characterized by (1) roughly comparable decreases in frequency for all of the core-size marker lines (including ν_{10}, ν_2, and ν_3) and (2) relatively smaller, but correlated, shifts in the oxidation-state marker line ν_4.

Figure 1 shows the Raman spectrum of ^{Ni}Hb obtained with 406.7- and 413.1-nm excitation and the spectrum of monomeric, four-coordinate Ni protoporphyrin in aqueous micellar solution (9). Excitation at 413.1 nm is at resonance with the red component of the split Soret band of Ni-reconstituted hemoglobin; at 406.7 nm the blue component of the Soret band is selectively probed. Comparison of the spectra shows that two sets of marker line frequencies exist. One set (labeled 4 in Figure 1) is enhanced by resonance with the blue Soret component; the other set (labeled 5) is enhanced by excitation of the red Soret component. Thus, the shifts in the core-size lines in going from set 4 → 5 are -39 cm^{-1} (ν_{10} at 1657 cm^{-1}), -20 cm^{-1} (ν_2 at 1593 cm^{-1}), and -34 cm^{-1} (ν_3 at 1519 cm^{-1}). At the same time the shift in the oxidation-state marker line is only -11 cm^{-1} (ν_4 at 1378 cm^{-1}). This pattern of shifts is consistent with axial ligation for the form whose Raman lines are labeled 5 in Figure 1.

Comparison with axial ligand complexes formed by Ni porphyrins in coordinating solvents (e. g. pyrrolidine, piperidine, and pyridine), however, shows a clear distinction between the frequencies of the "form 5" marker lines of the reconstituted proteins and those of the 6-coordinate models (9). The shifts of the 6-coordinate species relative to the 4-coordinate species are larger than for the form labeled 5 (Figure 1) of the proteins. The shifts for the 6-coordinate models are about -41 cm^{-1} (ν_{10}), -27 cm^{-1} (ν_2), and -41 cm^{-1} (ν_3) for the core-size markers and -12 cm^{-1} (ν_4) for the oxidation-state marker. No coordinating solvents give 6-coordinate complexes with frequencies close to those of form 5 of the protein. However, because the pattern of shifts for form 5, relative to the 4-coordinate form, is the pattern associated with axial ligation and because the marker frequencies of form 5 of the protein are distinct from those of known 6-coordinate Ni porphyrins in coordinating solvents, we propose that form 5 of the proteins is a 5-coordinate species.

Ligation-induced Core Expansion. Based on the empirical relationships (given by Spiro (24)) between core size and Raman frequencies for the lines ν_3, ν_2, and ν_{10} and using the observed center-to-nitrogen distance of 2.038 Å for the Ni tetramethylpyridylporphine bis-imidazole complex (26), the predicted frequencies are: ν_3, 1485; ν_2, 1560; ν_{10}, 1615 cm^{-1}. These predicted values compare favorably with experimental frequencies of 1475-1479 (ν_3), 1560-1568 (ν_2), and 1615-1618 cm^{-1} (ν_{10}) for a variety of 6-coordinate Ni-protoporphyrin complexes (9).

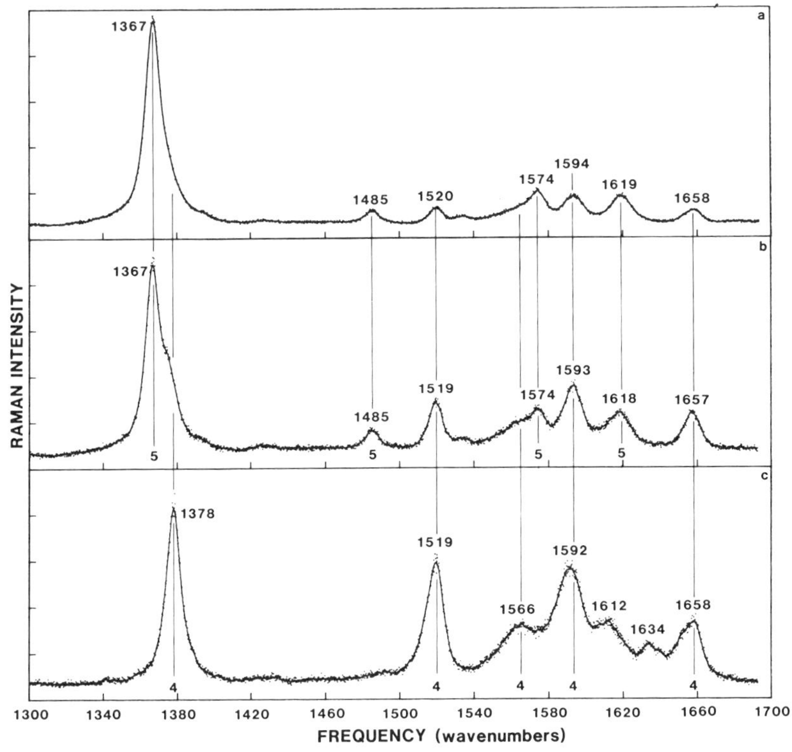

Figure 1. Raman spectra of nickel-reconstituted hemoglobin (pH 7.5) obtained with 413.1-nm excitation (a), with 406.7-nm excitation (b), and the spectrum of nickel protoporphyrin IX free acid in cetyltrimethylammonium bromide micelles in 0.1 M NaOH (c). Spectra in (b) and (c) were obtained simultaneously on a Raman difference spectrometer.

For the 5-coordinate species existing in the proteins a smaller core size of ~2.03 Å is predicted on the basis of the frequency of ν_3 and ν_{10}, but a still smaller core size ~2.00 Å is expected on the basis of ν_2. The discrepancy may result from distortion of the porphyrin macrocycle in the protein, nickel being out-of-plane, or from peripheral substituent orientational effects on ν_2.

Axial Ligand-Ni Stretching Mode. Isotopic substitution supports the identification of a 5-coordinate species in the proteins. Substitution of ^{64}Ni into ^{Ni}Hb results in a 7-cm^{-1} decrease in frequency of the line at 236 cm^{-1} (9). No other lines between 200 and 500 cm^{-1} shift by more than 1 cm^{-1}. A pure Ni-ligand mode would shift by 10 cm^{-1}. Thus, the mode at 236 cm^{-1} is about 70% pure.

The Ni-ligand mode can be identified as the Ni-axial ligand stretch because of the close analogy between the Ni isotopic substitution results and the Fe-isotope substitution data for native hemoglobin and myoglobin (27-30). For the native proteins more complete isotope substitution studies were done, including $^{14}N \rightarrow ^{15}N$ substitution of the pyrrole nitrogens, $^{57}Fe \rightarrow ^{54}Fe$ substitution, and partial deuteration of the protein (30). On the basis of this work and the direct correspondence between the Raman spectra of the Ni-reconstituted and native proteins, it is possible to assign the 236-cm^{-1} line to the Ni-histidine stretch. Histidine is indicated as the axial ligand by the frequency of the mode, which comes in the region of Fe-histidine stretching frequencies in heme proteins and model 2-methylimidazole-heme complexes (190-275 cm^{-1}) (27-31). The fact that ^{Ni}Hb has the normal T quaternary structure (9,13,32-36) also suggests that histidine is the axial ligand.

The low frequency Raman spectrum of the 6-coordinate models is distinctly different from the low frequency spectrum of the 5-coordinate sites of the Ni globins. In fact only one line, a very weak shoulder on the 300-cm^{-1} line, is observed in the expected metal-ligand stretching region below 300 cm^{-1}. Further, no predominately Ni stretches are found in the 100-500-cm^{-1} region. Instead, in both the 6- and 5-coordinate complexes many of the low frequency Raman lines show small (< ~1 cm^{-1}) ^{64}Ni isotopic shifts indicating some Ni-$N_{pyrrole}$ character.

Protein-Conformation Effects on the Metal-Protein Linkage. Since we have been unable to find model Ni-porphyrin complexes whose Raman frequencies come close to those of the Ni-reconstituted proteins, it is clear that the protein matrix has an impressive effect on axial ligation. We might also ask what is the effect of changes in protein conformation on axial ligation.

Native iron hemoglobin exists in a quaternary conformation which is distinguished by its low O_2 affinity. The low affinity protein conformation, called the T structure, has also been characterized by x-ray crystal structure (37,38) and many other physical techniques including resonance Raman spectroscopy (39,40). Specifically, the protein structure near the Fe porphyrin (heme) in the T structure is distinct from that of hemoglobins with bound exogenous ligands. These liganded hemoglobins have a different quaternary conformation called the R structure (38). Several deoxyhemoglobins and deoxymyoglobin also possess the R structure and they are characterized by high O_2 affinity. The Raman spectra of the R- and T-structure deoxyhemoglobins differ in the frequency of the Fe-histidine stretching vibration (39) and the π-electron density marker lines (40). Thus, a comparison of ^{Ni}Hb which has the T-conformation and ^{Ni}Mb which has an R-like structure can give some idea of the effect of a well defined protein conformational change at the Ni-porphyrin site.

^{Ni}Hb possesses both 4- and 5-coordinate forms as demonstrated by the Raman spectra (Figure 1) and the split Soret band of the absorption spectrum (9,36). In contrast, ^{Ni}Mb shows only the red Soret component and the Raman lines characteristic of the 5-coordinate form. Thus, myoglobin's R-like structure favors the 5-coordinate form. The R/T difference in affinity for histidine might also be expected to reveal itself in the strength of the Ni-histidine bond. In native Fe hemoglobin, the Fe-histidine bond increases in strength upon conversion from the T to R structure (31,39).

To determine the R/T difference in the Ni-histidine force constant we must locate the Ni-histidine mode in ^{Ni}Mb and compare the frequency with the 236-cm^{-1} value previously determined for ^{Ni}Hb (9). Figure 2 shows the Raman spectrum of natural abundance ^{Ni}Mb and ^{64}Ni-reconstituted Mb in the 100-500-cm^{-1} region. The spectra were obtained with 413.1-nm excitation. The Ni-histidine mode, which in ^{Ni}Mb is almost degenerate with a strong mode of the porphyrin ring at 243 cm^{-1}, cannot be resolved. However, the Raman difference spectrum indicates a decrease in intensity near 241 cm^{-1} and an increase near 233 cm^{-1} for the ^{64}Ni-reconstituted myoglobin relative to the natural abundance protein. For 415.1-nm excitation (spectrum not shown), the porphyrin mode is apparently weaker relative to the Ni-histidine mode. Even so, the Ni-histidine mode is not resolved in the spectrum of the natural abundance protein. However, the large isotopic shift in the ^{64}Ni protein reveals the mode as a clear low frequency shoulder at about 233 cm^{-1} in the 415.1-nm spectrum.

In addition, another weak line at ~270 cm^{-1} in natural abundance ^{Ni}Mb shifts to ~260 cm^{-1} for the ^{64}Ni protein (Figure 2). The weak 270-cm^{-1} line may represent a fraction of the protein with a different state of axial ligation. Alternatively, another almost pure Ni-porphyrin vibration is a possibility.

Using the 241-cm^{-1} value for the Ni-histidine frequency in ^{Ni}Mb, the T→R shift from the ^{Ni}Hb→^{Ni}Mb comparison is ~5 cm^{-1}. This value is consistent with the increase observed in a comparison of the corresponding Fe proteins and with the T→R shifts based on other Fe hemoglobins (3-8 cm^{-1}) (31). The similarity of the increases observed in the metal-histidine frequencies for the nickel and iron hemoglobins indicates that the effect of quaternary structure on the Ni-histidine bond is similar to the Fe case. Also, the effect of the protein conformation change is virturally independent of the particular metal in the porphyrin core.

Protein-Conformation Effects on Electronic Structure of the Porphyrin Ring. In spite of the similarity in the R/T changes at the metal-histidine bond for Ni and Fe hemoglobins, the effect of the R/T conformational change on the porphyrin's electronic structure in the two cases is quite different. In a wide variety of hemoglobins the T→R increase in ν(Fe-his) is associated with a

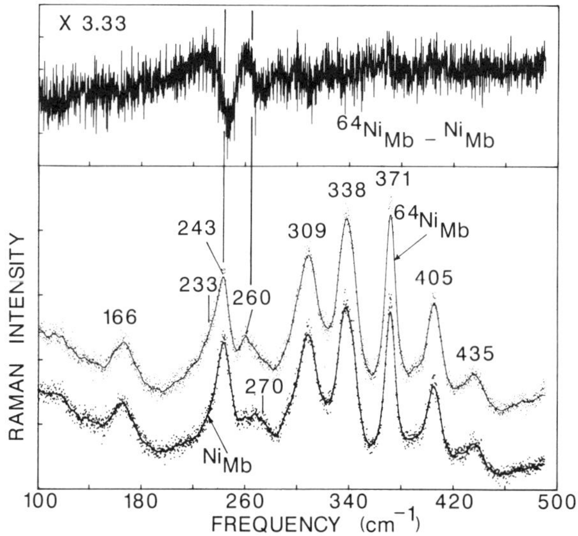

Figure 2. Raman spectra of ^{64}Ni-reconstituted myoglobin and natural abundance Ni-reconstituted myoglobin and the difference spectrum. The Raman difference spectrum shows clearly the isotope shift in the ~270-cm^{-1} line to ~260 cm^{-1} and the larger (intensity-wise) shift in the Ni-histidine mode at ~240 cm^{-1}.

decrease in ν_4. In contrast, ν_4 increases by 0.3 cm^{-1} for the Ni-reconstituted proteins (T→R) (9). The relationship between ν_4 and ν(Ni-his) is more like the linear correlation between these lines seen for peroxidases (41-43) than for the anti-correlation observed for Fe hemoglobins (31). The difference in the way the conformational change at the metal-histidine bond influences the porphyrin's electronic structure (i. e., the π-charge density, as measured by the oxidation-state marker line) for the Ni and Fe proteins, probably results from the differences in the way each metal interacts with the porphyrin ring. The differences in the metal-ring interaction is partly a result of the different electronic configurations of the metals (d^8 for Ni^{II}; d^6 for Fe^{II}) and the mixing of $e_g(d_\pi)$ metal orbitals with the $e_g(\pi^*)$ porphyrin orbitals in the case of Fe (44).

Excited State Axial Ligation Processes in Coordinating Solvents. When nickel porphyrins are irradiated with ~10-ns laser pulses in the 400-500-nm region, changes in axial coordination can be detected by changes in the Raman marker lines. The time evolution of axial ligand complexes of Ni porphyrins has been fully characterized by time-resolved absorption spectroscopy by Holten and coworkers (45-47). Recently, we have obtained (10) the first Raman spectra of the transient Ni-porphyrin species observed by Holten. The Raman spectra potentially contain considerably more structural information than the absorption spectra. On the other hand, because of the weakness of the Raman scattering it is more difficult to follow the time evolution of the transient species. The difficulty results from the neccessity of using an intense probe beam to excite the Raman spectrum as well as an intense pump pulse that generates the transient species at zero time. Unlike the weak probe pulse used in time-resolved absorption spectroscopy, the intense probe pulse needed to excite the Raman spectrum may interfere with the evolving transient species. Nevertheless, one can usually identify the spectrum of the transient species superimposed on the spectrum of the starting material by comparing the spectra obtained with high and low power pump/probe pulses (Figures 3 and 4).

Excitation wavelength dependence of the Raman lines presents another complication that is not a problem for transient absorption measurements. For example, in comparing spectra taken at two different excitation wavelengths, one must consider not only which and in what proportion each molecular species is pumped, but also keep in mind what differences in relative intensity of the Raman lines result from excitation at the two exciting wavelengths. The complication can usually be overcome, especially when the spectra of the two species are well resolved.

We have previously investigated ligand release in the 6-coordinate piperidine complexes of Ni octaethylporphyrin (10). For Ni(OEP), formation of the 6-coordinate complex is not complete, and so, the picture is complicated by the presence of both 4- and 6-coordinate species in the initial sample. However, upon excitation with the pulsed laser the relative proportions of the two Ni(OEP) species change as determined by changes in the relative intensities of the 4- and 6-coordinate sets of Raman marker lines. Thus, axial ligand release is observed in the excited state generated during the

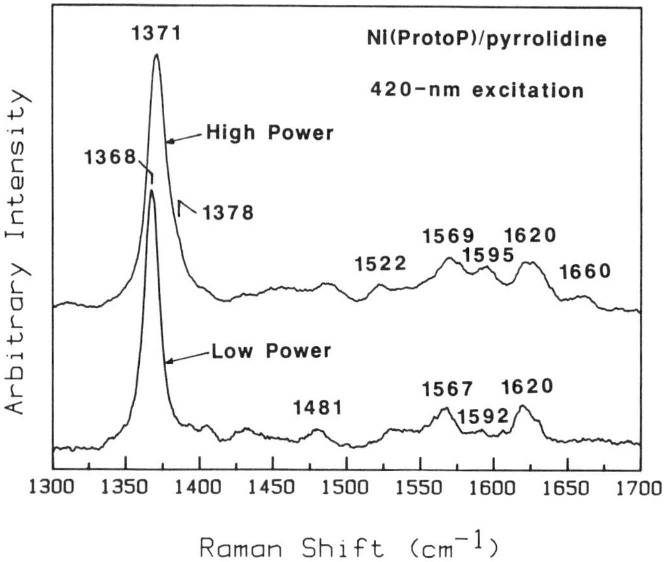

Figure 3. Transient Raman spectrum of nickel protoporphyrin IX in piperidine obtained with 10-ns laser pulses at 420 nm. (a) high power density; (b) low power density.

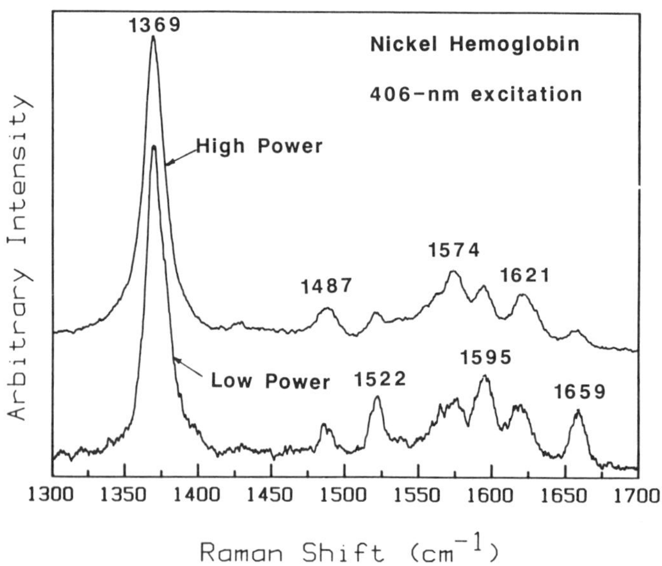

Figure 4. Transient Raman spectrum of nickel-reconstituted hemoglobin (pH 7.5) obtained with 406-nm, 10-ns laser pulses. (a) high power density; (b) low power density.

laser pulse. We also observe other excited state effects in the Raman spectra that, as yet, are not completely understood (10).

Figure 3 shows the 10-ns Raman spectra of Ni protoporphyrin in neat piperidine (>90% 6-coordinate form). The laser excitation wavelength of 420 nm preferentially excites the 6-coordinate form which happens to dominate the absorption at this wavelength (9-11). With respect to the low power spectrum, the spectrum obtained at high power shows increased intensity in the lines at 1660 (ν_{10}), 1595 (ν_2), and 1522 cm^{-1} (ν_3). These lines come from the 4-coordinate form. Also, ν_4 of the 6-coordinate form develops a high frequency shoulder near 1378 cm^{-1}, the expected frequency of the 4-coordinate species. The results clearly show that the piperidine ligands are lost in the excited state. Thus, we find that the Ni(ProtoP) case gives results similar to Ni(OEP) in piperidine.

Photoinduced ligand dissociation can be understood in terms of the ligand affinity properties of the excited states of the metalloporphyrin (47). Ni porphyrins in coordinating solvents display a ground-state equilibrium between the ligated $^3B_{1g}$ state with one electron each in the metal $d_{x^2-y^2}$ and d_{z^2} orbitals and the unligated $^1A_{1g}$ state with both electrons in d_{z^2} and none in $d_{x^2-y^2}$. (In the d^8 configuration of Ni(II) all other d orbitals are completely filled.) For the 6-coordinate species, absorption of light by the $\pi \rightarrow \pi^*$ transitions of the porphyrin ring generates the dissociative $^1A_{1g}$ d-d excited state (45-47). The ground state of the liganded species is the (d_{z^2}, $d_{x^2-y^2}$) $^3B_{1g}$ state; the lowest excited state $^1A_{1g}$ is rapidly (≤ 15 ps) reached by nonradiative decay. Because the filled d_{z^2} cannot act as a σ-acceptor orbital for the axial ligands, the ligands are lost until the excited state decays and ligands can be recovered. Rebinding of the ligands proceeds over an activation barrier and requires greater than 20 ns (45).

Photoinduced ligand uptake also occurs (45-47). Unligated Ni porphyrins in the equilibrium mixture of 4- and 6-coordinate species may be induced to bind ligands upon absorption of a photon. The 4-coordinate Ni-porphyrin species are in the $^1A_{1g}$ state initially. Upon photon absorption and decay of the $\pi \rightarrow \pi^*$ states the Ni porphyrin ends up in the lowest excited state which in this case is the unligated $^3B_{1g}$. The excited $^3B_{1g}$ state is associative; therefore, either ligands are acquired or the $^3B_{1g}$ state decays back to the unligated ground state.

We have observed the unligated $^3B_{1g}$ excited state in noncoordinating solvents (11,45-47), but not in ones that coordinate. Using transient absorption spectroscopy, Holten and coworkers found that the lifetime of the $^3B_{1g}$ state for Ni porphyrins in toluene is ~250 ps. During our 10-ns pulse, a sufficient concentration of the $^3B_{1g}$ excited state can be generated to obtain the Raman spectrum. In particular, we have observed the unligated $^3B_{1g}$ state for Ni(OEP) in toluene, Ni(ProtoPDME) in toluene and acetone, and Ni(ProtoP) in micelles (11). The $^3B_{1g}$ excited state is characterized by shifts in ν_4, ν_3, and ν_{10} to frequencies that are intermediate between the 4- and 5-coordinate values. For example, ν_{10} of the unligated $^3B_{1g}$ excited state of Ni(ProtoP) and Ni(ProtoPDME) is in the range 1625-1632 cm^{-1} for various solvent systems. For comparison, 5-coordinate Ni(ProtoP) is in the 1618-1621-cm^{-1} range, while 4-coordinate ν_{10} frequencies are in the 1652-1660-cm^{-1} range. Thus, about three-fourths of the core expansion that occurs on formation of the 6-coordinate complex is due to the d-d transition; the remainder of the expansion results from adding the axial ligands and the ligands' influence on the metal $4p_z$-$a_{2u}(\pi)$ interaction (19).

Ligation-Deligation Processes in Nickel-Reconstituted Heme Proteins. In contrast with the 6-coordinate models, 5-coordinate sites of ^{Ni}Hb and ^{Ni}Mb show virtually no difference between the Raman spectra obtained with low and high powers for excitation at 420 nm. The undetectably low quantum yield for dissociation for the nickel globins suggests that either (1) the dissociative $^1A_{1g}$ state is not reached (with significant yield) after absorptioon by the $\pi \rightarrow \pi^*$ transition or (2) the histidine ligand is regained rapidly on the nanosecond time scale of the Raman measurement because of cage (geminate) recombination. Rapid geminate recombination (i. e. much faster than the 10-ns probe pulse) would insure that the steady state concentration of 4-coordinate species does not increase for high pulse power.

For the 6-coordinate complexes we observe excited state ligand loss because the ligands are lost rapidly (~50 ps) and regained slowly with respect to our 10-ns pulse. Thus, during our pulse few deligated molecules regain ligands. Apparently, for the Ni globins the barrier to recombination is drastically lowered by the protein matrix and recombination of the ligand is much faster than in the coordinating solvents.

The absence of ligand release in the nickel globins is probably not the result of a change in the decay pathway. The same transitions are expected in the region between the $\pi \rightarrow \pi^*$ ring transitions for 5- and 6-coordinate Ni porphyrins (45,48); therefore, the low yield of photodissociation is not a result of additional decay pathways for the 5-coordinate species. Second, we also do not expect that the low lying d $\rightarrow$ d transition moves above the $\pi \rightarrow \pi^*$ states for the 5-coordinate form (48). Consequently, there are no alternative or new branching pathways for decay of $\pi \rightarrow \pi^*$ excited 5-coordinate Ni porphyrins and the d $\rightarrow$ d dissociative state remains the primary decay route. Thus, the most reasonable explanation for the lack of differences in the high and low power spectra of the Ni-reconstituted globins when the 5-coordinate sites are preferentially excited is based on rapid geminate recombination rather than faster or altered decay mechanisms.

The nickel proteins and the native iron globins show the same behavior in regard to fifth-ligand photodissociation but for totally different reasons. Heme proteins and other metal-reconstituted heme proteins have been investigated by transient absorption spectroscopy (49-53) and transient Raman spectroscopy (16,54-62). In none of the Fe proteins is loss of the histidine ligand observed even on a picosecond timescale. Soret excitation of carbonmonoxy and oxy Hb and Mb photolyzes CO or O_2, but not the histidine fifth ligand.

Although neither the Ni nor the Fe globins show evidence of photodissociation of the histidine ligand, the model compound data would, nevertheless, lead us to expect ligand ejection for Ni but not Fe. Time-resolved absorption (45-47) and transient Raman measurements (10,11) for 6-coordinate Ni-porphyrin complexes clearly predict that photodissociation of the fifth ligand will occur, since the transient species is identified as the 4-coordinate form. On the other hand, the transient species identified in the time-resolved absorption spectra of Fe-porphyrin 6-coordinate complexes is a 5-coordinate form (63-65). For example, Fe^{II}-porphyrin(1-methylimidazole)$_2$ complexes release only one of the imidazole ligands in the excited state (63-65). The lack of photodissociation of the fifth ligand for the Fe-porphyrin complexes results from the nature of the transient excited state generated. The excited state has only one electron in the d_{z^2} orbital not two as for Ni; the state is therefore less dissociative. Thus, on the basis of the results for the model complexes release of the histidine ligand is not predicted for the Fe globins (regardless of whether cage recombination occurs or not); ligand release is expected for the Ni-reconstituted proteins but not observed.

Copper-reconstituted cytochrome c (^{Cu}cyt-c) has also been investigated with transient absorption methods (49). No evidence of photoinduced ejection of the fifth ligand is observed in either ^{Cu}cyt-c or the model Cu-porphyrin 5-coordinate complexes (60-62). This is consistent with the likely transient state being a non-dissociative $\pi \rightarrow \pi^*$ or $a_{2u}(\pi) \rightarrow d_{x^2-y^2}$ charge-transfer state. Instead, excitation of 4-coordinate Cu-porphyrins in coordinating solvents results in weak binding of a fifth, but not a sixth, axial ligand. Ejection of the ligand acquired in the excited state occurs during or following deactivation through the charge-transfer state (62).

The lack of photodissociation in the Ni-reconstituted proteins argues for an impressive influence of the protein matrix on geminate recombination of the histidine ligand. The rebinding of the ligand is even much faster than in neat coordinating solvent where a ligand is always available for uptake. The lack of significant photodissociation may indicate the proteins maintain the histidine's orientation in the "bound" position even after photolysis has occurred.

Holten and coworkers have also shown that 4-coordinate Ni-porphyrin species existing in coordinating solvents are induced to pick up ligands by laser excitation (45-47). This ligand-uptake process can also be understood in terms of the d-electron configuration of the metal. For the Ni-porphyrin molecules that have no axial ligands, the $(d_{z^2}{}^2)$ $^1A_{1g}$ state is the ground state. Absorption of a photon by the $\pi \rightarrow \pi^*$ transition from the $^1A_{1g}$ results in a d → d transition moving one electron to the $d_{x^2-y^2}$ orbital. The resulting half empty d_{z^2} orbital then is free to accept charge from σ-donor ligands.

The photoinduced uptake of ligands is also observed in the case of ^{Ni}Hb. Unlike ^{Ni}Mb, ^{Ni}Hb is a mixture of 4- and 5-coordinate forms. As shown in Figure 4, excitation of the 4-coordinate sites in ^{Ni}Hb (excitation in the blue component of the Soret band) results in a decrease in the set of marker lines corresponding to the 4-coordinate form relative to the 5-coordinate form. Since we observe the same photoinduced ligation process in ^{Ni}Hb as observed for the model complexes it is reasonable to assume that the excited state levels of the 4-coordinate Ni-protoporphyrin system are not significantly altered by its protein environment. This further suggests that an alteration of decay pathways is not responsible for the deligation behavior of the 5-coordinate species but rather rapid geminate recombination. Most importantly, the rapid uptake of a fifth ligand on the 10-ns time scale suggests that the 4-coordinate Ni-porphyrin molecules are located in the usual heme pocket where a fifth ligand is readily accessible.

Summary

Resonance Raman scattering provides a valuable method of determining the state of axial ligation in nickel-reconstituted heme proteins and Ni-porphyrin complexes. A pattern of shifts in the Raman core-size and oxidation-state marker lines can be used to monitor changes in axial coordination. The shifts in the core-size lines (e.g. ν_{10}) indicate an expansion of the core from about 1.96 Å for the 4-coordinate Ni porphyrin to 2.04 Å for the 6-coordinate species, ~2.03 Å for the 5-coordinate species, and 2.01 Å for the unligated $^3B_{1g}$ transient excited state. In 5-coordinate complexes the axial ligand-Ni stretching mode has been identified. The mode can also be used to follow coordination changes. Five-coordinate Ni-porphyrin complexes have been observed, so far, only in the Ni-reconstituted proteins.

Transient Raman spectra in the 10-ns time regime indicate changes in ligation state for the Ni-porphyrins in aggreement with previous time-resolved absorption spectroscopic data. The 6-coordinate Ni-porphyrin complexes that exist in coordinating solvents are induced to release ligands upon photoexcitation of the dissociative $d_{x^2-y^2} \rightarrow d_{z^2}$ state. Excitation of 4-coordinate species that exist in coordinating solvents induces ligand binding via the associative $d_{z^2} \rightarrow d_{x^2-y^2}$ transition.

Although the same transition is expected for the 5-coordinate site in the Ni globins as for 6-coordinate models, ligand release is not observed, presumably, due to rapid geminate recombination in the protein. However, excitation of the 4-coordinate site of Ni hemoglobin does result in transient acquisition of a fifth axial ligand--most likely the proximal histidine.

Other excited-state effects besides coordination changes are observed in the transient Raman spectra (10,11). Further analysis of the excited states and dynamics of Ni-porphyrin complexes and Ni-reconstituted heme proteins should benefit from Raman spectroscopy's inherently rich structural information content. Transient Raman methods are now being applied to other metalloporphyrins and metalloporphyrin-based systems.

Acknowledgments

We thank Tomoko Yamamoto of the National Institutes of Health and Nai-Teng Yu of the Georgia Institute of Technology for helpful discussions. This work performed at Sandia National Laboratories and supported by the United States Department of Energy Contract DE-AC04-76DP-00789 and the Gas Research Institute Contract 5082-260-0767. Work at the University of New Mexico supported by the National Institutes of Health Grant GM33330, the donors of the Petroleum Research Fund administered by the American Chemical Society, and the Cottrell research fund of Research Corporation. EWF was supported by the graduate fellowship program of the Associated Western Universities.

Literature Cited

1. Baker, E. W.; Palmer, S. E. In "The Porphyrins"; Dolphin, D., Ed.; Academic: New York, 1979; Vol. 1, Part A, Chpt. 11.
2. Gunsalus, R. P.; Wolfe, R. S. FEMS Microbiol. Lett. 1979, 3, 191.
3. Ellefson, W. L.; Wolfe, R. S. In "Microbial Growth on C1 Compounds"; Heyden: London, 1981, p. 171.
4. Jaenchen, R.; Diekert, G.; Thauer, R. K. FEBS Lett. 1981, 130, 133.
5. Pfaltz, A.; Jaun, B.; Fassler, A.; Eschenmoser, A.; Jaechen, R.; Gilles, H-H.; Diekert, G.; Thauer, R. K. Helv. Chim. Acta 1982, 65, 825.
6. Keltjens, J. T.; Caerteling, C. G.; Van Kooten, A. M.; Van Dijk, H. F.; Vogels, G. D. Arch. Biochem. Biophys. 1983, 223, 235.
7. Ellefson, W. L.; Wolfe, R. S. J. Biol. Chem. 1981, 256, 4259.
8. Keltjens, J. T.; Whitman, W. B.; Caerteling, C. G.; van Kooten, A. M.; Wolfe, R. S.; Vogels, G. D. Biochem. Biophys. Res. Commun. 1982, 108, 595.
9. Shelnutt, J. A.; Alston, K.; Ho, J-Y.; Yu, N-T.; Yamamoto, T.; Rifkind, J. M. Biochemistry 1986, 25, 620.
10. Findsen, E. W.; Shelnutt, J. A.; Friedman, J. M.; Ondrias, M. R. Chem. Phys. Lett. 1986, 126, 465.
11. Findsen, E. W.; Shelnutt, J. A.; Alston, K.; Ondrias, M. R. J. Am. Chem. Soc. 1986, 108, 4009.
12. Wayland, B. B.; Del Rossi, K. J. J. Organomet. Chem. 1984, 276, C27.
13. Alston, K.; Schechter, A. N.; Arcoleo, J. P.; Greer, J.; Parr, G. R.; Friedman, F. K. Hemoglobin 1984, 8, 47.
14. Alston, K.; Storm, C. B. Biochemistry 1979, 18, 4292.
15. Shelnutt, J. A. J. Phys. Chem. 1983, 87, 605.
16. Findsen, E. W.; Ondrias, M. R. J. Am. Chem. Soc. 1984, 106, 5736.
17. Shelnutt, J. A. J. Am. Chem. Soc. 1983, 105, 774.
18. Shelnutt, J. A.; Dobry, M. M. J. Phys. Chem. 1983, 87, 3012.
19. Shelnutt, J. A.; Ondrias, M. R. Inorg. Chem. 1984, 23, 1175.
20. Shelnutt, J. A.; Straub, K. D.; Rentzepis, P. M.; Gouterman, M.; Davidson, E. R. Biochemistry 1984, 23, 3946.
21. Yamamoto, T.; Palmer, G.; Gill, D.; Salmeen, I. T.; Rimai, L. J. Biol. Chem. 1973, 248, 5211.
22. Spiro, T. G.; Strekas, T. C. J. Am. Chem. Soc. 1974, 96, 338.
23. Spiro, T. G.; Burke, J. M. J. Am. Chem. Soc. 1976, 98, 5482.
24. Spiro, T. G. In "Iron Porphyrins"; Lever, A. B. P.; Gray, H. B., Eds.; Addison-Wesley: Reading, 1982, Part II, p. 89.
25. Spaulding, L. D.; Chang, C. C.; Yu, N-T.; Felton, R. H. J. Am. Chem. Soc. 1975, 97, 2517.
26. Kirner, J. F.; Garofalo, Jr., J.; Scheidt, W. R. Inorg. Nucl. Chem. Lett. 1975, 11, 107.
27. Kitagawa, T.; Nagai, K.; Tsubaki, M. FEBS Lett. 1979, 104, 376.
28. Kincaid, J.; Stein, P.; Spiro, T. G. Proc. Natl. Acad. Sci. USA 1980, 76, 4156.
29. Nagai, K.; Kitagawa, T.; Morimoto, H. J. Mol. Biol. 1980, 136, 271.

30. Argade, P. V.; Sassaroli, M.; Rousseau, D. L.; Inubushi, T.; Ikeda-Saito, M.; Lapidot, A. J. Am. Chem. Soc. 1984, 106, 6593.
31. Ondrias, M. R.; Rousseau, D. L.; Shelnutt, J. A.; Simon, S. R. Biochemistry 1982, 21, 3428.
32. Alston, K.; Dean, A.; Schechter, A. N. Mol. Immunol. 1980, 17, 1475.
33. Alston, K.; Friedman, F. K.; Schecter, A. N. Hemoglobin 1982, 6, 15.
34. Alston, K.; Park, C. M.; Rodgers, D. W.; Edelstein, S. J.; Nagel, R. L. Blood, 1985, Vol. 64, Part II, 556.
35. Manoharan, P. T.; Alston, K.; Rifkind, J. M. Bull. Magn. Res. 1983, 5, 255.
36. Manoharan, P. T.; Alston, K.; Rifkind, J. M. Proc. Copper Coordination Conf. 1986, Albany, NY, in press.
37. Perutz, M. Nature(London) 1970, 228, 726.
38. Ladner, R. C.; Heidner, E. J.; Perutz, M. J. Mol. Biol. 1977, 114, 385.
39. Nagai, K.; Kitagawa, T. Proc. Natl. Acad. Sci. USA 1980, 77, 2033.
40. Shelnutt, J. A.; Rousseau, D. L.; Friedman, J. M.; Simon, S. R. Proc. Natl. Acad. Sci. USA 1979, 76, 4409.
41. Shelnutt, J. A.; Alden, R. G.; Ondrias, M. R. J. Biol. Chem. 1986, 261, 1720.
42. Shelnutt, J. A.; Satterlee, J. D.; Erman, J. E. J. Biol. Chem. 1983, 256, 2168.
43. Stump, R. F.; Deanin, G. G.; Oliver, J. M.; Shelnutt, J. A. Biophys. J. 1986, submitted.
44. Adar, F. In "The Porphyrins"; Dolphin, D., Ed.; Academic: New York, 1978, Vol. 3, Part A, Chpt. 2.
45. Kim, D.; Kirmaier, C.; Holten, D. Chem. Phys. 1983, 75, 305.
46. Kim. D.; Holten, D. Chem. Phys. Lett. 1983, 98, 584.
47. Holten, D.; Gouterman, M. Proc. Symp. Optical Properties and Structure of Tetrapyrroles 1984, Konstanz, Aug. 12-17.
48. Ake, R. L.; Gouterman, M. Theoret. Chim. Acta 1970, 17, 408.
49. Reynolds, A. H.; Straub, K. D.; Rentzepis, P. M. Biophys. J. 1982, 40, 27.
50. Hoffman, B. M.; Gibson, Q. H. Proc. Natl. Acad. Sci. USA 1978, 75, 21.
51. Greene, B. I.; Hochstrasser, R. M.; Weisman, R. B.; Eaton, W. A. Proc. Natl. Acad. Sci. USA 1978, 75, 5255.
52. Nagumo, M.; Nicol, M.; El-Sayed, M. J. Phys. Chem. 1981, 85, 2435.
53. Terner, J.; Voss, D. F.; Paddock, C.; Miles, R. B.; Spiro, T. G. J. Phys. Chem. 1982, 86, 859.
54. Friedman, J. M.; Rousseau, D. L.; Ondrias, M. R. Ann. Rev. Biophys. Bioeng. 1982, 11, 105.
55. Friedman, J. M., Ondrias, M. R., Rousseau, D. L. Science 1982, 218, 1244.
56. Scott, T. W., Friedman, J. M. J. Am. Chem. Soc. 1984, 106, 5677.
57. Ondrias, M. R., Friedman, J. M., Rousseau, D. L. Science 1983, 220, 614.
58. Findsen, E. W., Friedman, J. M., Scott, T. W., Chance, M. R., Ondrias, M. R. J. Am. Chem. Soc. 1985, 107, 3355.

59. Findsen, E. W., Friedman, J. M., Ondrias, M. R., Simon, S. R. Science 1985, 229, 661.
60. Kobayashi, T.; Huppert, D.; Straub, K. D.; Rentzepis, P. M. Photochem. Photobiol. 1979, 70, 1720.
61. Straub, K. D.; Rentzepis, P. M. Biophys. J. 1983, 41, 411a.
62. Kim, D.; Holten, D.; Gouterman, M. J. Am. Chem. Soc. 1984, 106, 2793.
63. Momenteau, M.; Lavalette, D. J. Am. Chem. Soc. 1978, 100, 4322.
64. Lavalette, D.; Tetreau, C.; Momenteau, M. J. Am. Chem. Soc. 1979, 101, 5395.
65. Dixon, D. W.; Kirmaier, C.; Holten, D. J. Am. Chem. Soc. 1985, 107, 808.

RECEIVED July 29, 1986

17

Resonance Raman Studies of Porphyrin Radical Cations, Excited States, and Ligation Photodynamics

Dongho Kim[1], Lisa Miller[2], Oliver Su[2], James Terner[3], and Thomas G. Spiro[2]

[1]Korea Standards Research Institute, Taedok Science Town, Taejon, Cungnam, Korea
[2]Department of Chemistry, Princeton University, Princeton, NJ 08544
[3]Department of Chemistry, Virginia Commonwealth University, Richmond, VA 23284

With a recirculating electrochemical cell it has been possible to obtain good quality resonance Raman spectra of the radical cations of several metallo-octaethyl porphyrins, with both B- and Q-band excitation. Characteristic frequency shifts are seen upon radical formation, whose patterns clearly distinguish a_{1u} from a_{2u} radicals. The directions of the shifts, however, do not accord with simple bonding arguments based on orbital symmetry. The first porphyrin triplet state RR spectra are reported, for zinc, magnesium, and palladium tetraphenylporphines, using a simple cw laser technique involving a comparison of spinning and stationary (photo-pumped) samples. The frequency shifts upon triplet formation are very small for the porphyrin skeletal modes. There is evidence, however, for extensive broadening of a phenyl mode, possibly reflecting a heterogeneous distribution of phenyl rotamers in the excited state. The same stationary sample cw laser technique has been used to monitor excited state deligation of nickel tetraphenylporphine (NiTPP) in pyridine. Saturation behavior is observed, from which the quantum yield ratio of photo-induced ligation of NiTPP to photo-deligation of $NiTPP(py)_2$ is estimated to be 0.05. NiTPP in piperidine gives evidence for the formation of a 5- as well as a 4-coordinate species via photodeligation of the initial 6-coordinate adduct. At equilibrium very little 5-coordinate adduct is formed at any piperidine concentration, but for nickel tetrakis(p-cyanophenyl)porphine a significant equilibrium fraction is present as 5-coordinate adduct in 1$\underline{M}$ piperidine. The ν_4 RR frequencies support a high-spin configuration for the 5-coordinate adduct, consistent with previous theoretical calculations.

Resonance Raman (RR) spectroscopy has been applied extensively to metalloporphyrins and heme proteins, and has provided useful information about molecular and electronic structure (1). The power of the technique which lies in the sensitivity of molecular vibrational

0097-6156/86/0321-0248$06.00/0

frequencies to changes in structure, and the selective enhancement of a limited number of vibrational modes via coupling with the electronic transition with which the laser excitation is in resonance. RR spectroscopy is applicable as well to excited states of porphyrins and to primary electron transfer products, provided that appropriate methods are used for the generation and sampling of these unstable species. The technique can provide structural information about these high energy molecules and can be applied as a monitor to probe their involvement in reacting systems.

We have examined RR spectra of a series of radical cations generated from metallo-octaethylporphyrins (OEP's). OEP's resemble physiological porphyrins in having aliphatic substituents at the pyrrole positions and hydrogen atoms at the methine positions. Their radical cations show an interesting dichotomy resulting from the near degeneracy of the highest filled a_{1u} and a_{2u} orbitals (2). Depending on the central metal and its ligands, the hole can reside on one or the other of these orbitals, giving rise to markedly different EPR (3) and absorption spectra (4). We find that the RR frequency shift patterns are similarly distinctive for these two classes of radical cations, but that the directions of the shifts are not what one would expect from simple orbital symmetry and bonding arguments (5).

Metalloporphyrin excited states have an extremely wide range of lifetimes. When d-orbital vacancies on a central metal lie at energies lower than the π-π^* states, the lifetimes become exceedingly short, and fast pulsed methods are needed for their detection. Downshifts and broadening in the picosecond RR spectrum of the oxy-hemoglobin photoproduct have been interpreted as arising from unrelaxed electronic excitation of the deoxy-heme (6). When d orbital vacancies are unavailable, however, the lifetime of the lowest triplet π-π^* state, formed by efficient intersystem crossing from the initially excited singlet levels, can be quite long (microseconds to milliseconds). We have found that, in favorable cases, RR spectra of metalloporphyrin triplet states can be obtained with cw laser excitation by the simple expedient of exciting stationary samples; the ground state RR spectrum, obtained by spinning the same sample, serves as a reference (7). Frequency shifts upon triplet formation for zinc, magnesium and palladium TPP's are found to be small, but there is evidence for a heterogeneous distribution of rotamers of the phenyl ring, consistent with increased phenyl conjugation in the excited states.

The same stationary sample cw technique has been used to study photodeligation dynamics for NiTPP in pyridine (8). The RR spectra show that the equilibrium between 6- and 4-coordinate species is shifted increasingly toward the latter with increasing laser power; the effect collapses when the sample is spun. The data indicate a much lower quantum yield for ligation than for deligation via the photoexcited species. For NiTPP in piperidine there is RR evidence that photodeligation produces a 5- as well as 4-coordinate product. The equilibrium fraction of a 5-coordinate adduct is very small at any piperidine concentration; but when para-cyano groups are substituted on the TPP phenyl rings thereby increasing the axial ligand affinity, the 5-coordinate fraction becomes readily detectable at 1M piperidine (9). The RR frequencies point to a high-spin configuration for the 5-coordinate adduct, consistent with theoretical expectations based on the calculations of Ake and Gouterman (10).

Experimental

Metalloporphyrins were purchased from Mid-Century Chemicals and purified by chromatography on alumina (A540) to remove fluorescent impurities. Radical cations were prepared *in situ* via controlled potential electrolysis in dichloromethane with tetrabutylammonium perchlorate supporting electrolyte. A standard photoelectrochemical cell was modified with connections for viton tubing connected to a glass capillary serving as a Raman cell, through which the solution was flowed via a peristaltic pump. The RR spectra were recorded at 90° from the capillary. The extent of oxidation was monitored by coulometry and UV-vis absorption spectroscopy. The triplet state and photodeligation studies were carried out with a standard NMR spinning cell in backscattering.

Raman spectra were recorded with a computer-controlled Spex 1401 double monochromator equipped with a cooled photomultiplier and photon counting electronics. Excitation was provided by Spectra Physics Kr^+ and Ar^+ cw lasers.

Results and Discussion

OEP Radical Cations. Figure 1 shows RR spectra, with 406.7nm excitation, of ZnOEP during its gradual conversion to radical cation via electrolysis (5). ZnOEP fluoresces from its lowest excited singlet state and RR spectra with excitation in the Q-band are obscured by the emission, but the background at 406.7nm is negligible. This wavelength is nearly coincident with the ZnOEP B absorption band (λ_{max}=404nm). For ZnOEP$^{+\cdot}$ this band is at 391nm and is only 0.6 times as strong; consequently the enhancement is considerably lower, as evidenced by the higher noise level of the spectrum. Nevertheless, the relative intensities of the bands are similar for the radical and the neutral porphyrin, and the frequencies are readily correlated. Large (19-24cm^{-1}) up- and downshifts are seen for ν_4 and ν_2, at 1373 and 1580cm^{-1} in ZnOEP (the porphyrin skeletal mode assignments are from reference (11), while smaller downshifts are seen for ν_3 (1483cm^{-1}) and ν_{10} (1614cm^{-1}); ν_{11}, at 1557cm^{-1}, becomes hidden by the downshifted ν_2 band of the radical cation.

The EPR spectrum of ZnOEP$^{+\cdot}$ shows it to have a hole in the a_{1u} orbital, while that of NiOEP$^{+\cdot}$ is assignable to an a_{2u} radical (3). Figure 2 compares the B-band excited spectra of NiOEP and NiOEP$^{+\cdot}$. Large shifts are again observed for the dominant bands, ν_4 (1382cm^{-1}) and ν_2 (1600cm^{-1}), but now they are in the opposite directions, down and up, respectively, to those shown by ZnOEP$^{+\cdot}$. Smaller downshifts are again seen for ν_3 (1519cm^{-1}) and ν_{10} (1655cm^{-1}). For NiOEP and its radical cation it was possible to obtain Q-band excited spectra, at 514.5 and 568.2nm, allowing identification and correlation of the non-totally symmetric A_{2g}, B_{1g} and B_{2g} modes (1) in the same frequency region. Their shifts were found to be quite small. RR spectra were obtained for several other OEP radical cations for which EPR and/or optical spectra had indicated an a_{1u} or a_{2u} assignment: (3,4) MgOEP$^{+\cdot}$ (a_{1u}), CuOEP$^{+\cdot}$ (a_{2u}), Co^{III}OEP^{++}$(Br^-)_2$ (a_{1u}) and Co^{III}OEP^{++}$(ClO_4^-)_2$ (a_{2u}). The frequencies and shifts on radical formation are given in Table I.

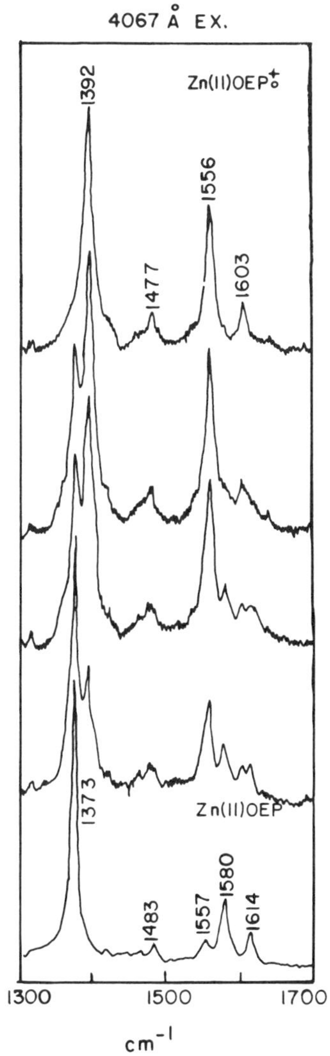

Figure 1. RR spectra with 406.7nm excitation for ZnOEP (0.5m$\underline{M}$, bottom) in $CH_2C\ell_2$ (50mW incident laser power) and for solutions with increasing amounts of monocation radical formed by electrolysis at 0.85V vs. SCE (15mW). Each scan was obtained with a fresh sample (Reproduced from Ref. 5. Copyright 1986 American Chemical Society.)

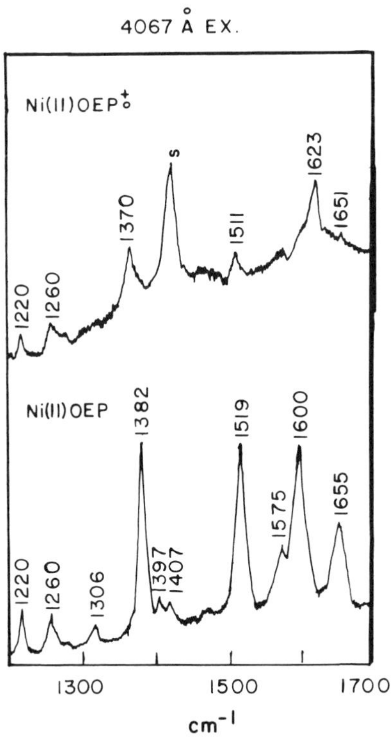

Figure 2. RR spectra with 406.7nm excitation for NiOEP (0.5m$\underline{M}$, bottom) in $CH_2C\ell_2$ (50mW) and for monocation radical formed by electrolysis at 1.25V vs. SCE (20mW).

TABLE I

RADICAL CATION VS NEUTRAL OEP FREQUENCY SHIFTS (cm^{-1})

a_{1u} Radicals

Assignment[a]	$[Co^{III}OEP]Br$	$\Delta^{+b}_{\bullet}$	$Zn^{II}OEP$	$\Delta^{+}_{\bullet}$	$Mg^{II}OEP$	$\Delta^{+}_{\bullet}$
$\nu_{10}(B_{1g})$ C_aC_m[c]	1641	- 4	1614	-11	1600	- 4
ν_2 (A_{1g}) C_bC_b	1592	-38	1580	-24	1571	-19
$\nu_{19}(A_{2g})$ C_aC_m	1595	- 2				
$\nu_{11}(B_{1g})$ C_bC_b	1572	- 4				
ν_3 (A_{1g}) C_aC_m	1510	- 3	1483	- 6		
ν_4 (A_{1g}) C_aN	1376	+14	1373	+19	1368	+22
$\nu_{21}(A_{2g})$ C_mH	1311	- 4				
$\nu_5+\nu_9(A_{1g})$ $\delta(C_mH)$	1256	+ 1				
ν_{13} (B_{1g})	1214	+ 4				

a_{2u} Radicals

Assignment[a]	$[Co^{III}OEP]ClO_4$	$\Delta^{+b}_{\bullet}$	NiOEP	$\Delta^{+}_{\bullet}$	CuOEP	$\Delta^{+}_{\bullet}$
$\nu_{10}(B_{1g})$ C_aC_m	1640	- 3	1655	- 4	1635	+ 2
ν_2 (A_{1g}) C_bC_b	1593	+20	1600	+23	1588	+23
$\nu_{19}(A_{2g})$ C_aC_m	1596	- 8	1609	- 5	1581	- 4
$\nu_{11}(B_{1g})$ C_bC_b	1572	- 1	1579	- 6	1570	- 6
ν_3 (A_{1g}) C_aC_m			1519	- 8	1500	- 5
ν_4 (A_{1g}) C_aN	1379	-22	1382	-12	1377	-15
$\nu_{21}(A_{2g})$ C_mH	1313	+ 5	1306	- 1	1313	+ 2
$\nu_5+\nu_9(A_{1g})$ $\delta(C_mH)$	1257	- 1	1260	0	1253	0
ν_{13} (B_{1g})	1214	0	1220	0	1212	0

[a] From ref. 11.

[b] $\Delta^{+}_{\bullet}$ frequency shift upon cation radical formation.

[c] Major bond stretching contributor to the potential energy distribution, as calculated in ref. 11b.

A clear and striking pattern emerges from these data. a_{1u} radicals consistently show large up and downshifts for ν_4 and ν_2 while a_{2u} radicals show large shifts in the opposite directions. Since these two bands are the dominant features of spectra obtained in resonance with the B absorption band, their shifts produce a clear indication of radical type, even when the spectra are relatively weak or poorly resolved. Thus, RR spectroscopy appears to be an apt tool for monitoring radical formation in biological-type porphyrins.

The observed shifts do not, however, accord with simple bonding and orbital symmetry arguments. Figure 3 is a diagram of the atomic coefficients and signs calculated for the a_{1u} and a_{2u} orbitals (2,12). It can be seen that the interaction between the C_b atoms of the pyrrole rings is bonding for the a_{2u} orbital and antibonding for the a_{1u} orbital, so that removal of an electron should decrease and increase the C_bC_b bonding in the a_{2u} and a_{1u} radicals, respectively. Although the internal coordinate contributions are mixed in the porphyrin normal modes, the primary contributor to ν_2 is C_bC_b stretching (11b). Yet, the observed ν_2 shifts are up for a_{2u} and down for a_{1u} radicals, opposite to the expected shifts. The primary contributor to ν_4 is C_a-N stretching (11b), and the observed shifts would lead us to expect that the C_a-N bond strength decreases for a_{2u} radicals and increases for a_{1u} radicals. Yet the interaction based on the orbital coefficients is antibonding in the a_{2u} orbital and non-bonding in the a_{1u} orbital. The discrepancies are not limited to metallo OEP's. Itoh and coworkers (13) have reported RR spectra for radical cations of metallo-TPP's, all of which form a_{2u} radicals, and have noted that the observed downshift of ν_4 is contrary to the orbital symmetry argument. In the case of TPP's, there appears to be no band which shifts up on radical formation, analogous to ν_2 of the OEP a_{2u} radicals.

Resolution of these discrepancies will require an actual calculation of the vibrational frequency shifts based on the estimated force constant changes upon radical formation. It is possible that both the normal mode analysis and electronic calculations will require modification to accommodate the data. An additional factor of importance is the possibility of a structural change in the porphyrin skeleton upon radical formation. The crystal structures of $ZnTPP^{\dot{+}}(C\ell O_4^-)$ (14) and $MgTPP^{\dot{+}}(C\ell O_4^-)$ (15) have revealed a distinct distortion of the porphyrin skeleton involving the ruffling of the pyrrole rings, relative to the more planar structure of the neutral porphyrin. Since both of the reported structures involve a_{2u} radicals, it is not clear whether the a_{1u} radicals undergo a similar or a different distortion.

Porphyrin Triplets. Figure 4 shows RR spectra of ZnTPP in dichloromethane, obtained with 476.5nm excitation (7). The bottom spectrum was obtained from the sample in an ordinary spinning cell and is the RR spectrum of ground state ZnTPP; frequencies and assignments are listed in Table II. The middle spectrum was obtained in the same cell, but without spinning, and with increased laser power (250mW). There are clear broadenings and shifts in the ZnTPP RR bands, and the solvent band at $1421cm^{-1}$ is greatly diminished in relative intensity. These effects are attributable to conversion of ZnTPP to its triplet state. Because the quantum yield for triplet formation is high (0.8)

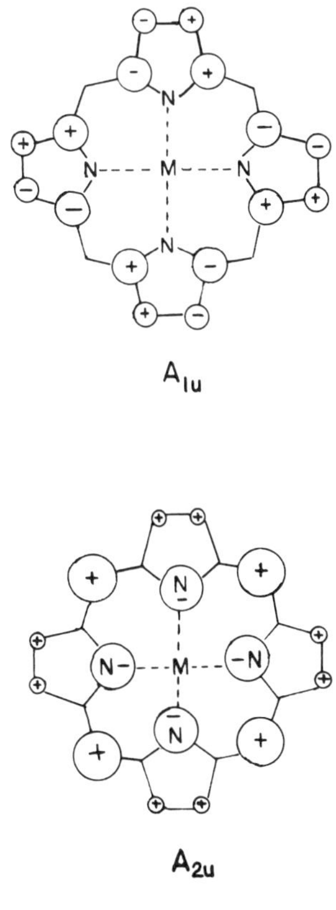

Figure 3. MO diagram for the a_{1u} and a_{2u} orbitals of porphine. The atomic orbital coefficients are proportional to the size of the circles; solid or dashed circles indicate sign.

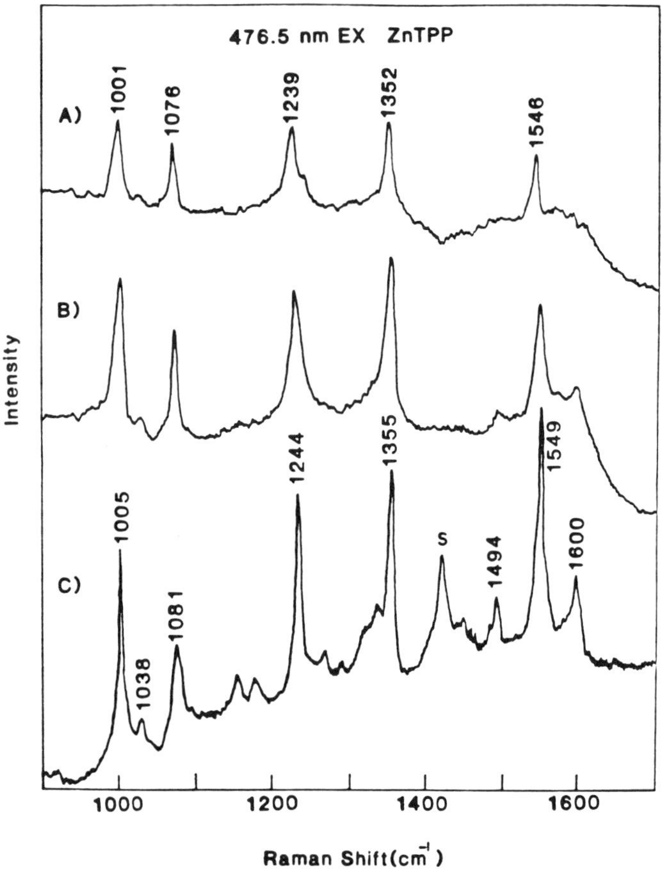

Figure 4. (A) Difference spectrum, B-C, attributed to ZnTPP*. (B) Stationary sample of ZnTPP (0.2m$\underline{M}$) in $CH_2C\ell_2$ in a NMR tube, excited at 135° with 476.5nm Ar^+ laser beam (250mW, focussed); the sample was cooled in a cold N_2 stream. The spectrum is the sum of five scans, each of a fresh sample from which O_2 was removed by freeze-pumping. (C) As in B, but the NMR tube was spun, and the laser power was lowered to 30mW; seven scans were added. (Reproduced from Ref. 7. Copyright 1986 American Chemical Society.)

TABLE II

RR Frequencies and Shifts (cm^{-1}) for MTPP Triplets and Radical Cations and Anions

Assignment	ZnTPP	Δ^{*a}	Δ^{+b}	Δ^{-c}	MgTPP	Δ^{*a}	Δ^{+b}	Δ^{-c}	PdTPP	Δ^{*}
A[d] phenyl[e]	1600	- 2	- 2	- 1	1602		- 2		1595	- 2
ν_2 C_bC_b	1549	- 3	- 7	+ 3	1550	- 4	-11	+ 5	1557	- 7
ν_{12} C_bC_b	1494	- 4	-19	+ 2	1496	- 3	- 9	+ 3		
ν_4 C_aN	1355	- 3	-12	- 6	1353	0	-20	+ 4	1364	- 1
C C_mPh	1244	- 5	0	- 6	1239	- 3	0	- 4	1235	- 1
ν_5 δ C_bH	1081	- 5	+ 8	-15	1080	- 5	-22	- 9	1080	- 3
E phenyl	1038				1033					
F phenyl	1005	- 4	0	- 2	1005	- 2	0	- 1	1013	0
ν_6 C_aC_m	1005	- 4	-22	-18	1005		-20	-13		

a Δ^i Frequency shift relative to MTPP for $MTPP^+$, $MTPP^-$, $MTPP^*$ ($i=+,-,^*$)

b From ref. 13a.

c From ref. 13b.

d Mode labels from ref. 18a.

e Principal internal coordinate contributor to the mode, from ref. 18b.

and the triplet lifetime is long (1.3ns) (16), pumping is effective even with the relatively low power cw laser excitation. Moreover, the excitation wavelength is in resonance with the strong triplet-triplet absorption (470nm, $\varepsilon \sim 71000M^{-1}$) (17). This absorption is responsible for the disappearance of the solvent band. The Soret band (416nm) of the ground state is at higher energy but is close enough to provide both resonance enhancement of the ground state and sufficient absorption ($\varepsilon \sim 4600$ at 476.5nm) for pumping. The top spectrum of Figure 4 is that of the triplet state, obtained by subtracting the bottom spectrum (ground state) from the middle spectrum (mixture); the weighting factor was gradually increased to the point where negative peaks began to be seen due to oversubtraction. Similar spectra were observed for MgTPP and PdTPP, with excitation at 488.0 and 457.8nm, respectively, for resonance with the triplet absorption bands, 485nm ($\varepsilon \sim 72000$) for MgTPP and 450nm ($\varepsilon \sim 47000$) for PdTPP (17). Triplet quantum yields (0.8 and 1.0 and lifetimes (1.3 and 0.4ns) are likewise suitably large for MgTPP and PdTPP (17).

The porphyrin skeletal mode frequencies are all slightly lower both for the triplet and for the ground state. Table II lists the triplet downshifts, 1-7cm^{-1}, along with the shifts observed for radical cations (13a) and anions (13b) for MgTPP and ZnTPP reported by Itoh and coworkers. Since the triplet excited states are formed by promoting an electron from the highest filled orbital to the lowest unoccupied orbital, the structural effects might be expected to be a superposition of those observed for radical cation and radical anion formation. Indeed, the opposing vibrational shifts which tend to be seen for electron addition and removal (Table II) help to explain the very small shifts seen for the triplet states.

An interesting feature of the triplet ZnTPP spectra (Figure 4) is the apparent broadening out of the ground state band at 1600cm^{-1}, assigned to a localized vibration of the phenyl subsituents (18). The enhancement of phenyl modes has been suggested (18a) to reflect an increase in phenyl conjugation in the excited state; conjugation is not significant in the ground state (19), since the phenyl orientation is nearly at right angles to the porphyrin ring (20). We suggest that the broadening of the 1600cm^{-1} band reflects an inhomogeneous distribution of phenyl orientations in the excited state. Such a distribution has a plausible physical basis, since the driving force for phenyl rotation into the porphyrin plane provided by the electronic excitation (the e_g orbital has particularly large coefficients at the meso carbon atoms (2)) encounters steric resistance from the non-bonded interactions between the protons at the ortho positions of the phenyl groups and those on the outer pyrrole carbon atoms (20). Consequently the phenyl torsion potential in the excited states may be relatively flat. Nevertheless, the vibrational frequencies are expected to be sensitive to the torsion angle for orientation close to co-planar because of the effect of conjugation.

Photodeligation of NiTPP. The same stationary versus spinning sample cw laser experiment was used to study photo-effects on nickel porphyrin ligation (8). Nickel porphyrins are 4-coordinate and low-spin in weakly-coordinating solvents, but 6-coordinate and high-spin in strongly coordinating solvents (21). Holten and coworkers (22) have shown with transient optical spectroscopy that the accessible excited state (low-spin) of the 6-coordinate species rapidly loses its axial ligands, while the excited state of the 4-coordinate nickel porphyrin (high-spin) becomes ligated, thanks to the hole created in the d_{z^2} orbital. It should therefore be possible to shift the equilibrium between 4- and 6-coordinate species with photons. Figure 5 shows precisely this effect for NiTPP in pyridine. Excitation of a stationary sample with 406.7nm laser radiation shifts the equilibrium from the 6- to the 4-coordinate species, as shown by the decrease and growth of the associated ν_4 RR bands (18a) at 1346 and 1369cm^{-1} respectively. The intensity ratio is plotted against laser power in Figure 6, where it is seen to increase monotonically from a starting value of 0.54 to a limiting value of $\sim$1.6. When the sample was spun through the laser, the spectral changes shown in Figure 5 were largely abolished, and the ratio of the bands was nearly independent of laser power (Figure 6).

The solid line in Figure 6 represents the equation

$$\frac{I_4}{I_6} = \frac{j_4(k_6 + I_o\sigma_6\phi_6)}{j_6(k_4 + I_o\sigma_4\phi_4)}$$

Where I_i, j_i, and σ_i are the observed Raman intensity, the molar scattering factor and the molar absorptivity of NiTPP (i=4) and $(py)_2$NiTPP (i=6), ϕ_4 and ϕ_6 are the quantum yields for converting $4 \rightarrow 6$ or $6 \rightarrow 4$ via light absorption, k_4 and k_6 are the thermal conversion rates, and I_o is the incident laser flux. This equation derives from a simple photokinetic scheme in which photoexcited NiTPP and $(py)_2$NiTPP are interconverted via capture or loss of ligands, or else return to their own ground state. From the limiting values of I_4/I_6 one can calculate $k_6\sigma_4\phi_4/k_4\sigma_6\phi_6 = 1.6/0.54 = 3.0$. From the optical absorption spectra (and the assumption that the $(py)_2$NiTPP spectrum is given by that of NiTPP in piperidine, for which $(pip)_2$NiTPP predominates) one can estimate $\sigma_4/\sigma_6 \simeq 90$ (at 406.7nm) and k_4/k_6 (the (4)/(6) equilibrium ratio) $\simeq 1.5$, whence $\phi_4/\phi_6 \simeq 0.05$. A low ratio of quantum yields is reasonable since ligation of the photoexcited NiTPP is a bimolecular process, while ligand loss from photoexcited $(py)_2$NiTPP is unimolecular.

Figure 7 shows laser induced changes in RR spectra of NiTPP in piperidine; again these changes are abolished by spinning the sample. The major species is the 6-coordinate complex, with ν_4=1346cm^{-1}. Another band is seen at 1356cm^{-1}, which increases with increasing laser power. In addition a band grows in at 1369cm^{-1}, the frequency of the 4-coordinate species. Thus photo-pumping of the deligation process is seen in piperidine as well as pyridine. The new species responsible for the 1356cm^{-1} band is suggested to be a 5-coordinate complex, with a single piperidine ligand.

The concentration of the 5-coordinate complex must be elevated in the photostationary state, since it is negligible at equilibrium; both the optical absorption (21) and RR spectra (9) display clean isosbestic points when NiTPP is titrated with piperidine in dichloromethane, showing that two piperidine ligands are added with no significant contribution from a mono-piperidine adduct. When, however, NiTPP is replaced with Ni[TPP(CN)$_4$] (TPP(CN)$_4$=tetrakis(p-cyanophenyl) porphine) the optical absorption spectra no longer display a clean isosbestic point, and the RR spectra (Figure 8) show a strong band

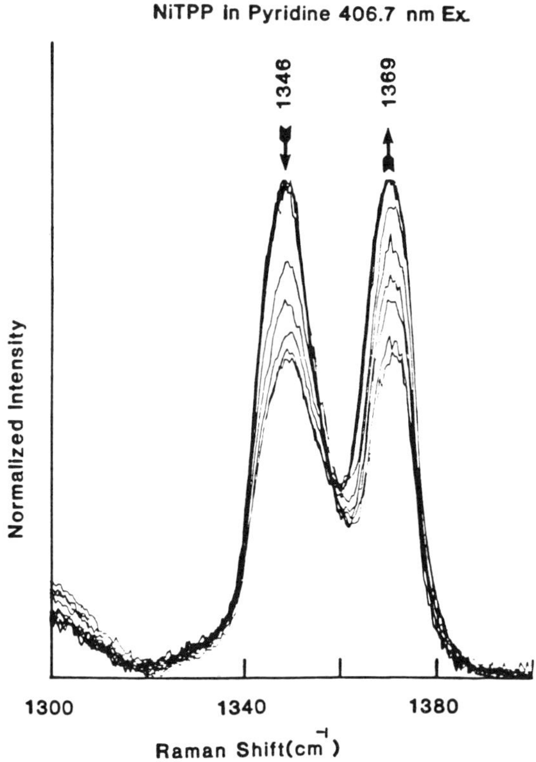

Figure 5. ν_4 Raman peaks for a stationary sample of NiTPP (0.2m$\underline{M}$) in pyridine, with 406.7nm excitation. Arrows indicate the decreasing ($1346 cm^{-1}$) and increasing ($1369 cm^{-1}$) relative band intensities at increasing laser power levels (1.5, 5, 10, 15, 20, 25, 50, 75, 90,125, and 140mW). In the absence of an internal standard (pyridine peaks are obscured by those of NiTPP) the spectra were scaled to whichever peak was stronger. (The intrinsic scattering factors are not the same for the two peaks.) Spectra were obtained via backscattering (135°) from a NMR sample tube, using a Spex 1401 double monochromator: $7 cm^{-1}$ slitwidth, 1sec time constant, $0.5 cm^{-1}$/sec scan rate. (Reproduced from Ref. 8. Copyright 1986 American Chemical Society.)

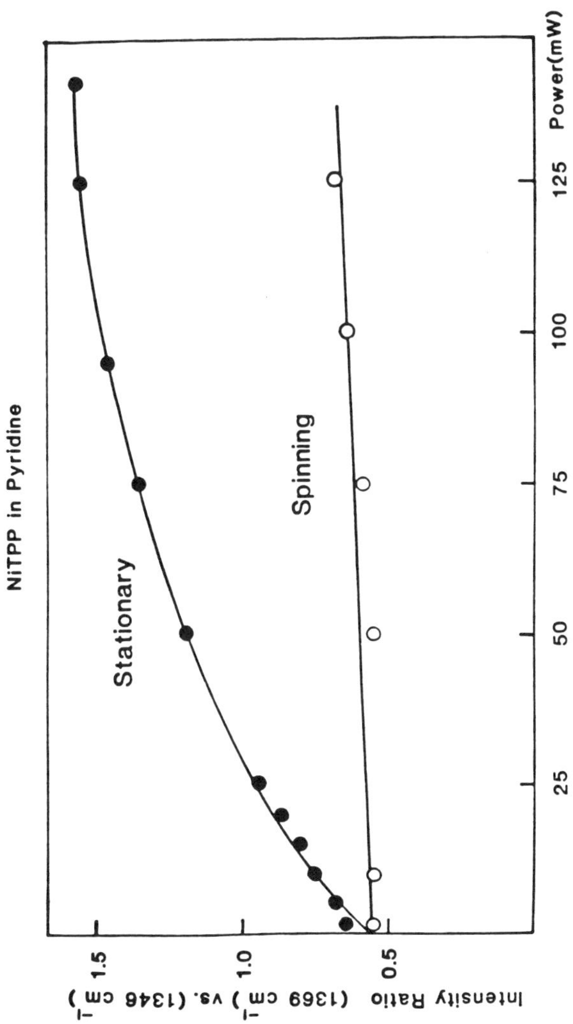

Figure 6. The 1369cm^{-1}/1346cm^{-1} Raman peak height ratio for NiTPP in pyridine vs. 406.7nm laser power for a stationary sample (●) (see spectra in Figure 1) and a spinning sample (O). (Reproduced from Ref. 8. Copyright 1986 American Chemical Society.)

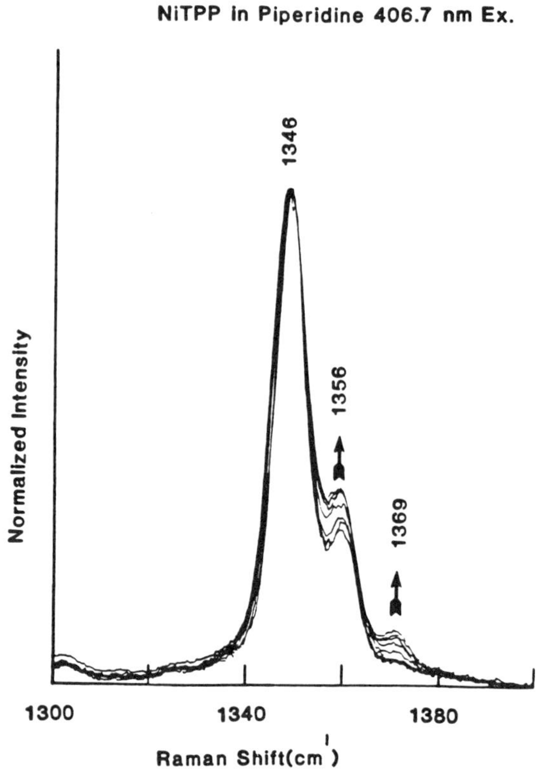

Figure 7. ν_4 Raman peaks for a stationary sample of NiTPP (0.5m$\underline{M}$) in piperidine, with 406.7nm excitation. Arrows indicate the increasing relative intensities of the 1356 and 1369cm^{-1} peaks with increasing laser power (5, 15, 25, 50, 75, 100, 125, and 150mW): 5cm^{-1} slitwidth, 1sec time constant, 0.5cm^{-1}/sec scan rate. The spectra were scaled to the 1346cm^{-1} peak. (Reproduced from Ref. 8. Copyright 1986 American Chemical Society.)

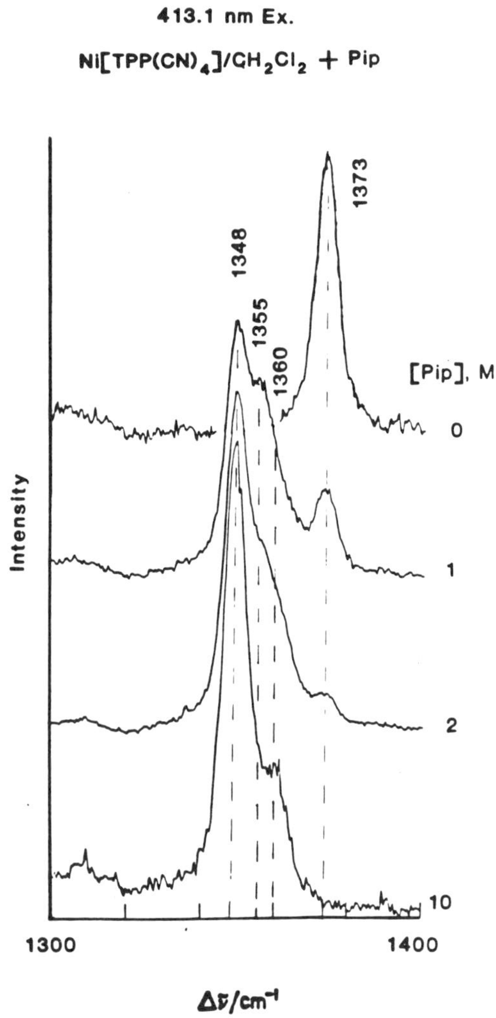

Figure 8. ν_4 region of the RR spectrum of Ni[TPP(CN)$_4$] (0.5m$\underline{M}$, spinning sample) in $CH_2C\ell_2$, with piperidine added to a concentration of 1 and 2$\underline{M}$, and in pure piperidine (10$\underline{M}$).

at $1355cm^{-1}$, in addition to the 1348 and $1373cm^{-1}$ ν_4 bands of the 6- and 4-coordinate species (9). The $1355cm^{-1}$ band reaches a maximum at 1M piperidine, and decreases again at higher piperidine concentrations, leaving a weaker band at $1360cm^{-1}$. The latter band, which is depolarized, is assigned to mode ν_{28} (B_{2g}) (18a) of the 6-coordinate complex. The superimposed $1355cm^{-1}$ band seen in 1M piperidine, which is polarized, is assigned to ν_4 of the 5-coordinate adduct. The frequency of this band is much closer to that of the 6-coordinate than of the 4-coordinate complex, as expected if the 5- as well as the 6-coordinate species has a high-spin configuration, consistent with the calculations of Ake and Gouterman (10). A large decrease in the ν_4 frequency upon adding piperidine ligands is believed to be associated with an expansion of porphyrin core due to promotion of an electron to the in-plane antibonding $d_{x^2-y^2}$ orbital in the high-spin configuration; porphyrin skeletal mode frequencies are known to be sensitive to the core-size, as well as to the electronic effects (23). The porphyrin center-to-pyrrole nitrogen distance (C_t-N) is ∿ 1.95Å in 4-coordinate Ni porphyrins (24), but 2.038Å in the bis-imidazole adduct of Ni-tetrakis(methylpyridyl)porphine (29). No structure is available for a 5-coordinate Ni porphyrin, but the $7cm^{-1}$ increase in ν_4 relative to that of the 6-coordinate $(pip)_2$NiTPP suggests partial relaxation of the core size due to the out-of-plane displacement of the Ni atom expected for a 5-coordinate high-spin adduct, by analogy with the relaxation seen for (2-MeImH)FeTPP (2.045Å C_t-N; Fe 0.5Å out-of-plane) (30) relative to high-spin $(THF)_2$FeTPP (2.057Å C_t-N, Fe in-plane) (31).

Acknowledgments

This work was supported by grant DE-AC02-81ER10861 from the U.S. Department of Energy. We thank Professor Dewey Holten for helpful discussions.

Literature Cited

1. Spiro, T. G. "Iron Porphyrins"; Lever, A. B. P., and Gray, H. B., Eds.; Addison-Wesley: Reading, MA, 1983, Part Two, pp. 85-159.
2. Gouterman, M. In "The Porphyrins"; Dolphin, D., Ed.; Academic Press, New York, 1978, Vol. III, Chapter 1.
3. Fajer, J.; Davis, M. S. In "The Porphyrins"; Dolphin, D., Ed.; Academic Press, New York, 1978, Vol. IV, pp. 197-256.

4a. Dolphin, D.; Muljiani, Z.; Rousseau, K.; Borg, D. C.; Fajer, J.; Felton, R. H. Ann. N.Y. Acad. Sci. 1973, 206, 177.

4b. Fuhrhop, J.-H.; Mauzerall, D. J. Am. Chem. Soc. 1969, 91, 4174.

5. Kim, D. H.; Miller, L. A.; Rakhit, G.; Spiro, T. G. J. Phys. Chem. 1986 (in press).
6. Terner, J.; Voss, D. S.; Paddock, C.; Miles, R. B.; Spiro, T. G. J. Phys. Chem. 1982, 86, 859.

7. Kim, D.; Terner, J.; Spiro, T. G. J. Am. Chem. Soc. 1986 (in press).
8. Kim, D.; Spiro, T. G. J. Am. Chem. Soc. 1986 (in press).
9. Kim, D.; Su, Y. O.,; Spiro, T. G. (submitted).
10. Ake, R. L.; Gouterman, M. Theoret. Chim. Acta 1970, 17, 408.
11a. Kitagawa, T.; Abe, M.; Ogoshi, H. J. Chem. Phys. 1978, 69, 4516.
11b. Abe, M.; Kitagawa, T.; Kyogoku, Y. J. Chem. Phys. 1978, 69, 4526.
12. Maggiora, G. M. J. Am. Chem. Soc. 1973, 95, 982.
13a. Yamaguchi, H.; Nakano, M.; Itoh, K. Chem. Lett. 1982, 1397.
13b. Yamaguchi, H.; Soeta, A.; Toeda, H.; Itoh, K. J. Electroanal. Chem. 1983, 159, 347.
14. Spaulding, L. D.; Eller, P. G.; Bertrand, J. A.; Felton, R. H. J. Am. Chem. Soc. 1974, 96, 982.
15. Barkigia, K. M.; Spaulding, L. D.; Fajer, J. Inorg. Chem. 1983, 22, 349.
16. Darwent, J. R.; Douglas, P.; Harriman, A.; Porter, G.; Richoux, M.-C., Coord. Chem. Rev. 1982, 44, 83.
17. Harriman, A. J. Chem. Soc., Faraday Trans. 2 1981, 97, 3660.
18a. Stein, P.; Ulman, A.; Spiro, T. G. J. Phys. Chem. 1984, 88, 369-374.
18b. Burke, J. M.; Kincaid, J. R.; Spiro, T. G. J. Am. Chem. Soc. 1978, 100, 6077.
19. LaMar, G. N.; Eaton, G. R.; Holm, R. H.; Walker, F. A. J. Am. Chem. Soc. 1973, 95, 63.
20. Eaton, S. S.; Eaton, G. R. J. Am. Chem. Soc. 1975, 97, 3660.
21. Walker, F. A.; Hui, E.; Walker, J. M. J. Am. Chem. Soc. 1975, 97, 2390.
22a. Kim, D.; Kirmaier, C.; Holten, D. Chem. Phys. 1983, 75, 305.
22b. Kim, D.; Holten, D. Chem. Phys. Lett. 1983, 98, 584.
23. Choi, S.; Spiro, T. G.; Langry, K. C.; Smith, K. M.; Budd, D. L.; LaMar, G. N. J. Am. Chem. Soc. 1982, 104, 4345.
24. 1.929Å in tetragonal (D_{2d}, ruffled) (25) and 1.958Å in triclinic (D_{4h}, planar) (26) Ni octaethylporphyrin; 1.960Å in Ni deuteroporphyrin (27), 1.957Å in Ni etioporphyrin (28).
25. Meyer, E. F., Jr. Acta Crystallogr. Sect. B. 1972, 28, 2162.
26. Gullen, D. L.; Meyer, E. F., Jr. J. Am. Chem. Soc. 1974, 96, 2095.
27. Hamor, T. A.; Caughey, W. S.; Hoard, J. L. J. Am. Chem. Soc. 1965, 87, 2305.
28. Fleischer, E. B. J. Am. Chem. Soc. 1963, 85, 146.
29. Kirner, J. F.; Garofalo, J., Jr.; Scheidt, W.R. Inorg. Nucl. Chem. Lett. 1975, 11, 107.
30. Hoard, J.L.; Scheidt, W. R. Proc. Natl. Acad. Sci., USA, 1973, 70, 3913; 1974, 71, 1578.
31. Reed, C. A.; Mashiko, T.; Scheidt, W. R.; Spartalian, K.; Lang, G. J. Am. Chem. Soc. 1980, 102, 2302.

RECEIVED April 21, 1986

18

Resonance Raman Investigation of Transient Photoinduced Ligation Changes in Nickel Porphyrin

E. W. Findsen[1], M. R. Ondrias[1], J. A. Shelnutt[2], and J. M. Friedman[3]

[1]Department of Chemistry, University of New Mexico, Albuquerque, NM 87131
[2]Sandia National Laboratories, Albuquerque, NM 87185
[3]AT&T Bell Laboratories, Murray Hill, NJ 07974

The first picosecond and nanosecond time-resolved Raman spectra of photoexcited metalloporphyrins in coordinating and noncoordinating solvents are presented. Our data demonstrate that resonance Raman spectra can be obtained from transient nickel porphyrin species and confirm the existence of photoinduced ligation changes in nickel proto and octaethyl porphyrin species on a subnanosecond timescale.

The metalloporphyrins form a diverse class of molecules exhibiting complex and varied photochemistries. Until recently time-resolved absorption and fluorescence spectroscopies were the only methods used to study metalloporphyrin excited state relaxation in a sub-microsecond regime. In this paper we present the first picosecond time-resolved resonance Raman spectra of excited state metalloporphyrins outside of a protein matrix. The inherent molecular specificity of resonance Raman scattering provides for a direct probe of bond strengths, geometries, and ligation states of photoexcited metalloporphyrins.

In the present investigation we use transient Raman spectroscopic techniques to study the picosecond and nanosecond ligation processes initiated by direct electronic excitation of nickel octaethyl porphyrin (Ni(OEP)) and nickel protoporphyrin IX (Ni(PP)). Similar time-resolved resonance Raman studies have successfully probed the dynamic processes involved in the photolysis and subsequent recombination of exogenous ligands in heme protein systems such as hemoglobin and myoglobin (1,2). Transient species of Ni(OEP) and Ni(PP) formed by photoexcitation in coordinating and noncoordinating solvents have been probed with ~25-psec and ~10-nsec time resolution and the resulting spectra are compared to those of equilibrium species. Recently the excited state kinetics in each of these time regimes was characterized by several studies using transient absorption spectroscopy (3-5). Our results are in general agreement with the molecular mechanisms inferred from these previous studies, but reveal well resolved vibrational information

0097-6156/86/0321-0266$06.00/0

about the excited electronic states and ligation behavior of the transient intermediates created after photoexcitation.

Materials and Methods

All Ni(OEP) and Ni(PP) samples (obtained from Porphyrin Products) were prepared (0.2-0.5 mM) in neat, spectral grade solvent (used without further purification) and were deoxygenated by purging with oxygen-free N_2 gas. No differences in the spectra were noted for aerobic and anaerobic samples.

The picosecond spectrum was generated using a Raman shift cell (H_2, 300 psi) which was pumped with the second Harmonic output of a Quantel active/passive modelocked picosecond laser. The first antistokes line (~436 nm) from the shift cell was then used to excite the sample and generate resonance Raman spectra of the species generated within 25 psec of excitation (pulsewidth measured by autocorrelation two photon measurements). The nanosecond spectra were obtained using a nitrogen pumped dye laser (Molectron) (10-nsec nominal pulse width) and a spectrometer system described elsewhere (6). The picosecond Raman data was collected using a 90° sampling geometry and a spherical lens to focus the laser into the sample while the nanosecond spectra were obtained using an ~180° backscattering geometry. The average power for the picosecond spectra was ~15 mW at 10 Hz (1.5 mJ/pulse). "Low power" nanosecond experiments at 406 and 440 nm were performed with the laser pulses (~.3 mJ/pulse) slightly focused with a cylindrical lens resulting in an energy density of ~5 mJ/cm^2 pulse. For "high power" experiments the beam was tightly focused with a spherical lens. We estimate the photon density at the sample to be approximately 100 times greater in the "high power" than in the "low power" experiments. All time-resolved spectra were obtained from samples placed in a static cuvette. The laser repetition rate in all transient experiments was fixed at 10 Hz. Sample temperatures in all cases were between 20-25°C. The equilibrium spectra were recorded from samples in a rapidly rotating cell using a Raman difference spectrometer (7) and 413.1 nm or 530.9 nm excitation from a krypton ion laser or 441.6 nm excitation from a He-Cd laser. Absorption spectra were recorded after each experiment to monitor sample integrity.

Results and Discussion

The spectra presented in Figures 1-3 demonstrate that high quality, transient resonance Raman spectra can be obtained for Ni(OEP) and Ni(PP) solutions using Soret excitation. These spectra can be interpreted on a molecular level by comparison with the extensive theoretical and experimental data base that exists for ground state nickel porphyrin species (8-16 and refs. therein). The coordination state of nickel porphyrins can easily be determined from the resonance Raman spectrum of the sample (10,12). Several modes in the Raman spectrum of porphyrins are quite sensitive to the state of axial ligation (10,12). In particular, the marker lines ν_4, ν_{11}, ν_2, ν_3, and ν_{10} (porphyrin skeletal mode designations follow those of Abe et al., (12a). The designation

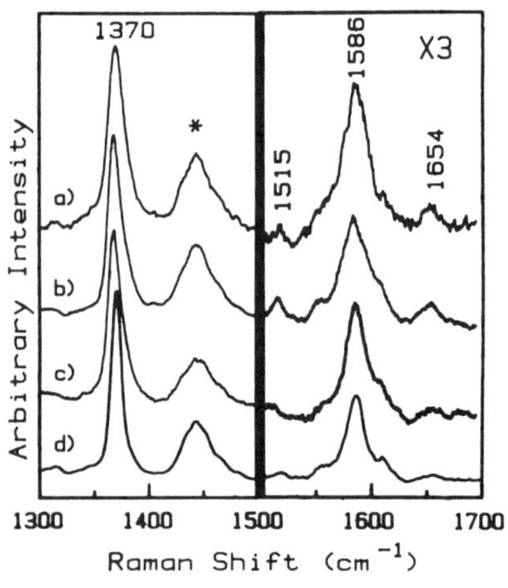

Figure 1. Resonance Raman spectra of photoexcited and equilibrium Ni(OEP) in neat piperidine obtained with 440 nm excitation. Trace a) displays the spectrum obtained within 25 psec of excitation. Traces b) and c) were obtained with 10 nsec excitation under high and low power conditions, respectively (see text). The spectrum of the equilibrium sample is shown in trace d) and was obtained with 441.6 nm excitation (~40 mW) from a cw He-Cd laser. Spectra bandpass was 7-9 cm^{-1} in all cases. Time-resolved spectra are the unsmoothed sum of 3-5 scans. Solvent bands are denoted by asterisks.

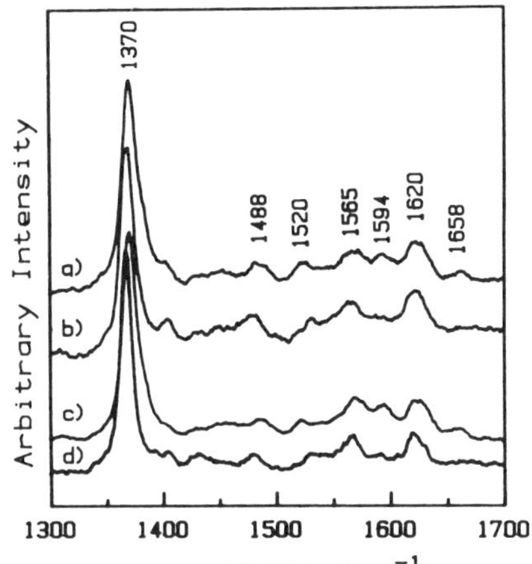

Figure 2. Transient Raman spectra of Ni(PP) (60-80 μM) in pyrrolidine. Traces a) and b) were obtained with 406 nm excitation at high and low power, respectively, while traces c) and d) were generated with 420 nm excitation (which is a compromise frequency which resonantly enhances both species to some extent) at high and low power, respectively. For low power spectra the average laser power (at 10 Hz) was .75-1.0 mW. The beam was only slightly focused onto the sample with a cylindrical lens. High power spectra were generated with 5-6 mW of average laser power sharply focused at the sample via a spherical lens. Spectra are the unsmoothed sum of 3-5 scans at 7-9 cm^{-1} spectral resolution.

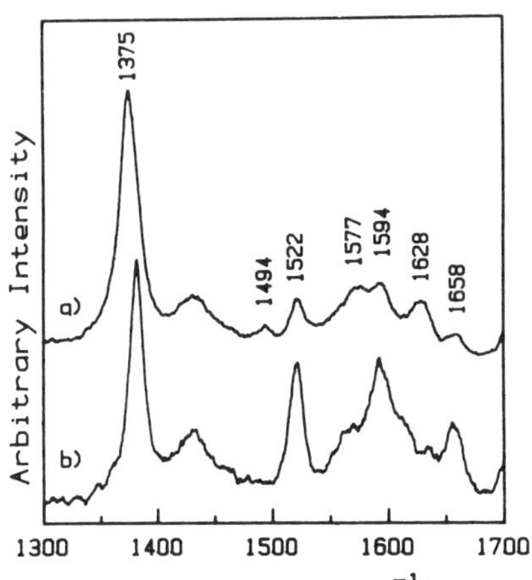

Figure 3. Transient Raman spectra of Ni(PP) in acetone/water (10% water by volume). Traces a) and b) were obtained under high and low power conditions, respectively, similar to those described for Figure 2. These spectra are representative of those obtained in noncoordinating sovent systems at high and low power.

$^n\nu_k$ refers to the kth normal mode in the nth axial ligation state.) show correlated shifts in frequency with ligation changes (10-12). For example, Ni(OEP) in piperidine at equilibrium exhibits the spectrum shown in Figure 1d. Two sets of marker lines are observed in the equilibrium piperidine solution; the set with low frequencies of these Raman lines (see Table I) are indicative of a 6-coordinate Ni(OEP) ligation state. These lines appear 10-40 cm^{-1} higher in frequency for Ni(OEP) in noncoordinating solvents (4-coordination) and are also significantly higher than for 5-coordinate nickel porphyrins (10). Comparison with the Raman spectra of Ni(OEP) in noncoordinating solvents shows that the majority of molecules in the piperidine solution are in the 6-coordinate form.

In Table I, the frequencies of various lines shown to be sensitive to the porphyrin's structure (8,10,15), are given for Ni(OEP) and Ni(PP) complexes in several time regimes and for several excitation wavelengths. As described below, the general results of the time-resolved spectra can be summarized as follows:

1. In coordinating solvents qualitative indications of an increase in formation of 4-coordinate Ni(OEP) are observed in the behavior of ν_{10} and ν_3 within 20-30 psec of excitation at high power,
2. High power nsec pulses also clearly produce an increase in the amount of 4-coordinate nickel porphyrin relative to either low power nsec pulses or an excitation of the ground state equilibrium as evidenced by the behavior of ν_2, ν_3, and ν_{10},
3. In noncoordinating solvent systems (toluene, methylene chloride, H_2O/Acetone, CTAB) power dependent shifts are observed for ν_4, ν_3, ν_2, ν_{10}, to positions near those observed for Ni(OEP) and Ni(PP) in coordinating solvents at equilibrium.

A number of elegant transient absorption studies have shown that nickel porphyrins possess a rich and complicated excited state photochemistry (3-5). In these studies it was concluded that in basic solvents such as piperidine and pyridine the state of ligation of the excited state porphyrin is different from that of the ground state system. In pyridine, ground state Ni(OEP) is mainly 4-coordinate ($\geq$75%). It was concluded that a change of ligation to predominantly a 6-coordinate system with solvent molecules as the axial ligands occurred after excitation from the ground state $^1A_{1g}$ to the $^3B_{1g}$ excited state. The $^1A_{1g}$ $\rightarrow$ $^3B_{1g}$ excitation is a d-d transition involving promotion of an electron from the doubly occupied d_{z^2} orbital to the empty $d_{x^2-y^2}$ orbital of the metal. The transition is not dipole-allowed but occurs upon de-excitation of the allowed $\pi\rightarrow\pi^*$ excitation of the macrocycle (16). In the solvents pyrrolidine and piperidine, however, the opposite situation exists. The ground state species is 6-coordinate $^3B_{1g}$, and upon promotion to the $^1A_{1g}$ excited state deligation occurs. Kim et al. (3,4) have explained the excited-state processes on the basis of σ-donation to the metal d_{z^2} orbital by the ligands. Thus an excited $^1A_{1g}$ state ($d_{z^2}^2$) is a ligand dissociative state because the acceptor orbital of the metal is

Table I

High Frequency Resonance Raman Bands of Ni(OEP) in Piperidine

	436 nm 25 psec	440 nm 10 nsec high power	440 nm 10 nsec low power	441.6 nm equilibrium
$^{4}\nu_{10}$	1654	1655	1654	1655
$^{6}\nu_{10}$	----	~1610 (sh)	~1610 (sh)	1610
$^{4,6}\nu_2$, $^{4}\nu_{11}$	1586	1585	1586	ν1587
$^{6}\nu_{11}$	~1565 (sh)	~1565 (sh)	~1565 (sh)	~1565 (sh)
$^{4}\nu_3$	~1515	1518	~1515	1517
$^{4,6}\nu_4$	1370	1368	1368	1370

Ni(PP) in Pyrrolidine
High frequency Raman bands (cm^{-1})

	406 nm Low Power	406 nm High Power	420 nm Low Power	420 nm High Power
$^{6}\nu_4$, $^{4}\nu_4$	1368	1370	1368	1370
$^{6}\nu_3$	1480	1488	1480	1484
$^{4}\nu_3$		1520		1521
$^{6}\nu_2$	1565	1565	1565	1569
$^{4}\nu_2$		1594		1594
$^{6}\nu_{10}$	1620	1620	1620	1620
$^{4}\nu_{10}$		1658		1658

Continued on next page

Table I. Continued

	Ni(PP) in H_2O/Acetone 406 nm excitation Low Power	High Power
$^{4}\nu_4$	1382	
$^{*}\nu_4$		1375
$^{4}\nu_3$	1522	1522
$^{*}\nu_3$		1494
$^{4}\nu_2$	1592	1594
$^{*}\nu_2$		1577
$^{4}\nu_{10}$	1662	1658
$^{*}\nu_{10}$		1628

completely filled. On the other hand, ligand bonding is favored by $^1B_{1g}$ and $^3B_{1g}$ (d_{z^2}, $d_{x^2-y^2}$) excited states that have a half filled d_{z^2} orbital that is free to accept charge from a filled ligand orbital. Intersystem crossing occurs readily in nickel porphyrin systems and rapid relaxation back to the ground electronic state ($^1A_{1g}$ for noncoordinating solvents, $^3B_{1g}$ for strongly coordinating solvents) occurs following excitation (3). Axial ligands added (or lost) in the excited state are, subsequently, lost (or regained) upon return to the equilibrium ground state.

Within ~25 psec of electronic excitation the resonance Raman spectrum of Ni(OEP) in piperidine indicates an increase in the amount of 4-coordinate Ni(OEP) and loss of the piperidine ligands. Evidence for this is found in the increase in intensity of the $^4\nu_{10}$ band for 4-coordinate Ni(OEP) (~1654 cm^{-1}) relative to that of the ligated species $^6\nu_{10}$ (~1610 cm^{-1}) in the picosecond spectrum (Figure 1a). Quantification of the extent of ligand loss within 25 psec of excitation is complicated by two factors: (1) a small amount of 4-coordinate Ni(OEP) is present in the equilibrium spectrum (Figures 1 and 2) scattering from the 6-coordinate species is more resonantly enhanced than that from 4-coordinate Ni(OEP) when 436-nm excitation is used. The resonance enhancement effect is quite evident in the behavior of $^6\nu_2$ (~1586 cm^{-1}) and $^6\nu_4$ (~1370 cm^{-1}) in the picosecond spectrum. Both modes appear at values characteristic of 6-coordinate Ni(OEP) and their relative intensities change dramatically compared to the equilibrium Raman spectrum. The changes in intensities of the modes observed within 25 psec relative to those observed with 10 nsec pulses may indicate additional excited state processes in the picosecond time regime. Despite these complications, we believe that the increase in $^4\nu_{10}$ (4-coordinate) intensity at ~1654 cm^{-1} relative to $^6\nu_{10}$ (6-coordinate) at ~1610 cm^{-1} qualitatively demonstrates an increase in the 4-coordinate species during the picosecond pulse.

The spectra generated with 10-nsec high power pulses also indicate a significant amount of photogenerated 4-coordinate for both Ni(OEP) and Ni(PP). Evidence of an increase in the amount of unliganded Ni(OEP) relative to equilibrium Ni(OEP) solution are is provided by in the increased relative intensities of 4-coordinate Ni(OEP) modes at ~1515 ($^4\nu_3$) and ~1655 ($^4\nu_{10}$) and a broadening to higher frequency of the band at ~1590 cm^{-1}. Spectra generated with low power 10 nsec pulses more closely resemble those of the equilibrium Ni(OEP) spectra. The increase in the amount of 4-coordinate Ni(OEP) evident in the high power spectra is apparently due to the creation of a significant concentration of the $^1A_{1g}$ state by the long lived (relative to d-d excited state lifetimes) pump/probe laser pulse. This effect is more clearly observed using Ni(PP) complexes (Figure 2).

The Raman spectra of Ni(PP) in coordinating solvents (Figures 2 and 3) clearly show that photoinduced ligation changes occur within the 10 nsec laser pulse width. The appearance of the modes $^4\nu_{10}$, $^4\nu_3$, and $^4\nu_2$ at 1658 cm^{-1}, 1520 cm^{-1}, and 1594 cm^{-1}, respectively, in the high power spectra of Ni(PP) in pyrrolidine indicates an appreciable amount of photodissociation of the 6-coordinate ground state species. The extent of photodissociation is both wavelength and power dependent. Photodissociation is observed using either 406 nm or 420 nm excitation. Even though the 4-coordinate modes are of similar relative intensity at both excitation wavelengths, the fraction of 4-coordinate molecules is appreciably less in the 406 nm spectra. This is expected since the absorptivity, and therefore the resonance enhancement, of the 6-coordinate species is considerably lower at 406 nm.

At either wavelength the proportion of 4-coordinate species increases directly with laser power density. Because the ligand recombination half-life (>20 nsec)) is longer than the laser pulse, a linear increase in the proportion of excited state species during the laser pulse, is expected using 440 nm excitation. This wavelength primarily excites the 6-coordinate species. Excitation at 406 nm produces a slightly more complicated situation since the photogenerated 4-coordinate molecules can, in principle, be excited by subsequent photons in the laser pulse. The proportions of 4- and 6-coordinate molecules will then be predicated upon the integrated solution (during the laser pulsewidth) of the differential rate equations involved in the excited state kinetics and should be quite sensitive to the ratio of the rate constants for the formation of the two species. Obviously, our data allows us to presently make only qualitative comparisons.

The close similarity between the photophysics of Ni(PP) and Ni(OEP) in coordinating solvents observed in this study argues strongly that the photodissociation (or photoassociation) of ligands is predicated upon metal-based excited states and that the decay pathways from porphyrin to metal d-orbital states are similar for the two molecules. Thus, our results corroborate the proposed photocycle of Holten and co-workers (3,4) and suggest that the initial excitation of the porphyrin π-system rapidly (<10's of psecs) decays to a d-d bottleneck excited state where ligation changes may occur.

The behavior of several modes of the photoexcited nickel porphyrin species is undoubtedly complicated by excited state influences other than the state of metalloporphyrin ligation. Four-coordinate Ni(OEP) and Ni(PP) displays shifts in ν_4, ν_2, ν_3, and ν_{10} to lower frequency as a function of increased photon density in noncoordinating solvents such as toluene, CH_2Cl_2, and H_2O/acetone which cannot be predicated upon changes in metalloporphyrin ligation.

Upon 10 nsec pulsed excitation of Ni(PP) in acetone/water (90:10) there is a decrease in relative intensity of ${}^4\nu_{10}$ (1656 cm^{-1}), ${}^4\nu_2$ (1593 cm^{-1}) and ${}^4\nu_3$ (1522 cm^{-1}) indicative of a decrease in the population of the 4-coordinate (ground state) species. Photoexcitation also produces a pronounced shift to lower frequency (~1374 cm^{-1}) and a broadening of ν_4. The new modes which appear in the spectra occur at frequencies that are distinct from those observed for equilibrium 4-, 5-, and 6-coordinate Ni(PP) species. In particular, (* indicates excited state species) ${}^*\nu_{10}$, ${}^*\nu_2$, and ${}^*\nu_3$ appear at 1627-1630 cm^{-1}, 1575-1580 cm^{-1} and 1492-1495 cm^{-1}, respectively. These frequencies are significantly higher than those observed for the ligated B_{1g} states of Ni(PP) in coordinating solvents (10). There are two possible explanations for the transient behavior of Ni(PP) in: 1) a very weak axial ligand (water) is photoassociating subsequent to Ni(PP) excitation, or 2) the observed spectra are those of the excited state ($^3B_{1g}$) Ni(PP) without bound axial ligands.

Water is a possible axial ligand for the transient Ni(PP) in these systems and has been shown to form weak complexes with other nickel porphyrin species (18). While we cannot unequivocally rule out weak, transient ligation, the observation of similar transient behavior in Ni(OEP) and Ni(PPDME) in noncoordinating, nonaqueous, solvents (toluene, methylene chloride (9, unpublished results)) leads us to conclude that the transient behavior of the Ni(PP) in acetone/water is not predicated upon ligand binding.

It is therefore likely that the 10 nsec laser pulses create a significant amount of the unligated $^3B_{1g}$ excited state. This is not unexpected since time-resolved absorption studies have demonstrated that excitation to this state from an unligated $^1A_{1g}$ ground state occurs in $\leq$15 psec, whereas decay from the $^3B_{1g}$ state requires $\approx$250 psec in noncoordinating solvents.

The spectral properties of this excited state species are of great interest since they represent the effects of the $(d_{z^2}^2) \rightarrow (d_{x^2-y^2}^1)$, $d_{z^2}^1$ metal centered transition upon the porphyrin. The most obvious effects are a substantial increase in porphyrin core-size. Previous studies (15) have established that the frequencies of ν_3 and ν_{10} are quite sensitive to porphyrin core-size. Their positions in the transient spectra indicate that the d-d transition alone engenders a core expansion from ~1.95 Å to ~2.02 Å. This is approximately two-thirds of the total expansion observed between equilibrium 4-coordinate Ni(PP) and equilibrium 6-coordinate Ni(PP). This is not unreasonable since the symmetry of the $d_{x^2-y^2}$ orbital allows it to interact directly with the porphyrin sigma system in an antibonding fashion.

The ~10 cm^{-1} shift to lower frequency of $^*\nu_4$ in the unligated B_{1g} state further indicates a substantial redistribution of electron density in the porphyrin orbitals. This mode has been traditionally viewed as a sensitive indicator of the oxidation state of the central metal presumably because of its sensitivity to the degree of π-backbonding between the metal and the porphyrin $e_g(\pi^*)$ orbitals (15). In this case, however, it appears that a substantial portion (~70%) of the frequency shift in ν_4 between equilibrium 4- and 6-coordinate Ni(PP) results from the net d-d transition. This sensitivity to the population of the metal $d_{x^2-y^2}$ orbital probably arises from the increased pyrrole nitrogen-nickel bond distances of the expanded metalloporphyrin which, in turn, leads to poorer overlap between the d_π and porphyrin $e_g(\pi^*)$ orbitals. It should be noted that the relationship between the frequency of ν_4 and the frequency of the core-size marker lines is similar to that observed for metal substitution (17). Metal substitution effects on marker line frequencies appear to be based on differences in covalency of the porphyrin-metal bond for different metals. Thus, π and σ orbitals in addition to $e_g(\pi^*)$ and metal d_π orbitals are probably involved.

In summary the behavior of Ni(OEP) in coordinating solvents observed within 25 picoseconds of photoexcitation appears to follow that reported from transient absorption studies (3,4). The Raman spectra conclusively show that there is an increase in the amount of 4-coordinate Ni(OEP) within ~25 psec of excitation. The spectra of Ni(OEP) and Ni(PP) in coordinating solvents observed with ~10 nsec pulses clearly show the loss of nitrogenous ligands in the excited state. In addition to corroborating the general nickel porphyrin photochemical scheme postulated by Kim et al. (3,4), we have demonstrated that transient Raman spectroscopy is a powerful means of directly observing the molecular basis of the photochemistry and photophysics of excited state metalloporphyrins on short timescales. The power dependencies of various nickel porphyrin modes in noncoordinating solvents suggest that the excited state formed has significant B_{1g} character. Further transient Raman studies using a variety of laser excitation frequencies, powers, and pump-probe protocols will be required to completely separate excited state effects from the effects of axial ligation. However, this preliminary study opens the door to a whole range of time-resolved resonance Raman studies of metalloporphyrin photochemistry, some of which are currently underway in our laboratories.

Acknowledgments

We gratefully acknowledge the support of the NIH (GM33330), Research Corporation, the donors of the Petroleum Research Fund as administered by the American Chemical Society (to MRO), The U.S. Department of Energy Contract DE-AC04-76-DP00789 and the Gas Research Institute Contract 5082-260-0767 (to JAS) and the graduate research scholarship fund of the Associated Western Universities (to EWF).

Literature Cited

1. Findsen, E. W.; Friedman, J. M.; Ondrias, M. R.; MacDonald, V. W. Science 1985, 229, 661.
2. Findsen, E. W.; Scott, T. W.; Chance, M. R., Friedman, J. M.; Ondrias, M. R. J. Am. Chem. Soc. 1985, 107, 3355.
3. Kim, D.; Kirmaier, C.; Holten, D. Chem. Phys. 1983, 75, 305.
4. Kim, D.; Holten, D. Chem. Phys. Letters 1983, 98, 584.
5. Kobayashi, T.; Straub, K. D.; Rentzepis, P. M. Photochem. Photobiol. 1979, 29, 925.
6. Findsen, E. W.; Ondrias, M. R. J. Am. Chem. Soc. 1984, 106, 5736.
7. Shelnutt, J. A. J. Phys. Chem. 1983, 87, 605.
8. Spaulding, L. D.; Chang, C. C.; Yu, N.-T.; Felton, R. H. J. Am. Chem. Soc. 1975, 97, 2517.
9. Cheung, L. D.; Yu, N.-T.; Felton, R. H. Chem. Phys. Letters 1978, 55, 527.
10. Shelnutt, J. A.; Alston, K.; Ho, J.-Y.; Yamamoto, T.; Rifkind, J. M. Biochemistry, 25, 620.
11. LaMar, G. N.; (Jensen) Walker, F. A. In "The Porphyrins"; Dolphin, D., Ed.; Academic: New York, 1979; Vol. 4, p. 61.

12. Shelnutt, J. A.; Straub, K. D.; Rentzepis, P. M.; Gouterman, M.; Davidson, E. R. Biochemistry 1984, 23, 3946.
12a. Abe, M.; Kitagawa, T.; Kyogoku, Y. J. Chem. Phys. 1978, 69, 4526.
13. Teroaka J.; Kitagawa, T. J. Phys. Chem. 1980, 84, 1928.
14. Kitagawa, T.; Ogoshi, H.; Watanabe, E.; Yoshida, Z. J. Phys. Chem. 1975, 79, 2629.
15. Spiro, T. G. In "Iron Porphyrins, Part 2;" Lever, A.P.B., Gray, H. B., Eds.; Addison-Wesley: New York; 1984.
16. Ake, R. L.; Gouterman, M. Theoret. Chim. Acta (Berl.) 1970, 17, 408.
17. Shelnutt, J.; Ondrias, M. R. Inorg. Chem. 1984, 23, 1175.
18. Pasternak, R. F.; Spiro, E. G.; Teach, M. J. Inorg. Nucl. Chem., 1974, 36, 599.
19. Shelnutt, J. A.; Dobry, M. M. J. Phys. Chem., 1983, 87, 3012.
20. Shelnutt, J. A. , J. Am. Chem. Soc. 1983, 105, 774.
21. Walker, F. A.; Hui, E.; Walker, J. M. J. Am. Chem. Soc. 1975, 97, 2390.

RECEIVED April 8, 1986

PHOTOCHEMISTRY

19

Mechanistic Studies of Thermal and Photoinduced Atropisomerization of Substituted Tetraphenyl Porphyrins in Solution and Organized Assemblies

Ruth A. Freitag, David C. Barber, Haruo Inoue, and David G. Whitten

Department of Chemistry, University of Rochester, Rochester, NY 14627

An investigation of thermal and photoinduced atropisomerization processes for substituted tetraphenylporphyrins is reported. The o-substituted "picket-fence" porphyrins exist as isolable atropisomers which can be interconverted by heating or by irradiation in degassed solution. The thermal reaction appears to involve a single one-bond rotation in which three of the four isomers must interconvert via stepwise formation of the 3,1 isomer; for free base and metalloporphyrins the equilibrium composition is fairly close to that anticipated on the basis of statistics. The diacids isomerize more rapidly than free base or metalloporphyrin to produce an equilibrium mixture somewhat enriched in the least hindered *trans*-2,2 isomer. The photochemical interconversion of atropisomers produces photostationary mixtures having nearly equilibrium composition for free base and metalloporphyrins. The diacid porphyrins also photoatropisomerize but in this case the photostationary mixture is strongly enriched in *trans*-2,2. There is a significant isotope effect on the quantum yield of atropisomerization for $H_2(D_2)$PF,THA which parallels the isotope effect on triplet lifetimes. The results are most consistently interpreted on the basis of an adiabatic mechanism for the photoatropisomerization in which the free energy of activation is lowered by *ca*. 12 kcal/mole as compared to the thermally induced process. Both thermal and photoatropisomerization reactions are retarded for 4,0 H_2PF,THA, the porphyrin with the longest side chains studied, in the presence of detergent micelles. For solutions containing micelles and vesicles it appears that at least two solubilization sites exist and that in one the porphyrins are unreactive while in the other "normal" solution-phase reactivity persists.

NOTE: This is number 48 in the series Photochemical Reactions in Organized Assemblies.

0097-6156/86/0321-0280$06.00/0

Excited states of the porphyrins and their metal complexes have been very well characterized through a number of experimental and theoretical investigations. The photophysics of these compounds as well as their ability to photosensitize through singlet and triplet energy transfer has been well documented (1-4). A large number of investigations have shown that porphyrins and their metal complexes can serve as both excited electron donors and excited electron acceptors in a number of photoredox processes (5-9). However, with the exception of ligand photoejection (10-12) and so-called "redox isomerism" (13) relatively little in the way of unimolecular photochemistry has been observed for the porphyrins or their metal complexes.

The occurrence of atropisomers, isomers which owe their stability to hindered rotation about a single bond, was first observed for ortho-substituted biphenyl compounds (14-16). The finding that resolvable 2,2',6-trisubstituted and 2,2',6,6'-tetrasubstituted biphenyls can be isolated and observed to racemize upon heating (14-16) has led to a number of studies of such isomers and their interconversion (17). Typically, in these compounds the two adjacent aromatic systems can not become coplanar without incurring severe steric repulsion. A relatively analogous situation occurs in the synthetic tetraphenylporphyrins and their metal complexes. Even for unsubstituted tetraphenylporphyrin (TPP), the steric repulsion between ortho phenyl and β-pyrrole hydrogens causes a preferred conformation of the phenyl rings to be close to perpendicular to the core (18,19). As would be anticipated TPP derivatives containing ortho substituents have rather severe steric barriers such that the substituent is essentially fixed on one side of the porphyrin plane at or near room temperature. In 1969 Gottwald and Ullman observed that tetra(o-hydroxyphenyl)porphyrin can be separated into four isomers, having identical absorption spectra, present in a ratio 1:2:4:1, that expected statistically for the four isomers schematically illustrated in Figure 1. In subsequent investigations a number of examples of isolable atropisomers among TPP derivatives and their thermal interconversion have been found (20-25). In a number of elegant investigations Collman and co-workers (26) were able to capitalize on the thermal stability of atropisomers with particularly bulky ortho substituents to obtain special properties for an isolated atropisomer, in this case the 4,0 or "picket fence" porphyrin. In a number of investigations we have found that the 4,0 isomer of tetrakis(o-hexadecylamidophenyl)porphyrin (H_2PF,THA) and related compounds are good surfactant molecules which can be used as their free base or metal complexes to form insoluble films at the air-water interface (27-29). Although the 4,0 isomer of H_2PF,THA was found to be quite stable in monolayer films or in transferred multilayers, subsequent investigations have shown that irradiation with visible light can produce photoatropisomerization of these picket fence porphyrins in solution (30,31)

The present paper reviews our study of the thermal and photochemical atropisomerization of a number of "picket fence" (substituted TPP) porphyrins. The study of the thermal atropisomerization with a number of different free-base and diacid "picket fence" porphyrins indicates that the thermal process likely involves a simple rotation around the porphyrin-phenyl bond in which ruffling of the porphyrin ring can assist but is not prerequisite;

the lack of a deuterium isotope effect when the N-H protons are substituted is consistent with this interpretation. Our study, including a determination of isotope effects on the triplet lifetime and photoatropisomerization quantum efficiencies indicates that the excited state process resembles the ground state reaction and in fact is probably best accounted for by an adiabatic interconversion of isomers in the excited triplet state. In addition to these studies on the mechanism of the thermal and photoatropisomerization in solution, the present paper also includes results of a study in a number of organized assemblies in which different substituted porphyrins are associated with different solubilization "sites" in surfactant solutions. These results suggest that the "site" occupied by the porphyrin can play a role in modifying its reactivity in both thermal and photochemical atropisomerization reactions.

Thermal Atropisomerization of Porphyrins and Metalloporphyrins in Homogeneous Solution

Results obtained in studies of the thermal atropisomerization of the various picket fence porphyrins used in these studies are generally parallel to those obtained in other studies of TPP derivatives with bulky ortho substituents (20-25). Thus for all the free base, diacid and metalloporphyrins studied a 1-bond isomerization process is observed. In each case the 4,0, *trans*-2,2 and *cis*-2,2 give only the 3,1 isomer upon heating (30,31). As data in Table I indicate,

Table I. Thermal Equilibration of "Picket-Fence" Porphyrins[a]

Compound	%4,0	3,1	trans-2,2	cis-2,2
[Statistical]	12.5	50	12.5	25
$H_2PF,TPro$	11	52	15	22
H_2PF,THA	4	47	20	29
H_4PF,THA^{2+}	3	37	33	27

[a]Hydrocarbon solvent

for the three different amide derivatives used, H_2PF,THA, $H_2PF,TPiv$, and $H_2PF,TPro$ the free base compounds equilibrate to produce mixtures differing only slightly from that expected on a statistical basis (31). A slightly lower than statistical amount of the 4,0 isomer and the enhanced over statistical value of *trans*-2,2 can be rationalized on the basis of lesser and greater stability of the two isomers, respectively, on steric grounds. A number of different metal complexes of the same porphyrins give very similar thermal equilibrium mixtures. In contrast the diacids equilibrate to give mixtures that have somewhat more *trans*-2,2 with relatively smaller amounts of 4,0 and 3,1 compared to that expected on the basis of statistics.

As has been elaborated elsewhere, activation energies measured for the thermal atropisomerization of these porphyrins and their metal complexes can be correlated with the degree of core rigidity and/or distortion from planarity for the different compounds (31). In general the metal complexes show higher activation energies than the free base porphyrins while the diacids show somewhat lower activation energies; representative values are listed in Table II.

Table II. Free Energies of Activation for Porphyrin Atropisomerization

Central Substituent	$\Delta G^{\ddagger}$, kcal/mole		
	TPro	THA	TPiv
H_2	29.1	29.1	30.6
H_4^{2+}	26.4	26.5	28.3
Zn	31.4	31.2	32.2
Pd	31.0	31.1	31.7

The thermal atropisomerization of the 4,0 isomers of H_2PF,THA and D_2PF,THA were compared in benzene solutions containing 1% (v/v) ethanol or ethanol-d; under these conditions it was found that there was no isotope effect within experimental error (k_H/k_D = 1.03 ± 0.04). This reinforces a view that the thermal isomerization, while perhaps assisted by deformations originating near the pyrrole nitrogens, does not require severe ruffling of the porphyrin macrocycle.

Photoatropisomerization in Solution

As mentioned above, several substituted biphenyls and binaphthyls exist as optically active atropisomers which can be interconverted or racemized by heating. In many cases the same compounds have been found to undergo photochemical interconversion (32-34). The occurrence of the photoatropisomerization has been attributed to an enhanced tendency for the biphenyl chromophore to achieve a planar conformation in its electronically excited states (33,35). Investigations of the photochemistry (32,33) and spectroscopy (36-38) of a number of biphenyl derivatives have led to the postulation of somewhat different excited state/ground state profiles which can account for the behavior of a variety of compounds (32-34). These are shown schematically in Figure 2 for biphenyl-type compounds in which the ortho substituents are not too large, it is likely that a minimum occurs in the excited state potential at nearly θ = 0°. In this case excitation of either twisted ground state isomer is followed by rapid rotation to the planar configuration from which radiative or nonradiative decay, concurrent with isomerization, can occur. The mechanism for the photoreaction in this case is thus a "diabatic" process quite analogous to the photoisomerization of olefins (39). On the other hand, as shown in Figure 2 part B for biphenyls with large ortho substituents it is probable that no minimum occurs at 0° but that the excited state shows a potential surface with two minima but a reduced energy barrier between them. In this case excitation produces an excited state in which interconversion between the two nonplanar configurations is much more rapid than in the ground state; thus, strictly speaking, in this situation the isomerization is an adiabatic process in which isomerization occurs prior to excited state decay (33,35). Several of the "picket fence" porphyrins and their metal complexes have bbeen found to undergo photoatropisomerization at room temperature under conditions where little or no thermal reaction occurs even over very long periods of time (21,24-26,31). In most cases the photostationary states resulting from prolonged irradiation (Table III)

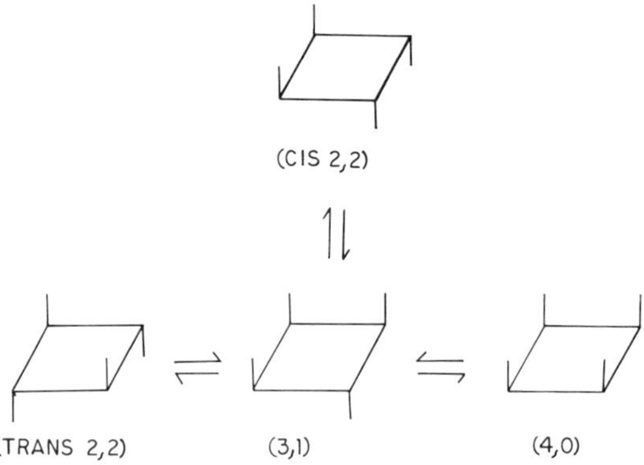

Figure 1. Schematic depiction of the four atropisomers of an *ortho* substituted tetraphenyl porphyrin.

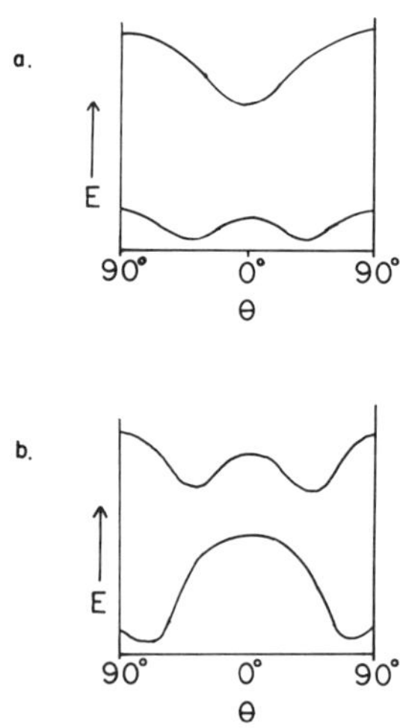

Figure 2. Ground state and excited state energy profiles for phenyl ring rotation (energy *vs.* dihedral angle, θ) in a) biphenyl, b) 2,2',6,6'-tetrasubstituted biphenyl (33).

Table III. Photostationary States Produced in Photoatropisomerization of "Picket-Fence" Porphyrins

Porphyrin	%4,0	3,1	trans-2,2	cis-2,2	ϕ^a
[Statistical]	12.5	50	12.5	25	
H_2PF,THA	5	48	20	27	1.3×10^{-3}
D_2PF,THA	4.5	47.4	19.5	28.6	2.3×10^{-3}
H_4PF,THA^{2+}	6	13	80	1	1.7×10^{-3}
H_2PF,TPro	9	50	16	25	2.3×10^{-3}
H_4PF,$TPro^{2+}$	0	22	78	0	

[a] Quantum yield for the conversion 4,0 → 3,1

were found to correspond quite closely to the equilibrium values obtained in the thermal atropisomerization at higher temperatures. It was found that photoatropisomerization could be initiated by irradiation throughout the visible and near ultraviolet transitions of the porphyrins. In all cases studied thus far, the isomerization has been found to originate from the excited triplet of the porphyrin or metalloporphyrin. Thus oxygen has been found to completely quench the isomerization even under conditions where there is negligible quenching of the porphyrin fluorescence by oxygen. Among the metal complexes, zinc and palladium porphyrins which are known to have relatively long triplet lifetimes (40,41) isomerize readily; in contrast copper and nickel porphyrins, whose triplet states are extremely short-lived, (41,42) are unreactive towards photoatropisomerization.

In conttrast to the situation with biphenyls or binaphthyls, photoatropisomerization of the porphyrins is inherently more complicated due both to the extended nature of the porphyrin chromophore and the presence of four sites within the "picket fence" porphyrins at which atropisomerization can occur. The simplest path would be a 1-bond labilization process analogous to that occurring in biphenyl; however it is plausible that due to extensive delocalization all four porphyrin-phenyl bonds might be equally labilized towards rotation or that extensive deformation of the macrocycle in the excited state could lead to specific labilization of an opposite or adjacent pair of porphyrin-phenyl bonds. Studies on initial products formed from 4,0 H_2PF,THA indicate that *trans*-2,2 and 3,1 are both produced promptly; *cis*-2,2 is only observed after somewhat prolonged irradiation. In contrast, with the other free-base and metalloporphyrins initial irradiation of 4,0 was found to produce only 3,1. The free-base "semi-capped" porphyrin H_2PF,DIP, in which two adjacent phenyl rings are bridged by a diisophthalamide linkage, was prepared for use in photoatropisomerization studies. This porphyrin, shown below, can exist in only two isomeric forms, analogs of the 4,0 and *cis*-2,2 atropisomers. Atropisomerization of H_2PF,DIP would require simultaneous rotation of two adjacent phenyl rings. In fact, no photoisomerization of 4,0 H_2PF,DIP was observed. At this point, it is unclear whether *trans*-2,2 H_2PF,THA is formed by simultaneous or sequential rotation (i.e., 4,0 → 3,1 → *trans*-2,2) of two phenyl rings.

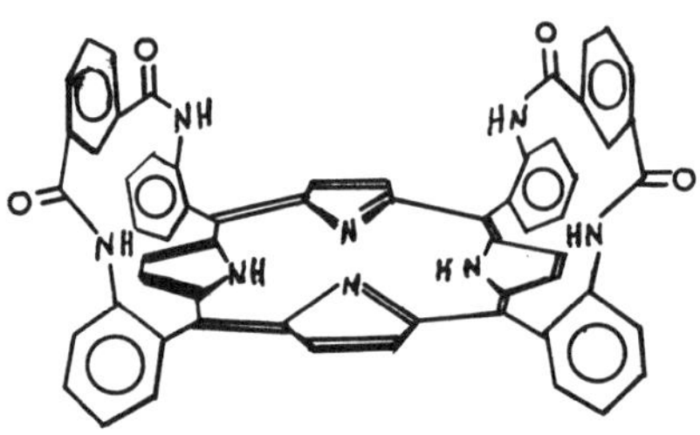

"4,0"

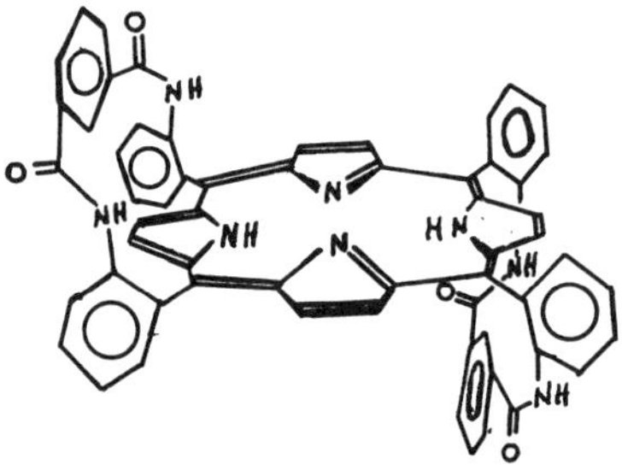

"CIS 2,2"

Although most of the free-base and metalloporphyrins studied give photostationary states very similar to the equilibrium content, the distribution of atropisomers produced upon prolonged irradiation of the porphyrin diacids differs markedly. In both cases studied the photostationary state of the diacid is strongly enriched in *trans*-2,2 compared to any other isomers. Although photoatropisomerization of the porphyrin diacids clearly originates from the triplet state of the macrocycle, a type of Förster cycle calculation based on the singlet-singlet absorption of the porphyrins suggest a reason for this behavior (43). Protonation of all four pyrrole nitrogens causes a steric repulsion which in turn causes the pyrrole rings to tilt alternately up and down with respect to the mean porphyrin plane. A crystal structure of the tetraphenyl porphyrin diacid (44) reveals the angle of this tilt to be 33°. The non-planar deformation of the porphyrin core disrupts the π-system and results in a loss of core conjugation. This results in a blue shift for the first and second visible bands for the diacids of octaalkylporphyrins (44-46). However for the tetraphenylporphyrin diacid the increase in conjugation resulting from interaction of the almost coplanar phenyl π-systems with the porphyrin π-system results in a red shift of the visible bands (44). For the individual atropisomers of H_4PF,THA^2 the 4,0 isomer exhibits a net blue shift upon protonation while the diacid of the *trans*-2,2 isomer is significantly red shifted. Small spectral shifts are seen for the *cis*-2,2 and 3,1 isomers. This has in turn been interpreted to mean that in the excited singlet state "stabilities" are shifted such that *trans*-2,2 > 3,1 > *cis*-2,2 > 4,0 for this diacid. A comparison of the spectral shifts observed upon protonation is given in Table IV.

The above mentioned deformation in the porphyrin diacid results in decreases in the free energy of activation for thermal atropisomerization, $\Delta G^{\ddagger}$, of 2-3 kcal/mole compared to the corresponding free bases (31). Even in the thermal isomerization the greater ground state stability of the *trans*-2,2 compared to the other isomers is apparent, although not nearly so much so as in the corresponding excited state.

Returning to the question of the mechanism of the photoatropisomerization, it is worthwhile to focus on the question of triplet lifetimes and their relation to the photoatropisomerization process. Table V compares triplet lifetimes obtained from transient absorption

Table IV. Comparison of Spectral Shifts Observed upon Protonation for H_2TPP and the Four Atropisomers of Each Picket-Fence Porphyrin[a]

Porphyrin	λ_{max} band I, nm free base	λ_{max} band I, nm diacid	λ_{max} band II, nm free base	λ_{max} band II, nm diacid	total shift, nm
H_2TPP	648	661	592	609	+30
H_2PF,THA					
(4,0)	647	643	591	592	-3
(*trans*-2,2)	648	654	591	600	+15
(3,1)	647	651	591	598	+11
(*cis*-2,2)	648	649	591	597	+7
H_2PF,TPiv					
(4,0)	647	638	590	589	-10
(*trans*-2,2)	648	651	591	598	+10
(3,1)	646	646	591	594	+3
(*cis*-2,2)	646	647	590	593	+4
H_2PF,TPro					
(4,0)	647	644	591	592	-2
(*trans*-2,2)	647	656	590	603	+22
(3,1)	647	650	590	597	+10
(*cis*-2,2)	647	650	590	598	+11

[a]All spectra were recorded in benzene.

Table V. First Order Decay Rate Constants and Triplet Lifetimes of Picket Fence Porphyrins

Compound	Transient Absorption (460 nm)		Ground State Bleaching (Soret)	
	k_1 (sec^{-1})	τ (μsec)	k_1 (sec^{-1})	τ (μsec)
H_2PF,THA[a]	935	1070	895	1120
H_2PF,TPiv[a]	1610	620	1690	590
	1390	720	1380	720
H_2PF,DIP[a]	1460	690	1380	720
H_2TPP[b]	740	1400		

[a]Measured in benzene
[b]Literature values, measured in toluene

and ground state bleaching measurements for several free-base and diacid porphyrins. It appears that the triplet lifetime decreases somewhat for the series H_2TPP > 4,0 H_2PF,THA > 4,0 H_2PF,TPiv and 4,0 H_2PF,DIP. The shortened lifetimes observed for H_2PF,TPiv and H_2PF,DIP, the two most constrained porphyrins, relative to those of H_2TPP and H_2PF,THA are probably significant. It is reasonable that this indicates that phenyl ring rotation is not an important mode of vibrational deactivation of the triplet excited state, since such an effect would lead to the opposite order of decay rates for this series of porphyrins. Interestingly the triplet lifetime of the diacid picket fence porphyrin is only 1/15th that of the corresponding free-base, the isomerization efficiency (Table III) is slightly greater.

The evidence presented thus far concerning the photoatropisomerization of the free-base and diacid porphyriins is most consistent with an adiabatic pathway in which the requisite energy for rotation about the pyrrole-phenyl bond is lowered in the excited triplet state. Although the energy is lowered enough to allow photoatropisomerization to be observed at room temperature, in most cases rotation is slow enough so that only rotation of a single group during the lifetime of an excited state is likely. The near coincidence of photostationary state and thermal equilibrium compositions suggest that the proposed adiabatic pathway occurs on a potential surface in which, for the free-base porphyrins at least, the relative energies of atropisomers differs little from the ground state. If we assume the occurrence of such an adiabatic process, a calculation of k^*_{isom}, the rate constant for atropisomerization in the excited state, is possible using the relationship:

$$\phi = \phi_t \frac{k^*_{isom}}{k_d}$$

where ϕ is the quantum yield of photoatropisomerization, ϕ_t is the quantum yield for production of the porphyrin triplet excited state and k_d is the excited state decay rate constant. This relationship allows the calculation of $\Delta G^{\ddagger *}$, the excited state energy barrier to isomer conversion. The results of these calculations for free base and diacid picket fence porphyrins are given in Table VI. For an adiabatic photoreaction the activation energy for atropisomerization is decreased 11-12 kcal/mole compared to the thermal process.

Deuteration has been previously shown to cause an increase in the lifetime of triplet free-base porphyrins (47). This has been attributed to the strong coupling of N-H tautomerism with nonradiative decay. In the case of mesoporphyrin IX the increase upon deuteration is approximately two-fold (48) As indicated in Table III deuteration of the picket fence porphyrin results in little change in the photostationary state composition but an almost two-fold increase in the quantum yield of 4,0 → 3,1. As stated above there is no measurable deuterium isotope effect on the thermal reaction; the proportionate increase in quantum yield and triplet lifetime upon deuteration of the picket fence porphyrin is thus completely consistent with the adiabatic mechanism described above. Although the evidence amassed does not completely rule out other possibilities, it seems that the photoatropisomerization is to date best described by the adiabatic pathway in which the porphyrin ground and excited state potential surfaces are modified much as illustrated in Figure 3.

Thermal and Photoatropisomerization of Free-Base Porphyrins in Organized Media

The long-chain "picket fence" porphyrin, H_2PF,THA, was originally synthesized for the express purpose of including the 4,0 isomer as a component in organized assemblies such as monolayer films at the air-water interface and surfactants in solution. It was anticipated that this molecule would have strong surfactant properties with the four long hydrophobic hydrocarbon chains oriented more or less perpendicular to the hydrophilic porphyrin plane and with all of the chains in one direction. Indeed, the 4,0 isomers of the free-base and several metal complexes are good surfactants which form stable, insoluble films at the air-water interface and can be incorporated into a variety of organized assemblies formed from surfactant molecules in solution (27-29). Monolayer films and supported multilayers formed by transfer of films containing the 4,0 isomer are stable at room temperature and show no evidence of photoatropisomerization (28,29). Although relatively little has been done in the way of investigation, it appears that isomers other than the 4,0 do not form stable films or transferred multilayers. Thus, as would seem reasonable, atropisomerization is eliminated when the porphyrin is incorporated into a highly organized and constrained environment. Since Langmuir-Blodgett films and assemblies made from these films represent a nearly crystalline or solid state-like environment, it appeared of interest to determine whether thermal and photoatropisomerization should be eliminated or hindered in more fluid organized assemblies formed by dispersion of surfactant molecules in aqueous solution.

An investigation of the effects of micellar environment on the thermal and photochemical atropisomerization of two picket fence porphyrins (4,0 isomers), one with long chains (C_{16}), H_2PF,THA, and one with short chains (C_3), H_2PF,TPro, has been carried out. As

Table VI. Calculations of Excited State Atropisomerization Rate Constants (k^*_{isom}) and Free Energies of Activation ($\Delta G^{\ddagger *}$)

Porphyrin	ϕ_{isom}	τ(μsec)	k_d(sec^{-1})	k^*_{isom}(sec^{-1})	$\Delta G^{\ddagger *}$(kcal/mol)
H_2PF,THA	$1.3x10^{-3}$	1100	910	1.4^a	17.2^b
H_4PF,THA^{2+}	$1.7x10^{-3}$	70	14,000	27^a	15.5^b

[a] Calculated assuming a quantum yield for production of the triplet state, $\phi_t = 0.87$ ($\phi_t = 1-\phi_f$; ϕ_f for H_2TPP in the absence of O_2 is reported to be 0.13).

[b] Calculated assuming a temperature of 25°C.

mentioned above these compounds have identical thermal atropisomerization rates at a given temperature and reasonably similar quantum yields for photoatropisomerization. In typical experiments the porphyrins were solubilized in aqueous sodium dodecyl sulfate (SDS) (0.05 M) or cetyltrimethylammoniumchloride (CTAC) (0.01 M); a porphyrin-surfactant solution was heated or irradiated and then subsequently extracted to remove the porphyrin from the surfactant for analysis by high pressure liquid chromatography (HPLC). Although the extraction-analysis procedure involves a number of difficulties which create some problems in precise quantitative analysis, it could be established that interconversion of atropisomers in this procedure does not occur so that the general patterns of thermal and photochemical atropisomerization for the two porphyrins could be established. Heating of an SDS solution of 4,0 H_2PF,TPro results in atropisomerization as verified by the appearance of the 3,1 isomer. At longer times the other isomers, cis-2,2 and trans-2,2 also can be detected. A rough estimate of the isomerization rate constant for 4,0 H_2PF,TPro indicates that the rate in SDS is of the same order of magnitude as that measured in homogeneous solution (xylenes). In contrast, 4,0 H_2PF,THA does not undergo thermal isomerization at all in SDS. Atropisomerization is not suppressed for 4,0 H_2PF,THA in a tetrahydrofuran/water solution; the suppressed rotation in the presence of micelles is likely to be a consequence of the presence of surfactant rather than the polarity of the environment. Thermal isomerization of trans-2,2 H_2PF,THA in an SDS micelle solution was also found to be prevented.

Similar results were obtained upon photolysis of SDS solutions of H_2PF,TPro and H_2PF,THA. All four isomers of H_2PF,TPro were present after irradiation in SDS micelles for 15 minutes. Photolysis of 4,0 H_2PF,THA in SDS did not result in production of the 3,1 isomer; photoatropisomerization of 3,1 H_2PF,THA in SDS micelles was also found to be prevented as none of the other isomers appeared after 30 minutes photolysis. The interesting observation that neither the 4,0 nor the 3,1 isomer undergoes atropisomerization indicates that it is not a special stabilization of one isomer in the micelle that prevents isomerization but suggests rather that the nature of solubilization for H_2PF,THA inhibits rotation about the porphyrin-phenyl bond.

Photoatropisomerization of H_2PF,TPro and H_2PF,THA was also investigated in CTAC micelles. As with the SDS solution, all four isomers of H_2PF,TPro appeared after irradiation for a short period. In the case of 4,0 H_2PF,THA it was found that irradiation produced a very slow conversion to the 3,1 isomer in contrast to what was observed in SDS.

The differences in photochemical and thermal atropisomerization behavior of the two "picket fence" porphyrins (4,0 isomers) in SDS micelles are accompanied by changes in the visible absorption bands of the two porphyrins. In all homogeneous solutions studied to date the two porphyrins, H_2PF,THA and H_2PF,TPro, have identical absorption spectra. In contrast as indicated in Figure 4 when the two porphyrins are solubilized in SDS there is a shift for H_2PF,TPro from 423 nm in benzene to 419 nm for the Soret band, while the Soret band for H_2PF,THA shifts from 423 nm in benzene to 435 nm in SDS. There are corresponding separations in the fluorescence spectra for the two porphyrins and the fluorescence from H_2PF,THA in SDS is of much lower

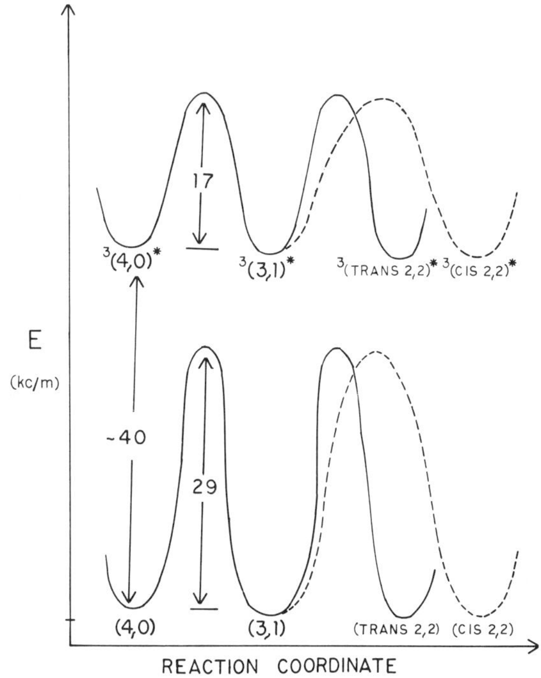

Figure 3. Proposed ground and triplet energy profiles for atropisomerization of H_2PF,THA.

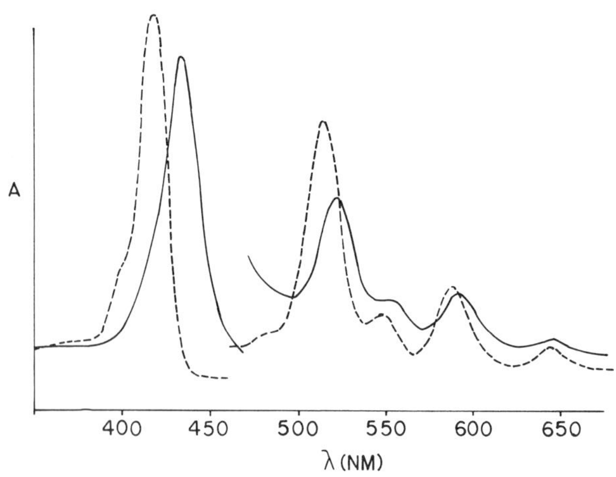

Figure 4. Absorption spectra of ——— H_2PF,THA and ------ H_2PF,TPro in 0.05 M SDS.

intensity. As mentioned above the two porphyrins show identical absorption spectra in all homogeneous solvents studied; however the spectra observed (Table VII) are quite solvent sensitive. From data in Table VII it is clear that the long chain H_2PF,THA experiences an environment in the SDS micelles similar to that in THF/H_2O while the shorter chain H_2PF,TPro experiences an environment similar to methanol. It appears reasonable that these differences, along with those in isomerization behavior arise from solubilization of these two porphyrins in two different environments in detergent solutions.

Preliminary studies of H_2PF,THA and H_2PF,TPro in SDS-water solutions by laser flash photolysis suggest that the differences in photoatropisomerization behavior for the two porphyrins may be accounted for by differences in triplet lifetimes. For H_2PF,TPro in SDS the lifetime is 660 μs, similar to that observed for this or other porphyrins in homogeneous solution. In contrast the triplet lifetime of H_2PF,THA in SDS was too short to be measured by laser flash photolysis (< 5 μs) and is much shorter than normally observed for unquenched porphyrin triplets. The reason for the reduced triplet lifetime of H_2PF,THA in aqueous SDS is not apparent but it is clear that the decrease could account for the lack of photoatropisomerization.

The difference in solubilization sites could be due to hydrophobicity and/or chain-length of the acyl residues attached to the macrocycle. Four porphyrins having intermediate-length chains -- H_2PF,THex; H_2PF,THept; H_2PF,TOct and H_2PF,TDec -- were prepared from the linear acid chlorides having 6-10 carbons, respectively. The four intermediate chain length porphyrins were solubilized in SDS and CTAC micelles. In both micellar solutions the two longer chain compounds H_2PF,TOct and H_2PF,TDec showed spectra very similar to that of H_2PF,THA; the Soret bands are characterized by a single peak at ~430 nm, somewhat blue shifted relative to the corresponding absorption for H_2PF,THA. For the two shorter chain compounds spectra showing two distinct Soret maxima were obtained; for H_2PF,THex the major Soret band is at 422 nm, slightly red shifted from the Soret band of H_2PF,TPro in micelles. However for H_2PF,THex there is also a shoulder at roughly 435 nm. For H_2PF,THept the shorter wavelength band is a shoulder while the longer wavelength band at 437 nm is the major peak. If we attribute, as above, the Soret bands at 420 and 435 nm to different solubilization sites in the detergent solution, it is clear that these porphyrins can occupy both sites to some extent. It is interesting that the Soret bands of the intermediate chain length porphyrins are characterized by the long wavelength absorption in concentrations of SDS below the critical micelle concentration. It is therefore possible but unproven at this time that the surfactant porphyrins associate with the detergent below the critical micelle concentration; pre-micellar aggregates have been proposed in other cases with cationic or surfactant dyes and other micelle-forming detergents (49-55).

Although more work is needed to clearly correlate the type of solubilization site occupied by different porphyrins with their reactivity in such sites towards atropisomerization, it is clear that different sites exist and that these sites show quite different reactivity in both thermal and photochemical processes. Preliminary studies have shown that related behavior probably occurs in other organized assemblies formed by dispersion of surfactant molecules in

Table VII. Absorption Maxima (nm) for H_2PF,THA and H_2PF,TPro in Different Solvents

Porphyrin and solvent	Soret	IV	III	II	I
H_2PF,THA					
benzene	423	515	549	590	647
methanol	418	517	550	592	647
acetone	420	516			
1:1 acetone:water[a]	429	519			
isopropanol	419	515			
1:1 isopropanol:water[a]	435	522			
ethyl acetate	420	515			
dioxane	422	516			
THF	421	516			
THF/H_2O[a]	431	522	~556	592	648
5×10^{-2}M SDS	435	522	~555	592	648
1×10^{-2}M CTAC	435	522	555	592	648
H_2PF,TPro					
benzene	423	515	548	589	646
methanol	417	515	548	590	646
THF/water[a]	440	524	557	593	647
5×10^{-2}M SDS	419	515	548	589	645
1×10^{-2}M CTAC	420	516	550	590	646

[a]Preliminary studies suggest that these solutions contain aggregated porphyrin.

aqueous or partially aqueous solutions. It has been found for example that in cationic dioctadecyldimethylammonium chloride (DODAC) vesicles H_2PF,TPro and H_2PF,THA exhibit different Soret band absorption maxima (420 nm and 437 nm, respectively) corresponding to the same porphyrins in SDS.

In contrast to the above results, all three "picket fence" porphyrins are solubilized in an oil-in-water microemulsion to yield a clear solution having a Soret band at 419-421 nm resembling that of H_2PF,TPro solubilized in micelles. In this case the microemulsion (composed of SDS, n-pentanol and dodecane) consists of oil "droplets" dissolved in bulk water; the radius of the droplet has been estimated to be 37 Å (56), well over twice that estimated for an SDS micelle (16 Å). Since the droplet in the microemulsion contains a much larger "interior", it is reasonable that it may be a better medium for solubilizing the porphyrin.

Summary

The studies reported in this paper show that thermal and photo-atropisomerization about the porphyrin-phenyl bond in tetraphenyl porphyrins is a fairly general process. The rates and efficiencies of both the thermal and photochemical reaction are influenced by a variety of factors including structural features within the porphyrin and the environment in which the porphyrin is solubilized. The thermal reaction appears well characterized by a simple rotation around the sterically constrained single bond; the photochemical reaction appears to parallel the thermal reaction very closely and is best described as an adiabatic process in which the triplet state has an activation energy lowered by 11-12 kcal/mol compared to the ground state. The reactivity, or lack thereof, exhibited by certain porphyrins in micelles and presumably other organized assemblies indicates the sensitivity of the atropisomerization process to environmental factors. Useful extensions of this study could involve the design and synthesis of other porphyrins in which atropisomerization (both thermal and photochemical) is more rapid or more efficient. For such porphyrins it would be anticipated that atropisomerization or lack thereof might provide a sensitive indication of interactions between the porphyrin and its environment.

Acknowledgments

We are grateful to the U. S. Army Research Office (contract numbers DAAG29-80-K-0066 and DAAG29-84-K-0011) for support of this work. This paper is taken in part from the Ph.D. dissertation of Ruth A. Freitag, University of North Carolina, Chapel Hill, 1983.

Literature Cited

1. Mercer-Smith, J. A.; Whitten, D. G. J. Am. Chem. Soc. 1978, 100, 2620.
2. Warden, J. T.; Bolton, J. R. Photochem. Photobiol. 1974, 20, 245.
3. Wildes, P. D.; Pacifici, J. G.; Irick, G.; Whitten, D. G. J. Am. Chem. Soc. 1971, 93, 2004.

4. Wyman, G. M.; Zarnegar, B. M. J. Phys. Chem. 1973, 77, 2584.
5. McLendon, G.; Miller, D. S. J. Chem. Soc. Chem. Commun. 1980, 533.
6. Harriman, A.; Richoux, M.-C. J. Photochem. 1981, 15, 335.
7. Inoue, H.; Chandrasekaran, K.; Whitten, D. G. J. Photochem. 1985, 30, 269.
8. Pileni, M.-P. Chem. Phys. Lett. 1980, 75, 540.
9. Mercer-Smith, J. A.; Mauzerall, D. Photochem. Photobiol. 1981, 34, 407.
10. Hopf, F. R.; O'Brien, T. P.; Scheidt, W. R.; Whitten, D. G. J. Am. Chem. Soc. 1975, 97, 277.
11. McLendon, G.; Smith, M. Inorg. Chem. 1982, 21, 847.
12. Hopf, F. R.; Whitten, D. G. J. Am. Chem. Soc. 1976, 98, 7422.
13. Whitten, D. G.; Meyer, T. J.; Hopf, F. R.; Ferguson, J. A.; Brown, G. Annals N. Y. Acad. Sci. 1973, 206, 516.
14. Turner, E. E.; LeFevre, R. I. W. Chemistry and Industry 1926, 45, 831, 883.
15. Bell, F.; Kenyon, J. Ibid. 1926, 45, 864.
16. Mills, W. H. Ibid. 1926, 45, 884, 905.
17. Testa, Principles of Organic Stereochemistry; Marcel Dekker, Inc.: New York, 1979.
18. Silvers, S.; Tulinsky, A. J. Am. Chem. Soc. 1964, 86, 927.
19. Hoard, J. L.; Hamor, M. J.; Hamor, T. A. J. Am. Chem. Soc. 1963, 85, 2334.
20. Gottwald, L. K.; Ullman, E. F. Tetrahedron Lett. 1969, 36, 3071.
21. Walker, F. A.; Avery, G. L. Tetrahedron Lett. 1971, 52, 4949.
22. Bonnet, J. J.; Eaton, S. S.; Eaton, G. R.; Holm, R. H.; Ibers, J. A. J. Am. Chem. Soc. 1973, 95, 2141.
23. Eaton, S. S.; Eaton, G. R. J. Am. Chem. Soc. 1975, 97, 3660.
24. Dirks, J. W.; Underwood, G.; Matheson, J. C.; Gust, D. J. Org. Chem. 1979, 44, 2551.
25. Hatano, K.; Anzai, K.; Kubo, T.; Tamai, S. Bull. Chem. Soc. Jpn. 1981, 54, 3518.
26. Collman, J. P.; Gagne, R. R.; Reed, C. A.; Halbert, T. R.; Lang, G.; Robinson, W. T. J. Am. Chem. Soc. 1975, 97, 1427, and references therein.
27. Schmehl, R. H.; Shaw, G. L.; Whitten, D. G. Chem. Phys. Lett. 1978, 58, 549.
28. Whitten, D. G.; Mercer-Smith, J. A.; Schmehl, R. H.; Worsham, P. R. Adv. Chem. Ser. 1980, 184, 47.
29. Chandrasekaran, K.; Giannotti, C.; Monserrat, K.; Whitten, D. G. J. Am. Chem. Soc. 1982, 104, 6200.
30. Freitag, R. A.; Mercer-Smith, J. A.; Whitten, D. G. J. Am. Chem. Soc. 1981, 103, 1226.
31. Freitag, R. A.; Whitten, D. G. J. Phys. Chem. 19833, 87, 3918.
32. Mislow, K.; Gordon, A. I. J. Am. Chem. Soc. 1963, 85, 3521.
33. Zimmerman, H. E.; Crumrine, D. S. J. Am. Chem. Soc. 1972, 94, 498.
34. Irie, M.; Yoshida, K.; Hayashi, K. J. Phys. Chem. 1977, 81, 969.
35. Imamura, A.; Hoffman, R. J. Am. Chem. Soc. 1968, 90, 5379.
36. Lewis, G. N.; Kasha, M. J. J. Am. Chem. Soc. 1944, 66, 2100.
37. Suzuki, H. Bull. Chem. Soc. Jpn. 1959, 32, 1340.
38. Wagner, P. J. J. Am. Chem. Soc. 1967, 89, 2820.

39. Saltiel, J.; Charlton, J. L. In Rearrangements in Ground and Excited States; Academic: New York, 1980; Vol. 3, Essay 14, 25.
40. Solov'ev, K. N.; Tsvirko, M. P.; Gradyushko, A. T.; Kozhich, D. T. Opt. Spectrosc. 1972, 33, 480.
41. Eastwood, D.; Gouterman, M. J. Mol. Spectrosc. 1970, 35, 359.
42. Eastwood, D.; Gouterman, M. J. Mol. Spectrosc. 1969, 30, 437.
43. Weller, A. Prog. in Rxn. Kinetics 1961, 1, 187.
44. Stone, A.; Fleischer, E. B. J. Am. Chem. Soc. 1968, 90, 2735.
45. Fleisher, E. B.; Stone, A. J.C.S. Chem. Comm. 1967, 332.
46. Meot-Ner, M.; Alder, A. D. J. Am. Chem. Soc. 1975, 97, 5107.
47. Gradyushko, A. T.; Tsvirko, M. P. Opt. Spektrosk. 1971, 31, 291.
48. Burgner, R. P.; Goncalves, A. M. J. Chem. Phys. 1974, 60, 2942.
49. Mukerjee, P.; Mysels, K. J. J. Am. Chem. Soc. 1955, 77, 2937.
50. Sato, H.; Kawasaki, M.; Kasatani, K.; Kusumoto, Y.; Nakashima, N.; Yoshihara, K. Chem. Lett. 1980, 1529.
51. Sato, H.; Kawasaki, M.; Kasatani, K. J. Photochem. 1981, 17, 243.
52. Kusumoto, Y.; Sato, H. Chem. Phys. Lett. 1979, 68, 13.
53. Sato, H.; Kusumoto, Y.; Nakashima, N.; Yoshihara, K. Chem. Phys. Lett. 1980, 71, 326.
54. Sato, H.; Kawasaki, M.; Kasatani, K.; Ban, T. Chem. Lett. 1982, 1139.
55. Law, K. Y. Photochem. Photobiol. 1981, 33, 799.
56. Almgren, M.; Greiser, F.; Thomas, J. F. J. Am. Chem. Soc. 1980, 102, 3188.

RECEIVED July 15, 1986

20

Quenching of Low-Lying Excited States in Porphyrins by Electron Acceptors in Rigid Matrices

Zbigniew Gasyna, William R. Browett, and Martin J. Stillman

Department of Chemistry and Centre for Chemical Physics, University of Western Ontario, London, Ontario, N6A 5B7 Canada

The photochemical formation of one-electron-oxidized, porphyrin π-cation radical species, in frozen solutions containing alkyl halides, has been characterized by optical absorption, magnetic circular dichroism (MCD) spectroscopies and the measurement of fluorescence intensity decay curves. The absorption and MCD spectra were analysed using a spectral envelope deconvolution computer program based on least squares and Simplex fitting procedures. The results of the deconvolution calculation of these spectra are consistent with the $^2A_{2u}$ and $^2A_{1u}$ models of the electronic ground state for $ZnTTP^{+\cdot}$ and $MgOEP^{+\cdot}$, respectively. The electron transfer process has been investigated by measuring fluorescence decay curves using a single-photon counting technique. The rate constant of the reaction has been determined for a number of the donor-acceptor systems, and was found to exhibit a strong dependence on exothermicity, reaching a maximum of 10^{13} s^{-1} in the most favorable reactions. These results are interpreted as indicating that there is an electron transfer reaction between the lowest excited singlet state of the porphyrin donor and the alkyl halide acceptor molecule.

Electron transfer reactions of photoexcited porphyrins and related compounds, have long attracted interest as model reactions for the photosynthetic apparatus of chloroplasts. In previous studies (1-3), we have shown that the products of photolysis in frozen solutions of a number of porphyrin-alkyl chloride donor-acceptor systems, are the one-electron oxidized radical species of the porphyrin and the reduced radical species of the acceptor. We now report a spectroscopic characterization of the porphyrin π-cation radical species formed, using absorption and magnetic circular dichroism (MCD) spectroscopies. In addition, we have carried out a complete spectral envelope deconvolution calculation that uses both absorption and MCD data to obtain band energy and intensity values for these complicated spectra. These results are compared with the spectral values predicted by theoretical calculations for both the $^2A_{2u}$ and $^2A_{1u}$ model species (4).

0097-6156/86/0321-0298$06.00/0

The study of the mechanism of the light-induced, intermolecular electron-transfer that occurs in frozen, rigid matrices between porphyrin electron donors, and quinone and alkyl halide electron acceptors, is an attractive experimental approach. Experiments carried out in the solid state provide a good environment in which to test the dependence of both the separation distance and excitation energy on the reaction rate, because the effects of diffusion become relatively insignificant. In such systems, electron transfer may be considered as a process that is competitive with the normal radiative decay of the excited state. Accordingly, it has been shown in a number of studies (5-7), that the efficiency of the "static" quenching of the donor or acceptor luminescence intensity correlates well with the calculated energetics for the electron transfer reaction itself. Analyses of pulsed excitation experiments (8-11) have demonstrated the presence of the predicted, but unusual, kinetics of the fluorescence decay measured from organic molecules which were studied in cryogenic glasses in the presence of electron acceptors. The photochemical oxidation of porphyrins in frozen solutions containing alkyl halides, is shown in this paper to involve the lowest excited singlet state of the porphyrin. We present results of deconvolution calculations carried out on the fluorescence intensity decay curves, from which we are able to show that the rate constant for the radical pair formation depends on the exothermicity of the electron transfer reaction.

Experimental

Materials and spectroscopic procedures. H_2TPP, ZnTPP and MgTPP were synthesized according to published procedures (12). MgOEP was kindly supplied by Dr. J. Fajer (Brookhaven National Laboratory). Reagent grade 2-chlorobutane, BuCl (BDH) was freshly distilled under nitrogen. CBr_4 (Kodak) was purified by recrystallization. Spectranalyzed CCl_4 (Fisher) was used without further purification. All other solvents were distilled under nitrogen before use. The quinones, p-benzoquinone, BQ (Fisher), tetrachloro-p-benzoquinone, p-Cl_4Q (Baker), tetrachloro-o-benzoquinone, o-Cl_4Q (Aldrich), and 2,3-dichloro-5,6-dicyanobenzoquinone, DDQ (Kodak) were purified by recrystallization. The porphyrins were dissolved in an appropriate solvent to which the acceptor was added. For the optical absorption measurements, the porphyrin solutions were placed in an optical cell, which was then plunged into liquid nitrogen in order to prepare the glass. The cell was quickly transferred into an Oxford Instrument CF204 cryostat, and the optical absorption spectra were recorded on a Cary 219 spectrophotometer. A 300 W tungsten-halogen Kodak projector lamp was used for the photolysis of the samples, which was carried out in the CF204 cryostat. The light was filtered through appropriate Corning filters. The MCD spectra were obtained with an Oxford Instruments SM-4 magnet which was mounted on a CD spectrometer built in this laboratory. All spectral data were stored digitally as recorded.

Spectral data analysis. The absorption and MCD spectra of ZnTPP and MgOEP were fitted with Gaussian lineshapes, and the band parameters were determined using the rigid-shift, Born-Oppenheimer, and Franck-Condon assumptions (13). Iterative, non-linear, least squares and SIMPLEX fitting procedures, used for the absorption and MCD band fitting, respectively, were developed in this laboratory (14). In order to enhance the quality of the fits calculated for each band envelope, these programs allowed the parameters calculated for the absorption spectrum to be used directly with the MCD and *vice versa*. The data presented here were calculated using the same parameters to fit both the absorption and the MCD spectrum measured for each complex.

Fluorescence and phosphorescence measurements. Steady-state fluorescence and phosphorescence measurements were made using a cylindrical Pyrex tube (4 mm o.d.) inserted into a quartz Dewar filled with liquid nitrogen. The fluorescence and phosphorescence spectra were determined on a Perkin-Elmer MPF-4 spectrofluorometer. Low-temperature fluorescence decay data were obtained using a Photochemical Research Associates (PRA, London, Ontario, Canada) Model 3000 Nanosecond Lifetime Fluorometer (15), which is based on the technique of time-correlated single-photon counting (16). A pulsed hydrogen arc lamp operating at ~30 kHz was used as the excitation source. Fluorescence lifetimes and decay curves were obtained by application of fast Fourier transform convolution and deconvolution techniques (17). Suitable programs were developed in this laboratory for use with the IBM Instruments S9001 computer (18).

Results and Discussion

Photolysis of H_2TPP. Figure 1 shows a series of optical absorption spectra measured at 79 K during the visible light irradiation of a glassy solution of H_2TPP in BuCl which contained 0.8 mol/L of CBr_4. The irradiation results in a decrease in the intensity of the porphyrin absorption bands located at 650, 600, 550, 525, and 420 nm (Figure 1, line *a*), while new, more prominent absorption bands grow in at about 690, 495, and 455 nm (Figure 1, lines *b*, *c* and *d*). Figure 1, line *e*, shows a spectrum which was calculated by subtracting, by computer, an estimated residual amount of the spectrum of the remaining unoxidized porphyrin (Figure 1, line *a*), from the spectrum recorded at the end of the irradiation (Figure 1, line *d*), on the basis of an estimated 85% photolytic conversion. The spectral features of the photochemical product displayed in Figure 1, line *e*, are characteristic of a porphyrin π-cation radical species (1,2), thus, we characterize the spectral changes shown in Figure 1 as arising from the formation of the $H_2TPP^{+\cdot}$ species.

Photolysis of ZnTPP and MgOEP. A number of metalloporphyrins have been photooxidized to produce the π-cation radical species at 77 K (1-3). The high yields of the photooxidized product obtained when alkyl halide glasses have been used, has resulted in the almost complete conversion to the π-cation radical species in several cases. Photolysis of ZnTPP and MgOEP can be carried out at low temperatures to yield nearly 100% of the photooxidized products. The solid lines in both the absorption (upper) and MCD (lower) panels of Figures 2 and 3, show the spectra recorded after photolysis of ZnTPP (Figure 2) and MgOEP (Figure 3) at 77 K, in the presence of CCl_4. The starred bands represent a small amount of remaining, unoxidized porphyrin.

Analysis of the absorption and MCD spectra of $ZnTPP^{+\cdot}$ and $MgOEP^{+\cdot}$. The optical absorption and MCD spectra of $ZnTPP^{+\cdot}$ and $MgOEP^{+\cdot}$, shown in Figures 2 and 3, are representative of the $^2A_{2u}$ and $^2A_{1u}$ ground states, respectively, that are found for porphyrin π-cation radical species. Results from deconvolution calculations carried out for both the absorption and MCD spectral envelopes of these π-cation radical spectra are also shown in Figures 2 and 3, plotted as the individual component bands that are

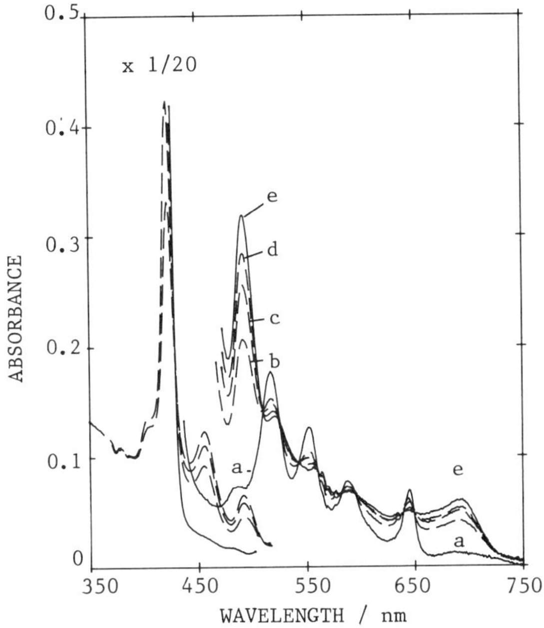

Figure 1. Absorption spectra measured at 79 K for (a) H_2TPP before irradiation, and (b-d) H_2TPP after increasing irradiation times, in a BuCl-CBr_4 (0.8 mol/L) glass. (e) The calculated absorption spectrum of 100% $H_2TPP^{+\cdot}$ in BuCL-CBr_4 (0.8 mol/L) at 79 K.

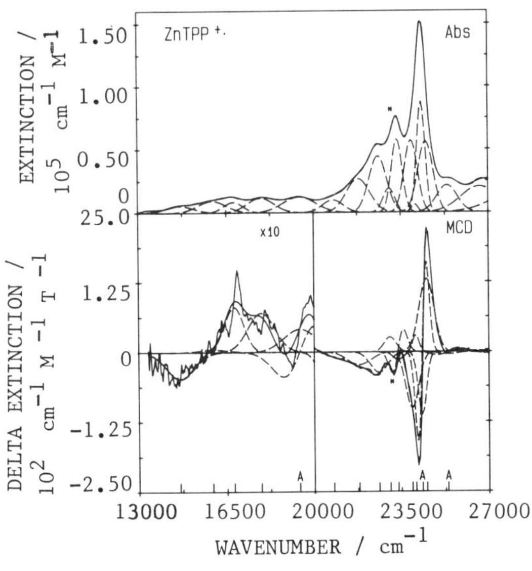

Figure 2. Computer deconvolutions of the absorption and MCD spectra of $ZnTPP^{+\cdot}$ in BuCl containing CCl_4 at 79 K. The starred band represents part of the spectrum of unoxidized ZnTPP.

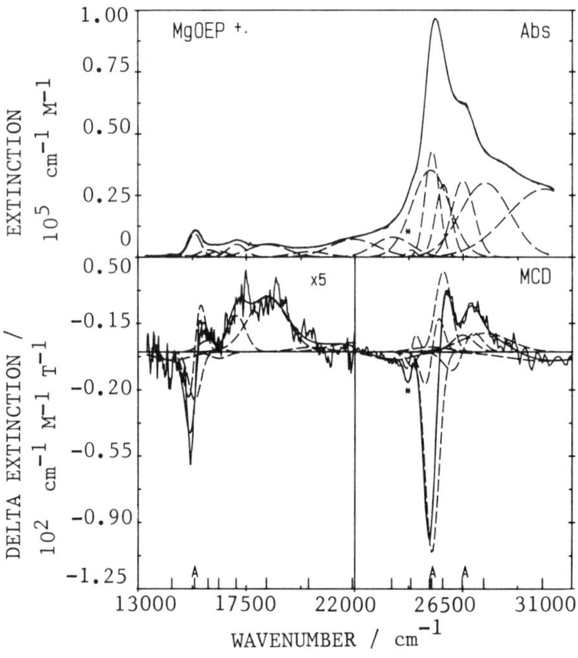

Figure 3. Computer deconvolutions of the absorption and MCD spectra of $MgOEP^{+\cdot}$ in BuCl containg CCl_4 at 79 K. The starred band represents part of the spectrum of unoxidized MgOEP.

responsible for the total, observed intensity. The parameters used for each of these component bands are tabulated in Table I. This table summarizes data from the theoretical calculations of Edwards and Zerner (4) for both $ZnTPP^{+\cdot}$, with either a $^2A_{2u}$ or $^2A_{1u}$ ground state and $MgOEP^{+\cdot}$, with a $^2A_{1u}$ ground state, giving the energy (in units of 1000 cm^{-1}) and oscillator strength (osc) for transitions to a number of excited states, identified by their excited state symmetry. The experimental data listed is for component bands used to fit the observed 77 K absorption and MCD spectra ("Calculated"), with the oscillator strength (osc) indicated for absorption bands and A/D values indicated for bands which were fitted with A terms in the MCD. In addition, for each absorption spectrum, a series of band centres were obtained ("Observed") by measuring the wavelengths of the band maxima directly from the spectra.

The deconvolution calculations were constrained to using the minimum number of Gaussian bands that would provide a "best fit" for *both* the absorption and MCD spectra using the *same* set of band centre and band width parameters. Our experience indicates that for spectra containing a large number of overlapping bands with similar intensity, fitting either the absorption or the MCD spectra separately produces ambiguous results. The predicted oscillator strengths ("osc" in Table I) of the absorption bands in the visible region of the theoretically-calculated spectra (4), roughly follow the trends actually observed in the absorption spectra, and quantified by the deconvolution calculation, Table I. Specifically, in the case of the $^2A_{2u}$ $ZnTPP^{+\cdot}$ species, a series of overlapping, weak bands is observed extending from 15 000 - 23 000 cm^{-1} (667 - 440 nm), with no well resolved features. By way of contrast, a well resolved band is seen as the lowest energy band at 15 400 cm^{-1} (649 nm) for the $^2A_{1u}$ $MgOEP^{+\cdot}$ species, this band is followed to higher energy by a series of overlapping bands that are similar to those observed for $ZnTPP^{+\cdot}$.

For both species, the deconvolution data show that in the UV region the absorption spectrum is dominated by several transitions, and that these transitions are all much more intense than the bands in the visible region. The deconvolution results clearly indicate that these π-cation radical spectra are far more complicated than the theoretical predictions have suggested (4), and that the band energies determined by the theoretical calculations (4), increasingly skew to higher energies as the energies of the transitions increase, relative to the band energies calculated directly from the spectral data.

The MCD data not only help to constrain the absorption spectrum fitting calculations, but also provide information about the degeneracy of the excited states involved, in the form of MCD A or B terms. The theoretical calculations (4) have indicated that many of the transitions which give rise to intensity in the absorption spectrum should be degenerate and this should result in A terms being observed in the MCD spectrum. However, few distinct A terms were observed in the spectra, Figure 2 and 3. No significant A term contribution was found in the visible region MCD spectrum of the $^2A_{2u}$ $ZnTPP^{+\cdot}$ species, although there appears to be a distict A term at 24 300 cm^{-1} (412 nm). In contrast, a resolved A term was found in the visible region of the $^2A_{1u}$ $MgOEP^{+\cdot}$ spectrum, at 15 300 cm^{-1} (653 nm), while to higher energies, the A terms appear to be much less intense, if present at all, than was found in the $ZnTPP^{+\cdot}$ spectrum.

Fluorescence intensity decay curves for H_2TPP. Figure 4A shows fluorescence decay profiles recorded for H_2TPP in BuCl and in BuCl-CBr_4 glassy solutions at 77 K. A single-exponential deconvolution was found to

Table I. Results of spectral envelope deconvolution calculations carried out on the absorption and MCD spectra of $ZnTPP^{+\cdot}$ and $MgOEP^{+\cdot}$.

$ZnTPP^{+\cdot}$	Theory[a]		Experiment		
Ex. State Symm.[a]	$^2A_{2u}$ Energy (osc)	$^2A_{1u}$ Energy (osc)	Observed[b] Energy	Calculated[c] Energy (osc)	MCD A/D
				14.6 (0.033)	
^2E_g	15.5 (0.01)	15.0 (0.04)		15.9 (0.056)	-
^2E_g	16.9 (0.00)	17.4 (0.00)	16.6	16.8 (0.035)	-
			17.9	17.9 (0.067)	-
^2E_g	19.8 (0.01)	18.6 (0.01)	19.4	19.4 (0.088)	0.863
				20.8 (0.045)	-
$^2A_{1u}$	21.3 (0.01)			21.8 (0.134)	-
			22.6	22.6 (0.169)	-
				23.4 (0.135)	-
				23.9 (0.159)	-
^2E_g	30.0 (0.01)	30.6 (0.02)		24.1 (0.008)	-
^2E_g	32.9 (4.42)	32.5 (3.85)	24.3	24.3 (0.157)	1.070
				24.5 (0.182)	-
			25.3	25.3 (0.095)	0.453
			26.7	26.6 (0.163)	-
^2E_g	36.6 (0.00)	37.0 (1.37)		30.1 (0.055)	-
^2E_g	42.2 (0.21)	40.5 (0.47)		31.2 (0.118)	-
^2E_g	43.5 (0.39)				

$MgOEP^{+\cdot}$	Theory[a]	Experiment		
Excited state symm.	$^2A_{1u}$ Energy (osc)	Observed[b] Energy	Calculated[c] Energy (osc)	MCD A/D
			14.4 (0.007)	
^2E_g	13.9 (0.077)	15.4	15.3 (0.027)	0.471
			15.9 (0.013)	-
^2E_g	16.1 (0.013)		16.4 (0.010)	-
		17.1	17.1 (0.021)	-
^2E_g	17.9 (0.002)	18.5	18.3 (0.044)	-
			20.1 (0.017)	-
$^2A_{2g}$	25.2 (0.032)		21.9 (0.077)	-
			23.5 (0.068)	-
^2E_g	29.7 (0.502)			
			25.1 (0.295)	-
^2E_g	30.7 (2.434)	25.3	25.2 (0.136)	0.409
			25.7 (0.103)	-
^2E_g	34.2 (0.405)	26.6	26.5 (0.154)	0.522
^2E_g	38.4 (0.850)		27.4 (0.330)	-
^2E_g	41.2 (0.264)		29.8 (0.468)	-

[a] From ref. (4). Band centres are given in units of 1000 cm^{-1}.
[b] Values obtained by measuring λ_{max} in the absorption spectrum only.
[c] Calculated with band shape fitting programs to provide acceptable fits in both the absorption and the associated MCD spectrum.

give a satisfactory fit for the fluorescence decay of H_2TPP in the neat BuCl glass at 77 K (line *a*). The fluorescence profiles for H_2TPP in the presence of CBr_4 were, however, found to deviate significantly from calculated, single-exponential decay curves (Figure 4A, lines *b* and *c*). This deviation was observed with excitation into either the B band (424 nm) or the Q band (560 nm) regions; both excitations led to identical quenching effects. Figure 4B shows the results obtained for MgTPP, clearly quite similar to the H_2TPP system. In both of these systems, we attribute the fluorescence quenching by CBr_4 to the electron transfer reaction: $MTPP^* + CBr_4 \rightarrow MTPP^{+\cdot} + CBr_4^{-\cdot}$ (1-3). The electron transfer process can be treated by formal analogy to the triplet-triplet energy transfer of Dexter's electron exchange mechanism (19). We have tested the function derived by Inokuti and Hirayama (20) for the decay of donor luminescence in a diffusion free, random ensemble of donor and acceptor molecules. The steady-state luminescence intensity of the donor is predicted, under these conditions, to depend exponentially on the concentration, C, of the electron acceptor,

$$I/I_o = e^{-4\pi R_q^3 C/3} \tag{1}$$

where, R_q is a "critical transfer distance".

A test of this dependence is shown in Figure 5 for the fluorescence quenching observed in a number of ZnTPP-acceptor solutions at 77 K. The time-dependence of the relative emission intensity, $I(t)/I_o$, may be given by

$$I(t)/I_o = e^{-\left[t/\tau_o + 4/3\ \pi(L/2)^3 C\ g(t) \right]} \tag{2}$$

In Equation 2, τ_o is the fluorescence lifetime in the absence of the acceptor, C is the acceptor concentration, and L is defined by the dependence of the electron transfer rate constant, k(R), on distance, R, as $k(R) = \nu e^{-2R/L}$, here ν is the electron-transfer rate constant for donor-acceptor pairs in contact. The function g(t) in Equation 2 is well approximated by $g(t)=(\ln\nu t)^3+1.732(\ln\nu t)^2+5.934(\ln\nu t)+5.445$, when $\nu t>10$ (20). The value of ν can be obtained by fitting this theoretical decay law into the experimental data.

The deconvolution of the data recorded in the presence of the electron acceptor, Figure 4A and 4B, lines *b* and *c*, represent best fits of the experimental decay profiles when ν is used as the adjustable parameter in Equation 2. The value of parameter L/2 was set to 0.83 Å, as recently determined by Miller et al. (21) for the reaction of radical anions of aromatic hydrocarbons with organic molecules, for which the assumption was made that this parameter should be invariant for reactions of a common electron donor with acceptors having different electron affinities. The magnitude of parameter L/2 is justified both by theoretical and experimental data obtained for the distance dependence of the electron-transfer rate. An alternative approach is to fit each experimental decay curve to Equation 2 allowing both L/2 and ν to vary. For example, fitting the data for the $MgTPP^* + CBr_4$ reaction system, gives L/2=0.5 Å and $\nu=1.2 \times 10^{13}\ s^{-1}$. However, even though adjusting L/2 can improve the quality of the fitting, the interpretation of the decay curves with such an adjustable L/2 may have no physical meaning since this parameter really depends on the nature of the electron donor alone, for example, on the binding energy of the electron on

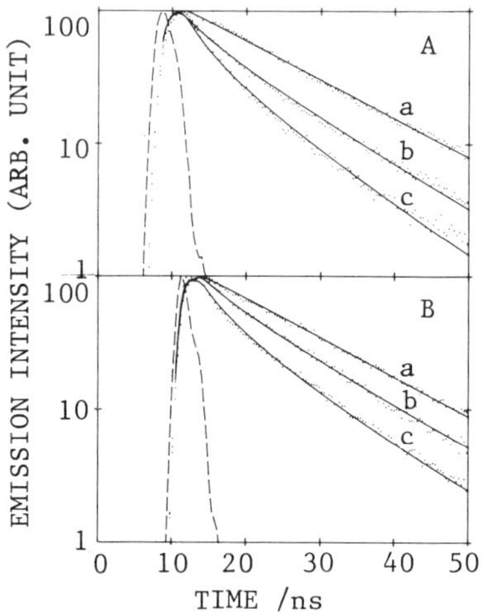

Figure 4. Observed fluorescence decay profiles (points) and theoretical fits (solid lines), calculated by the deconvolution program according to Equation 2, using ν as an adjustable parameter in the fit, for **(A)** H_2TPP and **(B)** MgTPP in BuCl containing CBr_4 at 77 K, following excitation at 515 nm and 424 nm, respectively. The concentration of CBr_4 was (a) 0, (b) 0.4, and (c) 0.8 mol/L. The parameters used in the fits were, for H_2TPP, τ_0=14.5 ns, L/2=0.83 Å and $\nu=1.4 \times 10^{13}\ s^{-1}$, and for MgTPP, τ_0=15.5 ns, L/2=0.83 Å and $\nu=6.5 \times 10^{12}\ s^{-1}$.

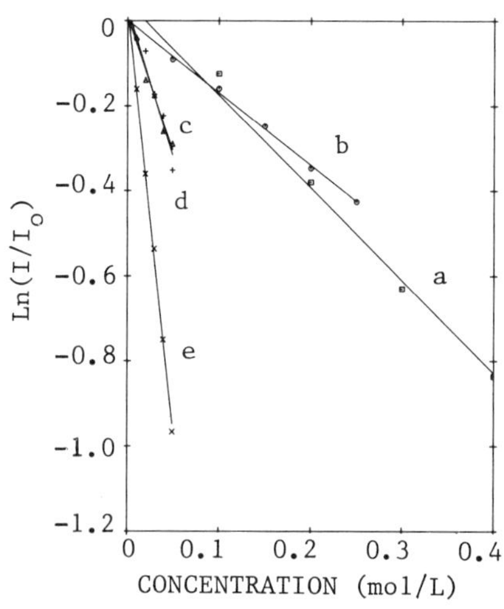

Figure 5. Relative fluorescence intensities at 77 K, plotted as $\ln(I/I_0)$, of ZnTPP excited at 555 nm vs. acceptor concentration. The acceptors solutions were: (a) CBr_4 in BuCl, and for (b), (c), (d), (e), BQ, o-Cl_4Q, p-Cl_4Q and DDQ, all in MTHF, respectively. The solid lines are least-squares fits to the experimental data.

the electron donor. It should be noted that the value of parameter L/2 for the electron-transfer quenching of an state may be actually different than assumed here, because there are differences in the electron binding energies between the aromatic radical anion used by Miller et al. (21) and the porphyrin excited singlet state that acts as the donor in this work. Thus measurements of fluorescence quenching by others have yielded L/2=0.7 Å (11) and 0.57 Å (10), for two different electron donors, the ruthenium phenanthroline complex and indole, respectively.

Table II presents the values of ν, the rate constant for the electron transfer reaction with the donor and acceptor in contact, calculated by deconvolution of the fluorescence decay curves for a number of excited porphyrin-alkyl halide systems. It appears that the rate parameter depends strongly on the calculated exothermicity for these reactions. Parameter ν contains information about the Franck-Condon factor of the electron-transfer reaction, which is in itself dependent on the reaction exothermicity and reorganization energy (22,23). Whether the rate constant for the electron-transfer reactions depends on the exothermicity in the manner predicted by theory, that is with a simple Gaussian dependence (22), cannot be answered at present because of the uncertainties in the energetics of the particular reactions studied here.

Table II. Energetics and rate constants for electron transfer quenching of the fluorescence of porphyrins by alkyl halides in BuCl at 77 K.

Reaction	$\Delta G(ET)/eV^{a}$	ν/s^{-1} [b]
$H_2TPP^* + CCl_4$	-0.13	5.0×10^8
$H_2TPP^* + CBr_4$	-0.61	1.4×10^{13}
$ZnTPP^* + CCl_4$	-0.56	1.0×10^{10}
$ZnTPP^* + CBr_4$	-1.04	2.5×10^{12}
$MgTPP^* + CHCl_3$	0.17	$< 3.0 \times 10^7$
$MgTPP^* + CCl_4$	-0.72	1.0×10^9
$MgTPP^* + CBr_4$	-1.20	6.5×10^{12}

[a] The thermodynamic potential change was calculated from data measured in fluids, assuming $\Delta G(ET)=\Delta G(D^+,A^-)-E(D^*)$, where $\Delta G(D^+,A^-)=E(D^+,D)-E(A,A^-)$, and $E(D^+,D)$ and $E(A,A^-)$ are redox potentials of the donor and acceptor, respectively, and $E(D^*)$ is the excited state energy. In the calculations, the redox potentials of 0.95 V, 0.71 V and 0.54 V (vs. SSCE) and energies of the singlet excited state of 1.86 eV, 2.05 eV and 2.03 eV were used for the donors H_2TPP, ZnTPP and MgTPP (24), respectively. The acceptor redox potentials were taken from the values of the irreversible half-wave reduction potentials which were equal to -1.67 V, -0.78 V and -0.30 V (vs. SSCE) for $CHCl_3$, CCl_4 and CBr_4 (25), respectively.

[b] The uncertainty in the ν values was estimated to be ±20%.

Acknowledgments

The authors gratefully acknowledge the financial support of this work by the NSERC of Canada through grants under the Operating, Equipment and Strategic (Energy) programs and from the Academic Development Fund at the UWO (to M.J.S.). The authors are associated with the Centre for Chemical Physics at the UWO.

Literature Cited

1. Gasyna, Z., Browett, W.R., Stillman, M.J., Inorg. Chim. Acta 1984, 92, 37.
2. Gasyna, Z., Browett, W.R., Stillman, M.J., Inorg. Chem. 1985, 24, 2440.
3. Gasyna, Z., Browett, W.R., Stillman, M.J., Inorg. Chem. 1984, 23, 382.
4. Edwards W.D., Zerner M.C., Can. J. Chem. 63, 1763 (1985).
5. Miller, J.R., Hartman, K.W., Abrash, S., J. Am. Chem. Soc. 1982, 104, 4296.
6. Miller, J.R., Peeples, J.A., Schmitt, M.J., Closs, G.L., J. Am. Chem. Soc. 1982, 104, 6488.
7. Guarr, T., McGuire, H., Strauch, S., McLendon, G., J. Am. Chem. Soc. 1983, 105, 618.
8. Khairutidinov, R.A., Sadovskii, N.A., Parmon, V.N., Kuzmin, M.G., Zamaraev, K.I., Dokl. Akad. Nauk SSSR 1975, 220, 888.
9. Sadovskii, N.A., Kuzmin, M.G., Dokl. Akad. Nauk. SSSR 1975, 222, 1380.
10. Namiki, A., Nakashima, N., Yoshihara, K., J. Chem. Phys. 1979, 71, 925.
11. Strauch, S., McLendon, G., McGuire, M., Guarr, T., J. Phys. Chem. 1983, 87, 3579.
12. Fuhrhop, J.H. In "Porphyrins and Metalloporphyrins"; Smith, K.M., Ed.; Elsevier: Amsterdam, 1975; pp 757-869.
13. Piepho S.B., Schatz P.N. (1983) in Group Theory In Spectroscopy, John Wiley and Sons, New York.
14. Browett W.R., Stillman M.J. unpublished.
15. Gudgin, E., Lopez-Delgado, R., Ware, W.R. Can. J. Chem. 1981, 59, 1037.
16. Ware, W.R. In "Creation and Detection of the Excited State"; Ware, W.R., Ed.; Marcel Dekker: New York, 1971; Vol. 1, pp 213-301.
17. Good, H.P., Kallir, A.J., Wild, U.P., J. Phys. Chem. 1984, 88, 5435.
18. Gasyna, Z., Stillman, M.J., to be published.
19. Dexter, D.L., J. Chem. Phys. 1953, 21, 836.
20. Inokuti, M., Hirayama, F., J. Chem. Phys. 1965, 43, 1978.
21. Miller, J.R., Beitz, J.V., Huddleston, R.K., J. Am. Chem. Soc. 1984, 106, 5057.
22. Marcus, R.A., J. Chem. Phys. 1956, 24, 966.
23. Ulstrup, J., Jortner, J., J. Chem. Phys. 1975, 63, 4358.
24. Darvent, J.R., Douglas, P., Harriman, A., Porter, G., Richoux, M.-C. Coord. Chem. Rev. 1982, 44, 83.
25. Stackelberg, M., Stracke, W., Z. Electrochem. 1949, 53, 118.

RECEIVED March 28, 1986

21

Photooxidation of Phthalocyanines

Photoinduced Reactivity of the Triplet State

Tebello Nyokong, Zbigniew Gasyna, and Martin J. Stillman

Department of Chemistry, University of Western Ontario, London, Ontario, N6A 5B7, Canada

Porphyrins and phthalocyanines can act as highly efficient electron donor molecules in linked donor-acceptor complexes that may be used in molecular solar cells. This paper describes the photochemical reaction that occurs in a series of metallophthalocyanines following irradiation into the phthalocyanine's Q band in the presence of electron acceptors which are irreversibly reduced. The products of these photooxidation reactions were characterized by optical absorption, magnetic circular dichroism (MCD) and electron paramagnetic resonance (EPR) spectroscopies. The MCD spectra, in particular, provide a starting point in the theoretical analysis of the origin of the electronic transitions in the spectra of the oxidized porphyrin species. These studies show that for zinc phthalocyanine, ruthenium phthalocyanine and cobalt phthalocyanine, one-electron oxidation of the π ring takes place. However, oxidation of the ferrous ion takes place before the π ring in iron phthalocyanine. The spectral data from photochemical experiments are compared with data from controlled potential electrolysis experiments. Transition energies calculated from band shape deconvolution programs that use both the absorption and MCD spectra are compared with the energies calculated theoretically.

Photochemical properties of transition metal phthalocyanines have been a subject of several investigations in recent years (1-5), due in part, to the prospect of their use in photovoltaic and photocatalytic processes. The ultra-violet photochemistry of these compounds has provided considerable information about the redox processes which can originate from π ring-centered $n\pi^*$ states; such photochemical reactions include, for example, hydrogen abstraction from solvent molecules (3-6). However, excitation into the lowest energy excited singlet state of the phthalocyanine π ring, where $\lambda_{max}> 640$ nm in, for example, ruthenium phthalocyanine (RuPc) and zinc phthalocyanine (ZnPc), populates only the lowest-lying triplet π^* excited state. This state can then be quenched by electron transfer to a number of electron acceptors (3-7). Spectral studies on the oxidation of RuPc in solution [8], and ZnPc in the solid state (9), show that the one-electron

0097-6156/86/0321-0309$06.00/0

oxidation is centered on the phthalocyanine ring rather than on the metal, yielding phthalocyanine complexes which involve a bivalent transition metal and a phthalocyanine π cation radical.

Although there have been several reports concerning the formation of the RuPc π cation radical in solution (8-10), there have been no reports that a stable ZnPc π cation radical species can be formed in solution. This is due mainly to the limited solubility of ZnPc in solvents commonly used in electrochemistry and photochemistry. In recent papers, we have reported that the photoinduced, one-electron oxidation of a variety of porphyrins (11-13) and of RuPc (8), takes place readily in the presence of alkyl halide acceptors. In the first part of this paper, we report that irradiation, using visible region light, of room temperature ZnPc solutions in methylene chloride containing CBr_4, leads to the one-electron photooxidation reaction. The resulting ZnPc π cation radical species has been characterized using the techniques of magnetic circular dichroism (MCD) and electron paramagnetic resonance (EPR) spectroscopies. The spectral properties of the photooxidized ZnPc species are compared with those of the electrochemically-oxidized ZnPc and with the spectra of the RuPc π cation radical species. We use the spectral data obtained from the ZnPc studies to aid in the assignment of the spectra of MPc π cation radicals species in general.

Unlike Zn(II)Pc and Ru(II)Pc, Co(II)Pc and Fe(II)Pc are known to undergo oxidation at the metal prior to ring oxidation (14,15). Electrochemical studies of CoPc have, however, shown that ring oxidation does occur before metal oxidation if weakly-coordinating ligands are employed (16). The products of the chemical oxidation of CoPc and FePc in solution, have been only poorly characterized in the past (17,18). Lever and coworkers (9,15) have reported the oxidation of CoPc and FePc in the solid state and characterized the reaction as metal oxidation, M(II)Pc(-2) $\rightarrow$ M(III)Pc(-2), followed by phthalocyanine π ring oxidation, M(III)Pc(-2) $\rightarrow$ M(III)Pc(-1). However, no electro- or photo-chemical oxidation studies of CoPc and FePc, carried out in solution, have been reported to date. We report here the first photochemical oxidation reaction that leads quantitatively, at short irradiation times, to stable products for both CoPc and FePc, in solution at room temperature. The development of a photochemical route for the quantitative formation of these MPc(-1) species, provides an important new synthetic technique for use in the production of such species.

Experimental

In order to produce phthalocyanine species which have a higher than normal solubility in solvents like dichloromethane the pyridine complexes were made. FePc (Eastman), CoPc (Kodak) and ZnPc (Eastman), were Soxhlet extracted with pyridine (py), evaporated to dryness and washed with hexanes to give the appropriate pyridine-coordinated metal phthalocyanine complex, $MPc(py)_n$. For CoPc, we characterize the complex as the diadduct, $CoPc(py)_2$, following the work of Cariati et al. (19). FePc forms diadducts in the presence of excess nitrogen binding ligands (20), so we characterize the pyridine complex used in this work as $FePc(py)_2$. ZnPc is known to bind only one axial ligand (21), hence the resulting pyridine complex is characterized as the ZnPc(py) species. The imidazolato- complexes, ($FePc(Im^-)_2$ and $ZnPc(Im^-)$), were prepared by refluxing FePc and ZnPc, respectively, in dimethylacetamide (DMA) in the presence of excess ligand. $ZnPc(CN^-)$ was prepared by refluxing ZnPc in DMA in the presence of excess of sodium cyanide. RuPc was prepared by literature methods (10). All solvents were freshly distilled. Ferrocene (fc), carbon tetrabromide (CBr_4), tetraethylammonium perchlorate ($[(Et)_4N]ClO_4$, Kodak), tetrapropylammonium perchlorate ($[(Pr)_4N]ClO_4$, Kodak), and tetrabutylammonium fluoroborate ($[(Bu)_4N]BF_4$, Fisher) were recrystallized before use.

Electrochemistry. Electrochemical measurements were carried out with the Princeton Applied Research Model 273 electrochemical system controlled by an IBM Instruments S9001 computer using the computer program ELECTRA (22). For cyclic voltammetry (CV) and differential pulse voltammetry (DPV), a platinum wire was used as an auxiliary electrode and a platinum bead was used as the working electrode. Ferrocene and a silver wire were used as internal standards, while the saturated calomel electrode (sce) was used as an external standard. For exhaustive electrolysis, an "H" cell was used in which the working electrode, consisting of a cylindrical platinum gauze (Johnson-Matthey, diameter 13 mm and height 40 mm), and the silver wire reference electrode, were housed in one compartment, while the auxiliary electrode, consisting of a 22 by 15 mm platinum plate, was housed in the other.

Photochemistry. For photochemical experiments, each MPc complex was dissolved in nitrogen-saturated dichloromethane, which contained 10^{-3} **M** CBr_4 as the electron acceptor. The solutions were irradiated with light from a 300 W tungsten-halogen projector lamp and excitation into the phthalocyanine's Q band (λ_{max}> 600 nm) was ensured by the use of a Corning CS 2-73 high energy (580 nm) cut-off filter. Absorption spectra were obtained with a Cary model 219 spectrophotometer, while MCD spectra were obtained with a JASCO J500C spectropolarimeter controlled by an IBM Instruments S9001 computer using the computer program CDSCAN5 (Kitchenham and Stillman, unpublished). EPR spectra were recorded on a Varian E-12 EPR spectrometer, equipped with a Varian E-257 temperature controller and interfaced to a Nicolet 1180 computer through a Nicholet Explorer III digital oscilloscope. For EPR measurements, the MPc solutions were transferred to 4 mm o.d tubes, glassed in liquid nitrogen and introduced into a dewar, filled with liquid nitrogen.

Quantum yield experiments, employing Reinecke's salt, $KCr(NH_3)_2(NCS)_4$, actinometry (23), were carried out with a PTI Quantacount, in conjunction with an LPS 200 power supply and a 100 W mercury lamp. For these measurements, the phthalocyanines were dissolved in nitrogen-purged dichloromethane containing CBr_4 (10^{-3} **M**). Irradiation took place in a 1 cm cell using wavelengths in the region of the phthalocyanine's Q band.

Results and Discussion

Both chemical and electrochemical oxidation techniques are well established as routes to the production of the π cation radical species of porphyrins and phthalocyanines (9,14,24). However, it is only recently that the formation of a stable porphyrin π cation radical species, at high yields at room temperatures, has been demonstrated using low energy light (11-13). The quenching of phthalocyanine excited states to give the π cation radical species is generally a more difficult task since only the lowest energy excited triplet states, with an energy of the order of 1.1 to 1.3 eV above the ground state (1), can be used, while the energies of the photoactive states in the porphyrins range from 1.4 to 2.3 eV (1), thus there is far less free energy available for reduction of an electron acceptor in the phthalocyanines when compared with the analogous porphyrin system, *vide infra*.

Ruthenium phthalocyanine. Photolysis of many complexes of RuPc in the presence of CBr_4, results in the formation of the ring oxidized phthalocyanine (8). Figure 1 shows absorption spectra recorded at various stages during photolysis of RuPc(CO)(DMF). The most significant changes observed are (i) the loss in intensity of the Q band at 641 nm, and (ii) the concurrent growth in intensity of bands near 400, 510 and 700 nm. The photochemically-induced quenching of intensity in the Q band, can be reproduced using electrochemical oxidation if controlled potential coulometry is carried out at the oxidation couple potential of $E_{1/2}$ = 0.77 V vs. the saturated calomel electrode (sce), Table I. EPR spectra measured at 79 K for RuPc(-1)(CO)(DMF) formed photochemically in a frozen glass, exhibited a free electron-like band centred at 2.0048, Table II. Therefore, we have characterized the final spectrum shown in Figure 1 as arising from the Ru(II)Pc(-1) π cation radical species.

Zinc phthalocyanine. The visible region spectra recorded during the photolysis of a range of ZnPc(L) complexes, where L=CN^-,Im^- and py, revealed several changes connected by good isosbestic points, Figure 2. These changes are similar to the changes observed when solutions of the pyridine and imidazolate complexes of ZnPc are exhaustively electrolysed near the potential of the positive peak in the cyclic voltammetry trace ($E_{1/2}$ = ~0.70 V vs. sce (25), Table I). In addition, these spectral changes are similar to those observed in the photochemical (Figure 1 and reference 8) and electrochemical (10) oxidations of complexes of RuPc. The spectrum of the oxidized ZnPc(py) species exhibits four major bands, centered at 440, 500, 720 and 830nm. Insert (a) of Figure 2 shows the last spectrum recorded during the photolysis in Figure 2, from which the residual, unoxidized fraction of ZnPc(-2)(py) has been subtracted by computer; in essence, Insert (a) shows a spectrum of 100% of the π cation radical complex, $[ZnPc(-1)(py)]^{+\cdot}$. Insert (b) shows the time dependence of the spectral changes observed during the irradiation. The appearance of isosbestic points in the series of spectra shown in Figure 2, is a good indication that the four new bands are due to the formation of the same species, which we characterize as the ZnPc(-1) π cation radical. The general broadening of each of the bands in the absorption spectrum, together with a decrease in intensity, seems to be typical behaviour of ligand oxidation of porphyrins and phthalocyanines (8,11). Since the electron is removed from the a_{1u} orbital of the phthalocyanine ring (26), the original Q band transition (identified as $1a_{1u}(\pi) \longrightarrow 1e_g(\pi^*)$, following the four-orbital notation of Gouterman et al. (26)) is drastically affected.

Figure 3 shows the absorption and MCD spectra recorded under the same conditions during a photolysis of ZnPc(-2)(py). In the upper panel (ABS) are shown the initial absorption spectrum recorded before photolysis (dashed line), identified by the intense Q band, and the final spectrum recorded at the completion of the photooxidation (solid line). The lower panel shows the MCD spectrum recorded for the photooxidized product, the $[ZnPc(-1)(py)]^{+\cdot}$, between 300 and 860 nm (solid line), and the UV portion of the MCD spectrum of the unoxidized complex (dashed line) for comparision purposes. There was still some residual unoxidized ZnPc present when the final spectra were recorded, and this results in the appearance of

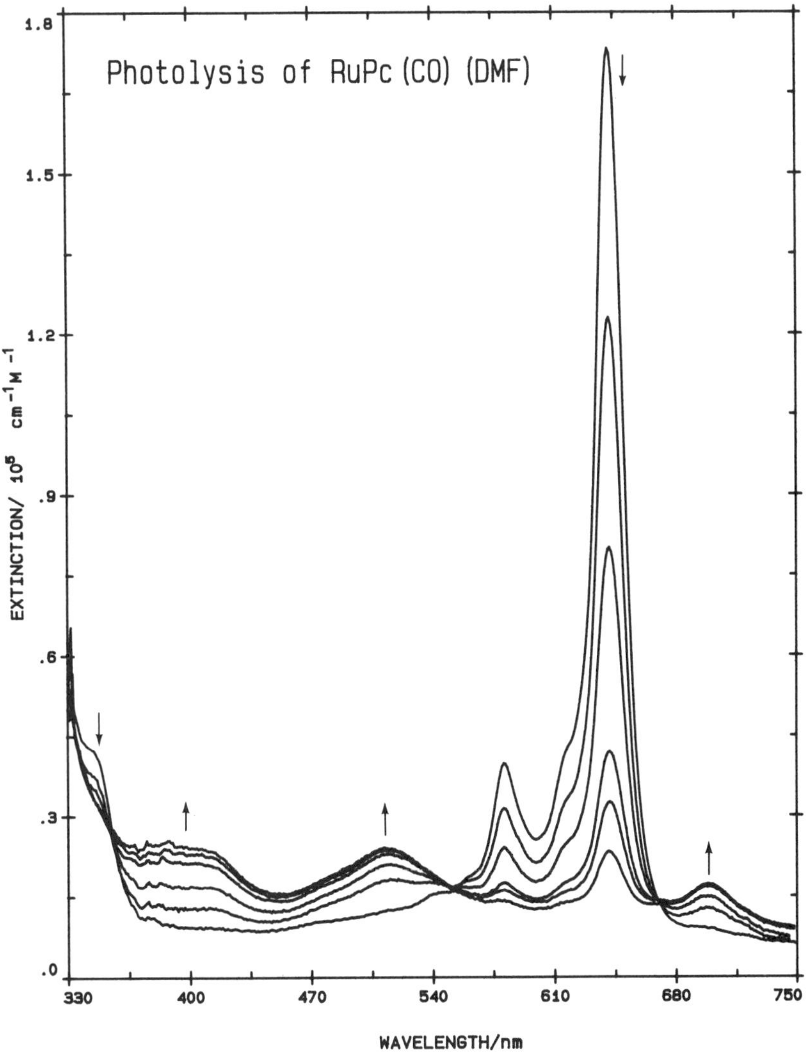

Figure 1. Absorption changes during the photolysis of RuPc(CO)(DMF) in CH_2Cl_2 in the presence of 10^{-3} **M** CBr_4.

TABLE I

The half-wave potentials of the M(II)Pc(-2)(L) complexes.

COMPLEX	SOLVENT[a]	$E_{1/2}$ (V)[b] (fc$^+$/fc)	$E_{1/2}$ (V)[c] SCE	ASSIGNMENT
ZnPc(py)	DMA	0.23 -1.43 -1.75	0.70 -0.96 -1.28	Pc(-1) Pc(-3) Pc(-4)
ZnPc(Im$^-$)	DMA	0.24 -1.45 -2.00	0.71 -0.98 -1.53	Pc(-1) Pc(-3) Pc(-4)
RuPc(DMSO)$_2$	DCM	0.30	0.77	Pc(-1)
RuPc(CO)(DMF)	DCM	0.30	0.77	Pc(-1)
CoPc(py)$_2$	PY	0.65 0.43 -0.29 -1.08 -1.40 -1.75 -2.07	1.12 0.90 0.18 -0.61 -0.93 -1.28 -1.60	Pc(0) Pc(-1) M(III) M(I) Pc(-3) Pc(-4) Pc(-5)
CoPc	DCB	0.45 -0.86 -1.50 -1.74	0.92 -0.39 -1.03 -1.27	Pc(-1) M(I) Pc(-3) Pc(-4)
FePc(py)$_2$	DMA	0.41 0.22 -1.46 -1.76	0.88 0.69 -0.99 -1.29	Pc(-1) M(III) Pc(-3) Pc(-4)
FePc(Im$^-$)$_2$	DMA	0.72 0.13 -0.96 -1.29	1.19 0.60 -0.49 -0.82	Pc(-1) M(III) M(I) Pc(-3)

a) 0.1 M [$(Pr_4)N]ClO_4$ dissolved in an appropriate solvent was used as an electrolyte. DMA = dimethylacetamide; DCM = dichloromethane; py = pyridine; DCB = dichlorobenzene.

b) fc = ferrocene

c) $E_{1/2}$ values versus sce were obtained by adding $E_{1/2}$ (fc$^+$/fc) = 0.47 V to the $E_{1/2}$ vs. fc$^+$/fc values listed in the table, following the method outlined in reference 25.

TABLE II

The triplet state redox potentials of the MPcL complexes, and the quantum yields and the EPR parameters for the photooxidation products of these species.

COMPLEX	Q_{redox}[a] (10^{-3})	Triplet redox[b] potentials (V)	g value[a]	Band width[a] (gauss)
ZnPc		-0.44		
ZnPc(Im -)	10.7		2.0055	3.8
ZnPc(py)	1.0		2.0059	4.0
ZnPc(CN^-)	5.7		2.0068	4.0
RuPc(py)		-0.61		
RuPc(DMSO)	0.2		2.0055	14.4
RuPc(CO)(4-Mepy)	0.3		2.0048	7.2
RuPc(CO)(DMF)	0.3		2.0048	7.2
FePc		-0.54		
FePc$(py)_2$	3.2			
FePc(Im^-)	1.5			
CoPc$(py)_2$	1.0			

a) Quantum yields and EPR parameters were obtained in dichloromethane containing 10^{-3} M CBr_4, using the Q band excitation wavelength. For ZnPc, RuPc and CoPc, the photooxidation results in the formation of the π cation radical, while metal oxidation occurs in the case of the FePc species. The EPR spectra were obtained at 79 K.

b) The triplet state redox potentials of MPc are calculated as:

$$E(M(II)Pc(-1)/M(II)Pc(-2)_T^*) = E_{1/2}(M(II)Pc(-1)/M(II)Pc(-2)) - E_T(M(II)Pc(-2))$$

for RuPc and ZnPc,

and as $E(M(III)Pc(-2)/M(II)Pc(-2)_T^*$

$$= E_{1/2}(M(III)Pc(-2)/M(II)Pc(-2)) - E_T(M(II)Pc(-2)$$ for FePc.

The value of the triplet state energy of CoPc species is not known.

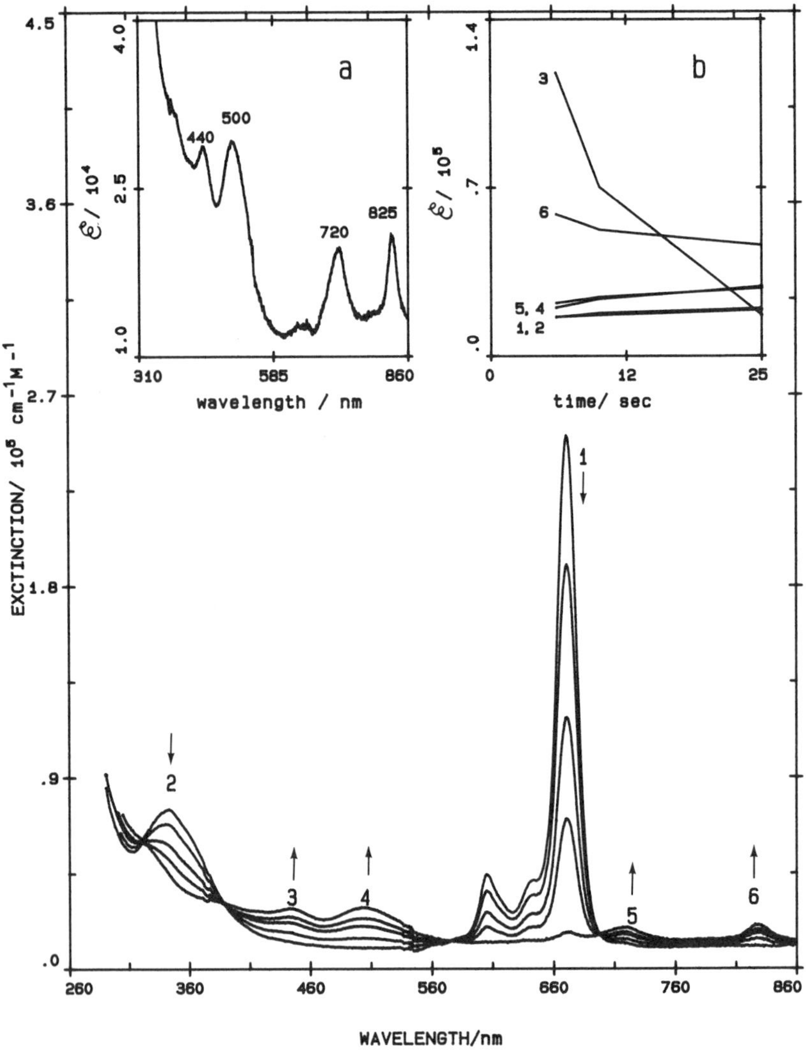

Figure 2. Absorption spectrum changes observed during the photolysis of ZnPc(py) in dichloromethane in the presence of 10^{-3} **M** CBr_4.

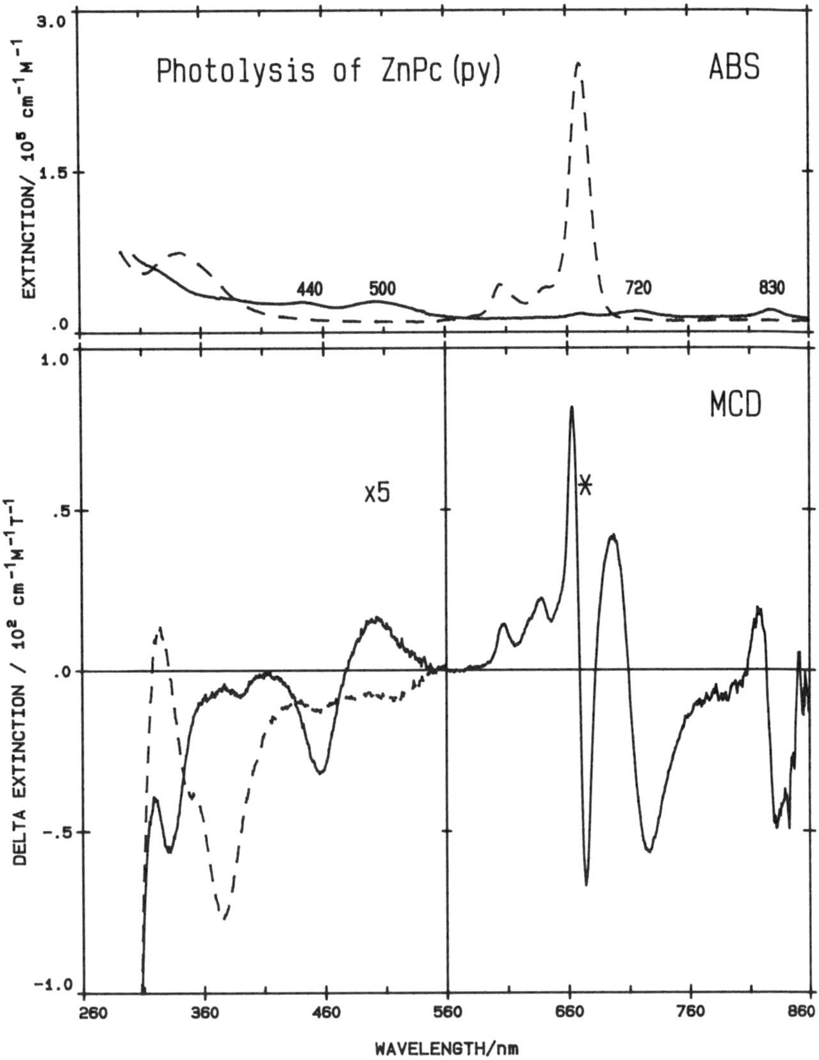

Figure 3. The MCD spectra of the neutral (dashed line), and photolysed (solid line), ZnPc(py) in CH_2Cl_2 in the presence of 10^{-3} **M** CBr_4. * = the remaining unoxidized species.

the narrow A term at 670 nm, indicated by a star. Despite the apparent intensity in the MCD spectrum, it is clear from the absorption spectrum that there is less than 5% unoxidized impurity. The Q band A term is the only feature that is intense enough to change the MCD spectrum of the ZnPc(-1) species significantly. While the MCD spectrum in the visible region is relatively well resolved, the spectral changes below 400 nm are far more complicated, as there is significant absorption due to either the electron acceptor in the photochemical experiments, or the electrolyte in the electrochemical experiments. For this region, full analysis of the MCD spectra using deconvolution procedures will be essential.

Comparison between experimental and calculated band energies. Theoretical calculations on the electronic transitions of neutral metallophthalocyanines are quite numerous (27-29). In Figure 4, we have plotted both the theoretically-predicted band energies (replotted from references 26 and 28, and given the heading "Calc") and the band energies (and degeneracies) obtained by deconvolution of the absorption and MCD experimental data (shown with headings "Exp") for the oxidized and neutral ZnPc species discussed in this paper. In addition, for comparison purposes, we have included the analogous data for the zinc porphyrin system. The theoretically-predicted porphyrin data were taken from reference 30, and the experimentally-observed porphyrin data were taken from reference 31. In Figure 4, the longer horizontal lines represent transitions to degenerate states as determined by the appearance of MCD A terms, while the shorter horizontal lines represent transitions to nondegenerate states.

In comparing the experimental and calculated energies for the transitions in the neutral ZnPc species, we find that there is a good agreement for the Q band with an energy of 15 000 cm^{-1} observed and 14 950 cm^{-1} calculated. However, below 500 nm (above 20 000 cm^{-1}) the number of theoretically-predicted transitions does not match the number of bands observed in the spectrum. In particular, the MCD spectrum of ZnPc in dimethylacetamide, with an axial ligand of cyanide (32), clearly exhibits a degenerate transition at ~26 000 cm^{-1} (385 nm). This feature can also be observed in the MCD spectra of many other ZnPc(L) species, but with not such good resolution. Using the band-fitting programs we have obtained good estimates of the band centres for these two degenerate transitions in a number of complexes of ZnPc (32). We suggest that this transition corresponds to the B band for the zinc phthalocyanine complexes. The degenerate transitions at 32 000 cm^{-1} (313 nm) and 39 000 cm^{-1} (256 nm) are then assigned as the N and L bands, respectively. Another degenerate transition is observed at 41 000 cm^{-1} (244 nm) in the methylene chloride solutions used in this study, which are significantly more transparent in the UV than the more commonly-used solvents such as DMA, DMSO and pyridine. A comparable band has not been observed in the porphyrin systems (29). We associate this 244 nm band with the C band of the Gouterman model (28).

Figure 4 shows that the visible region of the ZnPc π cation radical consists of many more degenerate transitions than were observed in the neutral species. This is quite unlike the situation found for the porphyrin π cation radical species (31). Very few detailed absorption spectra have been published for phthalocyanine π cation radical species, thus we will compare the data reported in this paper with data obtained from a theoretical calculation previously carried out for the porphyrin π cation radical species (30). These calculations do predict that a larger number of transitions should be seen in the visible region than are expected for the neutral

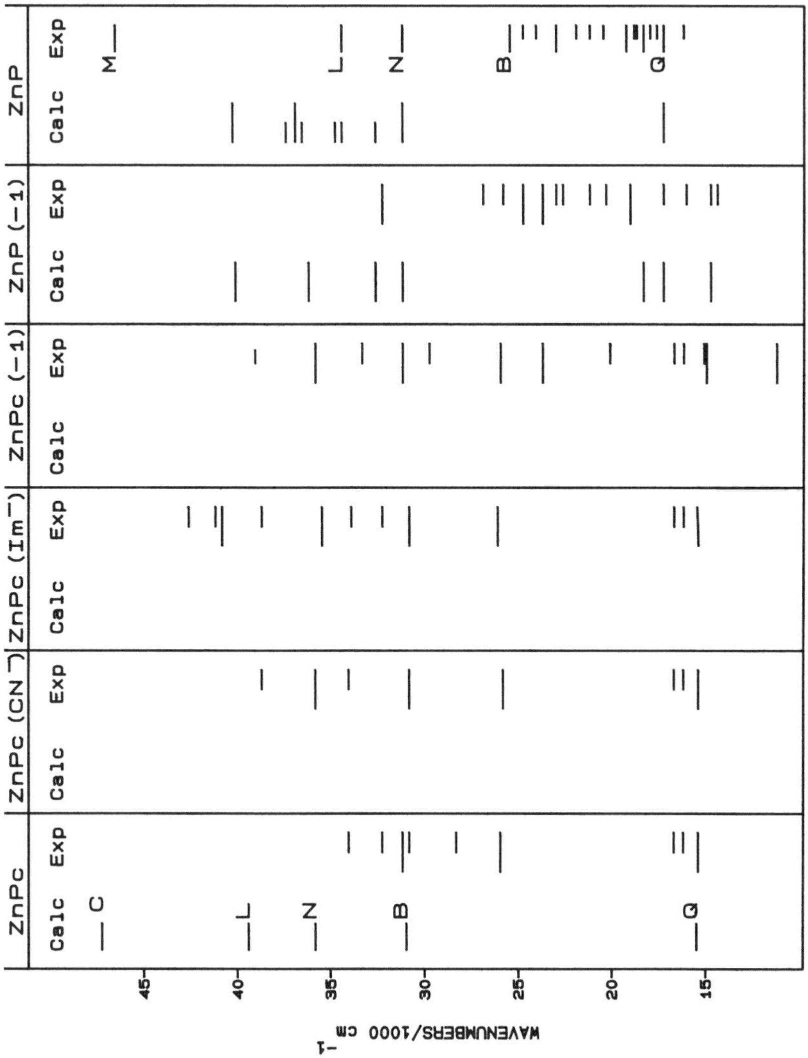

Figure 4. Comparison between the energies of transitions observed in the absorption and MCD spectra of neutral and oxidized ZnPc and zinc porphyrin species.

species (28). However, it is only with the availability of spectral envelope deconvolution calculations carried out on both absorption and MCD spectra, that accurate band energies and transition degeneracies can be obtained for these complex spectral systems. We have begun such deconvolution calculations for the π cation radical species of both porphyrin (31) and phthalocyanines (32).

The first attempts to rationalize the spectral patterns of the π cation radicals of the phthalocyanines have only recently been presented by Lever and coworkers (26). They associate the 440 and 720 nm bands of ZnPc with two degenerate transitions, both $^2E_g \longrightarrow {}^2A_{1u}$, which arise from transitions from low-lying, filled molecular orbitals into the partially-filled, highest occupied molecular orbitals. The MCD spectrum reported in Figure 3, shows that both of these bands exhibit positive MCD A terms. These features identify degenerate transitions, which means that the MCD bands at 440 and 720 nm are consistent with the theoretical assignment of Lever et al. (26). Although the A term under the 440 nm band is not clearly resolved, the degenerate nature of the transition involved has been confirmed by band shape fitting (32). The 500 nm band has been assigned (26) to a transition from the partially-filled, highest occupied MO to the lowest unoccupied MO ($^2A_{1u} \longrightarrow {}^2E_g$), the Q band transition of the π cation radical. However, the MCD spectrum (Figure 3) shows that this band has MCD B term character, signifying that it is a non-degenerate transition; this is inconsistent with the theoretical assignment. Similar MCD spectra have also been observed for the RuPc π cation radical species (8), which suggests that the non-degenerate nature of the 500 nm transition may be general for MPc(-1) species. Finally, we turn to the 830 nm band. The MCD spectrum clearly shows that this is a degenerate transition, however, at present there is no predicted transition to account for its presence. Because the Zn(II) ion in ZnPc is a d^{10} ion, we do not expect charge transfer transitions in the visible/near-IR region for the Zn(II)Pc π cation radical spectra, and as such all transitions observed for ZnPc(-1)L should be $\pi \longrightarrow \pi$ or $\pi \longrightarrow \pi^*$ in nature.

Photolysis of CoPc. Changes during the photolysis of $CoPc(py)_2$ dissolved in dichloromethane in the presence of CBr_4 are shown in Figure 5. Insert **(a)** shows a spectrum calculated by subtracting the spectrum due to the residual neutral species from the final spectrum recorded after the photolysis, while Insert **(b)** shows the time dependence of the intensity of individual bands in the absorption spectrum observed during the irradiation. Following photooxidation of $CoPc(py)_2$, new bands are formed at 410, 450, 510 and 730 nm; the 450 nm band is very much less clearly resolved than the other bands. The shapes and positions of the 450, 510 and 730 nm bands are similar to those observed during the formation of the π cation radical species of both ZnPc and RuPc. However, the intensity in the 410 nm band has a quite different time dependence following irradiation when compared with the other bands (Insert **(b)** of Figure 5, curve 1). At long irradiation times, the intensity of the 410 nm band continues to increase, while the intensity in the 510 and 730 nm bands (curves 2 and 4, respectively, of Insert **(b)**) begin to decrease in intensity. This indicates that the 410 nm band arises from a species that is different from the species responsible for the 510 and 730 nm bands. The growth of the 410 nm band at long irradiation times suggests that the new species arises from a photoreaction of the π cation radical. The original neutral species could be regenerated, either before or after substantial formation of the 410 nm band, by addition to the oxidized species of sodium dithionite.

The general broadening and decreased intensity of the absorption

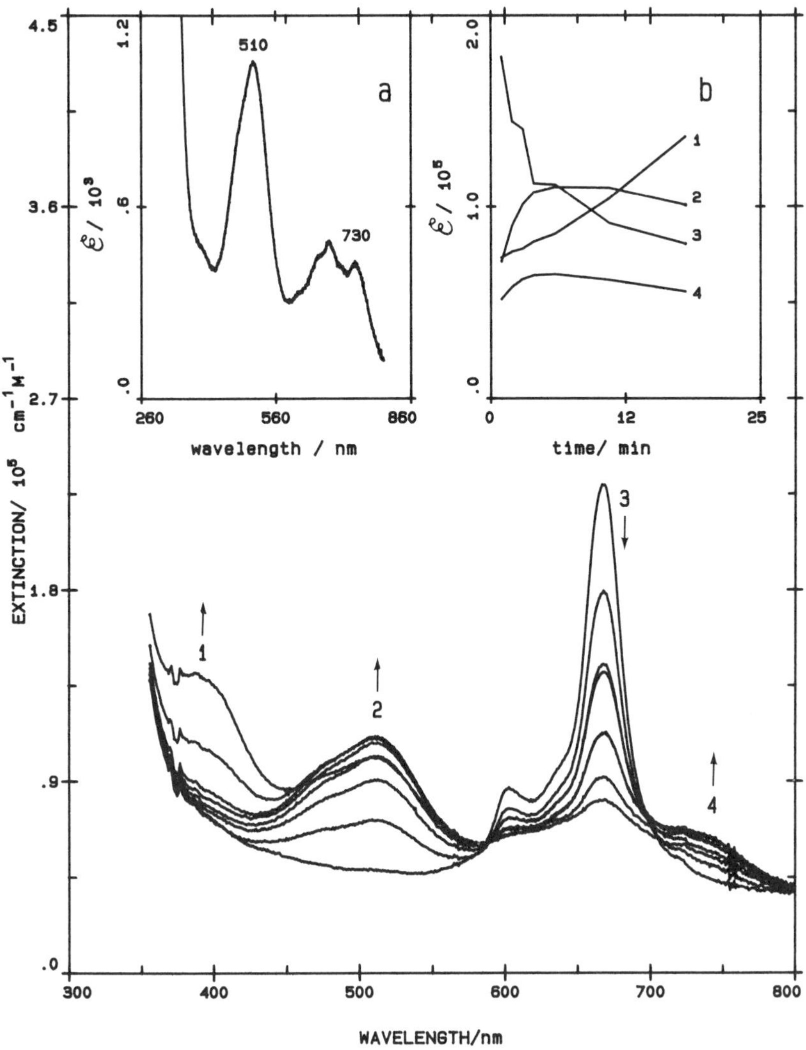

Figure 5. Absorption changes during the photochemical oxidation of $CoPc(py)_2$ in CH_2Cl_2 in the presence of 10^{-3} **M** CBr_4.

bands, and the loss of the main MCD A term under the Q band, all demonstrate that the photolysis of $CoPc(py)_2$ results in the generation of a π cation radical. The spectral changes (Figure 5) show no evidence of metal oxidation prior to the formation of the CoPc π cation radical. In the presence of strongly coordinating ligands like pyridine, it is expected that Co(II)Pc(-2) will be oxidized first to Co(III)Pc(-2), then to Co(III)Pc(-1) (16); while in weakly coordinating solvents, the Co(III) species does not form (16). The cyclic voltammetry and differential pulse voltammetry results (Table I) for CoPc, do indeed show that potentials attributable to metal oxidation are only observed in the presence of strongly coordinating solvents like pyridine. When weakly coordinating solvents, like dichlorobenzene, are employed, metal oxidation potentials are not observed (Table I). Although $CoPc(py)_2$ readily dissolves in weakly coordinating solvents, Cariati and coworkers (19) have shown that the coordinated pyridine molecules are lost after a few seconds in solution, resulting in the unligated CoPc, which then exhibits π ring oxidation prior to metal oxidation. We have found that π ring oxidation takes place initially when $CoPc(py)_2$ is dissolved in dichloromethane. The observation of an EPR signal by Cariati et al. (24) for their CoPc complex was good evidence that the compound should be formulated as $Co(II)Pc(-2)(py)_2$, and not the Co(III) complex.

The excited states used as the photoreductant in the CoPc are difficult to determine. No long-lived excited state is known for CoPc, and, therefore, we are unable to identify the states involved in the electron transfer reaction. The longest living excited state in CoPc is expected to have a life time on the nanosecond time scale by analogy with other first row (open shell) transition metal phthalocyanines (33).

Photolysis of FePc. The photochemical oxidation of $FePc(Im^-)_2$ or $FePc(py)_2$ in dichloromethane, in the presence of CBr_4 as an electron acceptor, results in similar isosbestic absorption changes. Figure 6 shows the spectral data obtained during the photolysis of $FePc(Im^-)_2$. Insert **(a)** shows the difference spectrum obtained when the spectrum due to residual, unoxidized neutral species was subtracted from the spectrum recorded at the end of the photolysis. Insert **(b)** shows the time dependence of changes in the intensity of individual bands in the absorption spectrum during the irradiation. The absorption changes are quite different from those observed when ZnPc, RuPc or CoPc are photolysed. Upon photolysis, two new bands are formed, at 554 and 684 nm. The 684 nm band has an intensity close to that of the original Q band, suggesting that the electron has not been removed from the phthalocyanine ring. The 554 nm band (curve 3 of Insert (**b**), in Figure 6) follows a similar time dependence on irradiation as the 684 nm band (curve 2 of Insert (**b**)), indicating that both bands are a result of the formation of the same species. The absorption spectra of the new species is similar to the spectra observed for the 5-coordinate, low spin $[Fe(III)Pc(Im)]^+$ complex in dichloromethane (34), which suggests that photolysis in the presence of CBr_4, results in oxidation at the metal centre, Fe(II) to Fe(III), rather than at the π ring ligand. These photochemically-induced spectral changes are similar to those observed during the electrochemical oxidation (at 0.69 V vs. sce, Table I) of $FePc(py)_2$, Figure 7, curve 2.

The lowest energy excited triplet state of FePc is known to have a life time of about 45 nanoseconds (33) and we expect the life time of the lowest energy excited triplet state of $FePc(Im^-)_2$ also to be in the same time scale. Electrochemical oxidation at the phthalocyanine ligand oxidation

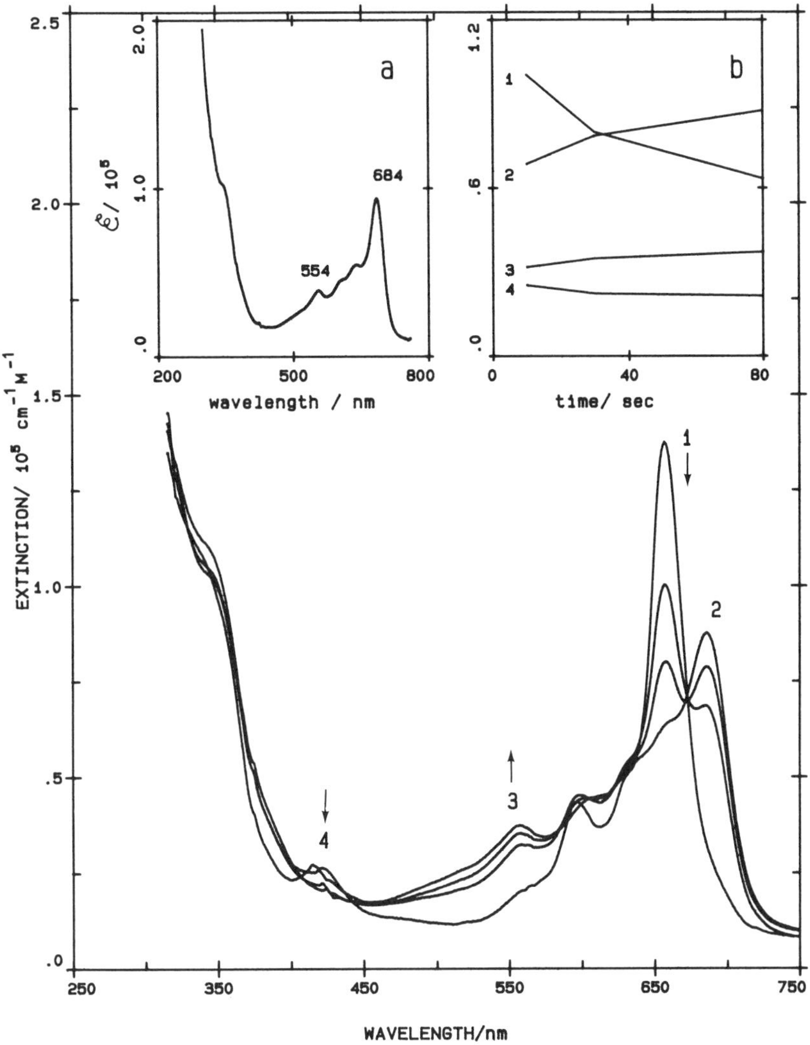

Figure 6. Absorption changes during the photolysis of $FePc(Im^{-})_2$ in CH_2Cl_2 in the presence of 10^{-3} **M** CBr_4.

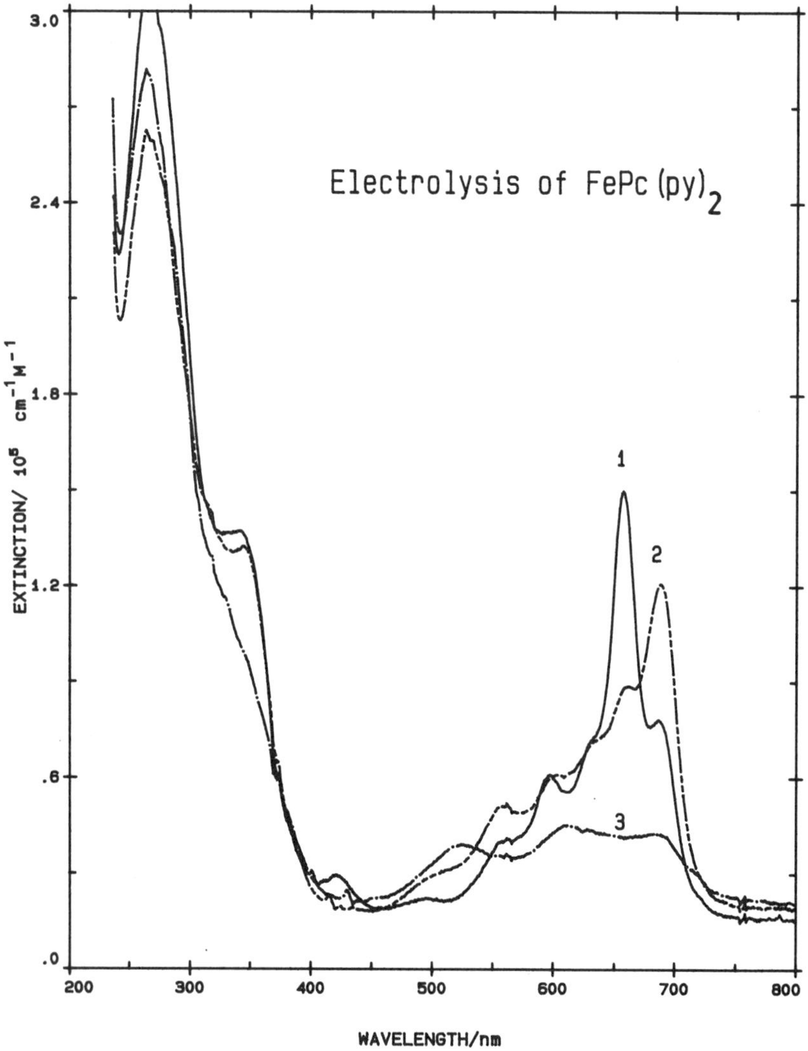

Figure 7. Absorption changes during the electrochemical oxidation of $FePc(py)_2$ in CH_2Cl_2 in the presence of 0.05 **M** $[(Et)_4N]ClO_4$. (1) The unoxidized, neutral species. (2) The spectrum recorded following metal oxidation at 0.69 V vs. sce. (3) The spectrum recorded following oxidation of the ring at 0.88 V vs. sce

potential (0.88 V vs sce, for $FePc(py)_2$), results in the formation of the new band at 520 nm (curve 3, in Figure 7) and the disappearance of the 684 nm band (curve 2). The general broadening and low intensity of each of the new bands is typical of π ring oxidation in phthalocyanines (8,10).

A mechanism to account for the observed photochemistry. Based on a scheme proposed for porphyrins (16), the photooxidation of ZnPc and RuPc can be explained as follows:

$$M(II)Pc(-2) \xrightarrow{h\nu} M(II)Pc(-2)^*$$

$$M(II)Pc(-2)^* + CBr_4 \longrightarrow [M(II)Pc(-1)]^{+\cdot} + CBr_3{}^{\cdot} + Br^-$$

where MPc^* represents the phthalocyanine in an excited triplet state. The $CBr_3{}^{\cdot}$ species is expected to dimerize rapidly following the reaction. The quantum yields of the resulting phthalocyanine π cation radical are shown in Table II. The EPR spectra obtained at 79 K for the ZnPc and RuPc π cation radicals give isotropic g values (Table II), which are very close to the free electron value of 2.0023 and are characteristic of the π cation radicals of phthalocyanines (10,15).

If the mechanism for the photoredox reaction of the metallophthalocyanines is based on the assumption that it is the lowest energy triplet state of the phthalocyanine that is involved in the reduction of electron acceptors such as CBr_4, then the reducing power of the metallophthalocyanine triplet state must be more exothermic than the oxidation potential of the acceptor, in order for the photooxidation to occur. As seen in Table II, the redox potentials of the $M(II)Pc(-1)]^{+\cdot}/M(II)Pc(-2)^*$ and $M(III)Pc(-2)/M(II)Pc(-2)^*$ couples are more negative than the redox potential of CBr_4 ($E_{1/2}$ = -0.3 V vs sce [35]), which means that the thermodynamic requirement for the reaction to occur is fulfilled.

In conclusion, photolysis of ZnPc, RuPc and CoPc results in the formation of the π cation radical species through oxidation of the ring, while in the case of FePc, metal oxidation occurs prior to ring oxidation, but π ring oxidation is demonstrated to take place following metal oxidation.

Acknowledgments

We wish to thank NSERC of Canada through the Operating, Equipment and Strategic (Energy) programs, and the Academic Development fund at the University of Western Ontario (UWO) for financial support (to MJS). We also wish to thank the Canadian International Development Agency (CIDA) and the National University of Lesotho for a research scholarship (to TN). We wish to thank Dr. William Browett for providing the deconvolution data for the porphyrins. The authors are associated with the Centre for Chemical Physics at the UWO.

Literature Cited

1. Darwent, J.; Douglas, P.; Harriman, A.; Porter, G.; Richoux, M.C. Coord. Chem. Rev. 1982, 44, 83.
2. Darwent, J.; McCubbin, I.; Phillips, D. J. Chem. Soc. Faraday Trans. 2, 1982, 78, 347.
3. Prasad, D.R.; Ferraudi, G. J. Phys. Chem. 1982, 86, 40.40,(1982).

4. Prasad, D.R.; Ferraudi, G. Inorg. Chem. 1983, 78, 1672.
5. Ferraudi, G.; Prasad, D.R. J. Chem. Soc. Dalton Trans. 1984, 2317.
6. Prasad, D.R.; Ferraudi, G. Inorg. Chem. 1982, 21, 4241.
7. Ohno, T.; Kato, S.; Lightin, N.N. Bull. Chem. Soc. Japan. 1982, 55, 2753.
8. Nyokong, T.; Gasyna, Z.; Stillman, M.J. Inorg. Chim. Acta. 1986, 112, 11.
9. Myers, J.F.; Rayner Canham, G.W.; Lever, A.B.P. Inorg. Chem. 1975, 14, 461.
10. Dolphin, D.; James, B.R.; Murray, A.J.; Thornback, J.R. Can. J. Chem. 1980, 58, 1125.
11. Gasyna, Z.; Browett, W.R.; Stillman, M.J. Inorg. Chem. 1985, 24, 2440.
12. Gasyna, Z.; Browett, W.R.; Stillman, M.J. Inorg. Chem. 1984, 23, 382.
13. Gasyna, Z.; Browett, W.R.; Stillman, M.J. Inorg. Chim. Acta. 1984, 93, 37.
14. Wolberg, A.; Manassen, J. J. Am. Chem. Soc. 1970, 92, 2982.
15. Rayner Canham, G.W.; Myers, J.; Lever, A.B.P. J. Chem. Soc. Chem. Comm. 1973, 483.
16. Lever, A.B.P.; Licoccia, S.; Magnell, K.; Minor, P.C.; Ramaswany, B.S. Adv. Chem. Ser. 1982, 201, 237.
17. Ogata, Y.; Murumo, K.; Kwan, T. Chem. Pharm. Bull 1969, 17, 1194.
18. George, P.; Ingram, D.J.E.; Bennett, J.E. J. Am. Chem. Soc. 1957, 70, 1870.
19. Cariati, F.; Galizzioi, D.; Marazzoni, F.; Busetto, C. J. Chem. Soc.Dalton Trans. 1974, 556.
20. Ouedraogo, G.; More, C.; Richard, Y.; Benlian, D. Inorg. Chem. 1981, 20, 438.
21. Taube, R. Pure and Appl. Chem. 1974, 38, 427.
22. Nyokong, T.; Stillman, M. J. J. Automatic Chemistry 1986, in press.
23. Wegner, E.; Adamson, A.W. J. Am. Chem. Soc. 1972, 88, 394.
24. Dolphin, D.; James, B.R.; Murray, A.J.; Thornback, J.R. Can. J. Chem. 1980, 58, 1125.
25. The electrochemical half-wave potentials ($E_{1/2}$) are reported in this paper versus the fc^+/fc couple. These values were calculated by subtracting the measured $E_{1/2}$ (ferrocene) versus silver wire, from the measured $E_{1/2}$ (MPc complex) versus silver wire. The half-wave potentials ($E_{1/2}$) that are reported versus the sce, were calculated by adding $E_{1/2}$ (ferrocene) versus sce (0.47 V), to each $E_{1/2}$ (MPc complex) versus the fc^+/fc couple.
26. Minor, P.C.; Gouterman, M.; Lever, A.B.P. Inorg. Chem. 1985, 24, 1894.
27. Henriksson, A.; Roos, B.; Sundbom, M. Theoret. Chim. Acta. (Berlin) 27,1972, 303.
28. McHugh, A.J.; Gouterman, M.; Weiss, C. Theoret. Chim. Acta. 1972, 24, 346.
29. Schaffer, A.; Gouterman, M.; Davidson, E.R. Theoret. Chim. Acta. 1973, 30, 9.
30. Edwards, D.; Zerner, M.C. Can. J. Chem. 1985, 63, 1963..
31. Gasyna, Z.; Browett, W. R.; Stillman, M. J., 1986, this volume, p000.

32. Nyokong, T.; Gasyna, Z.; Stillman, M. J., submitted to Inorganic Chemistry.
33. Ohtani, H.; Kobayashi, T.; Ohno, T.; Kato, S.; Tanno, T.; Yamada, A. J. Phys. Chem. 1984, 88, 4432.
34. Kobayashi, N.; Koshiyama, M.; Funayama, K.; Osa, T.; Shirai, H.; Hanabusa, K. J. Chem. Soc. Chem. Comm. 1983, 913.
35. Stackelberg, M.; Stracke, W. Z. Electrochem. 1949, 53, 118.

RECEIVED March 28, 1986

22

Cation Radical Formation with Exposure of Magnesium Porphyrin Thin Films to Light and Oxygen

Robert M. Burgess[1], Martin Gouterman, Gamal-Eddin Khalil, and James van Zee

Department of Chemistry, University of Washington, Seattle, WA 98195

When thin films of Mg tetraphenylporphin (MgTPP) are exposed to oxygen, radical cations $MgTPP^+$ form in the dark; this dark reaction obeys a Langmuir absorption isotherm, is largely reversible, and is attributed to oxygen molecules on the surface. On exposure to light more $MgTPP^+$ forms, some of it irreversibly and some reversibly. We attribute the irreversible reaction to oxygen in the bulk. The $MgTPP^+$ is monitored in the visible adsorption, in growth of an EPR signal, in the appearance of new bands in the IR (some of which are assigned to O_2^-), and by reduction in the film fluorescence. Mg octaethylporphyrin shows optical absorption changes but no EPR. Zinc tetra(perfluorophenyl)-porphyrin shows no cation formation under these conditions.

Twenty years ago Harrison and coworkers reported that oxygen increased the conductivity of copper phthalocyanine films (1). Later, Assour and Harrison reported observations of EPR radicals in free base phthalocyanines, which increased with exposure to oxygen and decreased with exposure to hydrogen (2). The effect of oxygen on organic semiconductors has been widely researched since these early studies (3,4). It has been found that photoconductivity increases on oxygen exposure for phthalocyanines (5,6), xanthine dyes (7), and porphyrins (8). An oxygen/water mixture was shown to be necessary to produce large band bending in the photovoltage studies of aluminum/phthalocyanine interfaces (9) and was shown to increase the photovoltage of cells based on palladium tetraphenylporphyrin (PdTPP) deposited on n-doped SnO_2 electrodes (10). The latter study also showed that the effect of oxygen could be partially reversed by heating the device in a hydrogen atmosphere.

[1]Current address: Boeing Electronics Company, P.O. Box 3999 MIS 88-43, Seattle, WA 98124

0097-6156/86/0321-0328$06.00/0

In another study, Bree and Lyons (11) with anthracene thin films showed that the surface photo current increased non-linearly with partial pressure of oxygen. Their results were consistant with the photocurrent change being a function of a partial monolayer coverage of oxygen. For the high pressure dependency, greater than 10 torr, they fit their data to a Langmuir adsorption isotherm, and the "intermediate" pressures followed a Freudlich isotherm.

Recent work in our own laboratory (12) on photovoltaic devices [quartz/Al/AlO_x/MgTPP/Ag] showed that exposure to both oxygen and water was needed for these devices to show photovoltaic activity. The fact that diamagnetic porphyrins are p-type semiconductors (4), may also be related to the requirement of oxygen; i.e. oxygen diffuses into the porphyrin providing electron traps, thus generating holes in the valence band.

To obtain more direct information on the role of O_2 in the electrical properties of porphyrin films, we undertook some spectroscopic studies, namely FTIR, EPR, electronic absorption and emission. Absorption and emission data were mainly used to confirm the presence of porphyrin cation species and the absence of additional oxidation products. Absorption data was also used to estimate the ratio of cation species to neutral species by measuring the change in the neutral species' absorbance. FTIR data was used to probe the existence of any bonded oxygen, any neutral or reduced oxygen along with any porphyrin cation ring modes. The EPR data probed the oxygen pressure and light dependancy of the cation radical formation along with an attempt to observe a reduced oxygen radical species, which has been observed by EPR for oxygen adsorbed on various metal substrates (13,14).

We first review the spectroscopy of porphyrin cations and oxygen anions before presenting our study of their formation in films.

Porphyrin Cations, Oxygen Anions: Review

In this paper we confine our studies to tetraphenylporphyrin (TPP), etioporporphyrin (Etio) [i.e. tetramethyltetraethylporphyrin] and tetraperfluorophenylporphyrin (TFPP), Zn and Mg complexes. MgTPP and MgEtio have two of the lowest oxidation potentials of porphyrin complexes, 0.54 and 0.4 volts vs. SCE, respectively (15), while ZnTFPP has an oxidation potential of 1.3 V (16). On oxidation these molecules form pi-cation radicals (i.e. an electron is lost from the highest filled ring orbital). The major differences between MgTPP and MgEtio are their solid state structures and the unoccupied orbital of the cation radical. The differences in their solid state structure arises by virtue of the steric hindrance of the phenyl groups on TPP. The crystal structure of both show a triclinic arrangement with one molecule per unit cell (17,18). The crystal cell angles for both molecules are not far from 90°; therefore, the crystal distances, a, b and c, can be used as intermolecular distances. For the ZnTPP complex the phenyl groups are nonplanar with respect to the porphyrin ring system: a = 14.8, b = 17.2 and c = 14.6Å, and for NiEtio a = 14.6, b = 12.3 and c = 12.3Å.

By observing the formation of the porphyrin cation radicals of MgTPP and MgOEP, the differences between their ground states were elucidated using EPR and visible absorption data (15,19). Absorption spectra, obtained at various stages in the oxidation process, showed that the cation has bands to the red and blue of the Soret absorption (350–450 nm) and to the red of the visible bands (600–900 nm). These data were obtained electrochemically in solution using as counter ions Br^- or ClO_4^-, although it was noted that I_2 could also be used as an oxidant. EPR solution data showed that the cation radicals of MgTPP and MgOEP both exhibit hyperfine structure with measured g-values of 2.0025 and 2.0031, respectively. Upon exposure to oxygen the hyperfine structure of the MgTPP cation radical was shown to disappear. By selectively deuterating different positions of the porphyrin ring systems, Fajer et. al. (19), have shown that the ground state of $MgTPP^+$ is a $^2A_{2u}$ and that of $MgOEP^+$ is a $^2A_{1u}$, as predicted by semiempirical MO theory (20).

Cation radicals of several metallated TPP and OEP complexes were reported to have unique IR active ring modes (21). These absorptions were observed in the region 1250–1290 cm^{-1} for TPP complexes and 1520–1570 cm^{-1} for OEP complexes. IR and raman active modes have also been observed for several different reduced oxygen species adsorbed on various metal substrates. Observations of the adsorption of oxygen on silver by surface enhanced raman spectroscopy (SERS) (22), show superoxide molecules (O_2^-) have a frequency of 1053 cm^{-1}, and peroxide (O_2^{2-}) of 697 cm^{-1}. This same study reported additional peaks were observed at 815 cm^{-1} and 1286 cm^{-1} if the oxygen exposure took place while the getter ion pump was operating. The 815 cm^{-1} line was identified as an adsorbed ozonide species, but they had no explanation for the line observed at 1286 cm^{-1}

Experimental

Film preparation for optical/IR studies. These porphyrin thin films were prepared by vacuum sublimation using radiation shielded quartz crucibles wrapped with tungsten wire. Samples were heated slowly by passing an electric current through the tungsten wire. Sample degassing was accomplished by increasing the current through the wire slowly with a variac. When fully degassed, the temperature increases to the sublimation point and deposition occurs. Films were usually grown to their final thicknesses, on the order of 300–500 nm, over a 12–24 hour period. The substrates used for the absorption and emission studies were standard 1"x3"x1/16" glass slides and the FTIR substrates were 1/4" thick NaCl optical flats. For film deposition, a low vacuum (10^{-6} torr) bell jar was used with an oil rotary pump for the roughing stage, and an oil diffusion pump with a cryostatic trap for the lowest pressure range. Film thicknesses for samples prepared using the bell jar system were calculated from absorption data assuming the extinction coefficient from solution data.

Sample chamber for optical and FTIR studies. For optical and FTIR studies the ambient atmosphere was controlled by placing the sample in vacuum tight chambers described elsewhere (23). These chambers were fitted with optical flats as viewing windows to allow straight throughput for absorption data acquisition and perpendicular observation of emission. The optical flats were sealed by use of O-rings.

Transfer from the preparation bell jar to the sample chamber was accomplished by first backfilling the system with dry nitrogen or argon for several minutes. When the bell jar was removed, the sample was quickly moved to the chamber while continually blowing the inert gas across it. While this process does involve brief exposure to air, our results show that the effect of exposure to oxygen in the dark was largely reversible. Thus, we believe that any effects due to this transfer are small.

Exposure of the sample chamber to various ambients was controlled by use of a manometer system with evacuation obtained using a rotary pump. The lowest pressure obtainable using this system was 20 mtorr (base pressure of the rotary pump) with the highest pressure being 800 torr (maximum allowable on the manometer).

Optical and FTIR instrumentation. Absorption spectra were taken on three instruments. A Hewlett Packard 8450 UV/Vis Spectrophometer was used for the visible region of the spectrum. This instrument was computer controlled and used to measure standard absorption spectra as well as absorption difference spectra. A Cary 14 Spectrophotometer was used for the near IR region, and an IBM IR/90 Series

FTIR Spectrometer was utilized to measure absorbance as well as absorbance difference spectra in the region 600–4000 cm^{-1}. Emission spectra were taken on a commercially available Perkin Elmer 650-10S Fluorimeter.

The light source used to form the radical cations was a 100 Watt tungsten bulb. Absorbance difference spectra were calculated by using the relationship $\Delta A = \log(I_o/I)$, where I_o and I are the stored absorbance spectra before and after the exposure of the films to oxygen and/or light, respectively.

EPR sample preparation. For EPR samples, either the bell jar system was used or the following preparation procedure was followed. A small amount of sample (10–50 mg) was placed inside an EPR tube, attached to a vacuum apparatus and evacuated to ~10^{-5} torr with an oil diffusion pump. A cylindrical oven, consisting of a quartz tube wrapped with tungsten wire, was placed over the EPR tube. Heat was supplied by passing a current through the wire with a variac. Samples consisted of porphyrin sublimed on the walls of the EPR tube. The film morphology of samples prepared in this manner are quite different from those prepared using the bell jar. These samples consist of needle-like crystals entending towards the center of the EPR tube. The placement of the tube in the EPR cavity insured that only the sublimed porphyrin was observed.

EPR Instrumentation. The EPR instrument used in this study was a commercially available Varian E-3 EPR Spectrometer. The apparatus used to measure the lifetime of the photoassisted radical formation is shown, schematically, in Figure 1. The light source was a water-cooled 1000 watt tungsten lamp. The light was pulsed using an electronic shutter with a rise time of under 2 ms. The shutter, as well as the timing for the experiment, was controlled by special "FOCAL" functions available on the PDP 8/e computer (24). Data was acquired by first setting the magnetic field to the differential absorption maxima. After the buffer memory was activated, the sample was pulsed with white light. The output signal from the EPR instrument passed through a preamplifier with a time constant of 10 ms. The time varying signal then passed through a null amplifier to eliminate the DC offset signal. Finally, the data was integrated by the computer for a preselected number of light pulses.

The Varian E-3 was also used for the study of radical cation formation as a function of oxygen pressure in the dark. Oxygen was supplied by a high pressure cylinder, and the lower pressures were controlled by a manometer assembly. The initial studies were performed by setting the magnetic field to the differential absorption maxima and observing the signal as various pressures and light levels perturbed the sample. Using an internal standard, it was realized that increasing oxygen pressures could perturb the baseline of these scans resulting in a DC offset mistaken for a true oxygen pressure dependancy. Therefore, all subsequent scans were made by scanning through the peak and using the peak height as a measure ofthe signal strength. Copper tetraphenylporphyrin (CuTPP), which is unlikely to form a cation in the presence of oxygen (ring oxidation is 0.9 V vs SCE (15), was used as the internal standard.

Source of materials. MgTPP was obtained from Strem chemicals and was purified by elution through a dry-packed column of Al_2O_3 (Woelm, Activity Grade I, neutral) with dichloromethane and acetone. ZnTFPP was synthesized in our laboratory by first making the free base (25), followed by metallation using Zn acetate (26). MgEtio was synthesized using a procedure from Fuhrhop et al. (27).

Observations

Optical absorbance changes with oxygen and light. The first observation to be presented for the solid state oxidation of porphyrins by oxygen is by visible absorption spectroscopy. The data presented in Figure 2 are absorption spectra for thin film samples of MgTPP, MgEtio and ZnTFPP before (b) and after (a) exposure to both oxygen and light. Exposure to oxygen alone or light alone has no effect on the film absorption spectrum (data not shown). The MgTPP absorption data shows clearly the reduction of the visible bands of the normal ground state and the rise of bands at 500 and 800 nm that correspond to the absorbance of the cation radical (28). The MgTPP films show the largest percentage of converted neutral species, approximately 10% reduction in the visible absorption bands. The MgEtio and ZnTFPP films of Figure 2 were thicker than that of MgTPP, and their spectra extend only down to 590 nm. On exposure to light and oxygen, the MgEtio shows a clear absorbance increase in the region 700 to 900 nm while the ZnTFPP shows essentially no change. The MgEtio data can be compared to changes reported by Fajer et al. (28) for the conversion MgOEP → $MgOEP^+$ in solution. Our Q(1,0) and Q(0,0) bands are red shifted, ~600 cm^{-1} and ~200 cm^{-1}, with respect to theirs. They observed a relatively sharp band at 720 nm in the cation, while we observe a very broad band with a maximum of 800 nm. The absorbance rise we observe in the near IR is roughly comparable to the intensity loss in the visible, which is the same for the

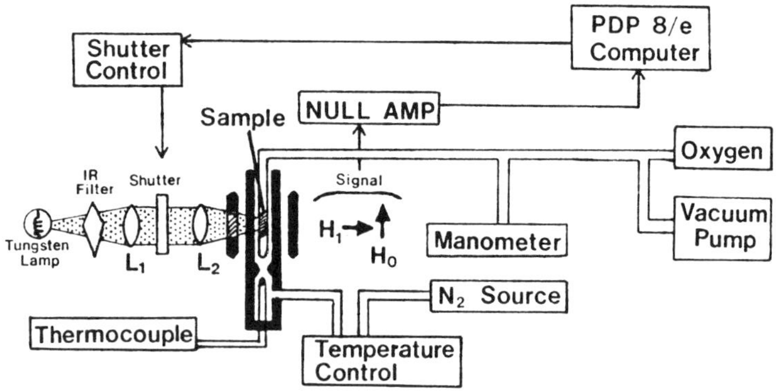

Figure 1. Block diagram of EPR apparatus used to measure the lifetime of photoexcited porphyrin cation radicals.

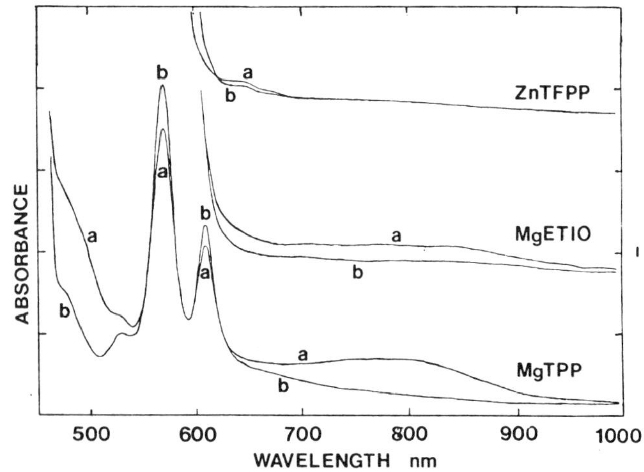

Figure 2. Absorption spectra obtained on MgTPP, MgEtio and ZnTFPP thin films (b) before and (a) after exposure to oxygen and light.

$MgOEP \rightarrow MgOEP^{+}$ conversion. Thus, although our near IR band is broader and more red shifted than that reported for the isolated molecule oxidation in CH_2Cl_2, the assumption that we are observing a similar oxidation is reasonable.

EPR spectra produced by oxygen in the dark. The EPR signal strength was measured on MgTPP thin films upon exposure to various pressures of oxygen in the dark. The oxygen pressure varied from 20 mtorr, the base pressure obtained using a rotary pump, to 760 torr. The initial EPR scan shows a small porphyrin cation radical peak at 3357 gauss, with a g-value of 2.0031 (Figure 3a) possibly due to the residual oxygen present at the base pressure. In addition to this peak, several other sharp peaks, 1-2 gauss wide and of varying intensities, are observed over a 1000 gauss range. These peaks were reported to be the result of the coupling of the rotational levels of gaseous oxygen to the spins of the unpaired electrons (29). These oxygen peaks diminish and gradually disappear as the ambient pressure is increased (Figure 3b). Also shown in Figure 3b is the increase of the porphyrin cation absorbance with an increase in oxygen pressure. Note also, that no change occurs with the CuTPP absorbance peaks.

Room temperature spectra were taken over the range of pressures 1-760 torr. Typical results obtained are plotted in Figure 4 along with the calculated fit of the data points to a Langmuir adsorption isotherm,

$$\text{EPR signal} = k_a P/(1+k_a P) \quad (1)$$

Least square fits were obtained by the gradient-expansion algorithm developed by Marquardt (30). The DC offset (i.e. EPR signal at 20 mtorr) was subtracted from each data point before performing the least squares fit. The inverse of the fitting parameter, k_a, is the pressure at which the EPR signal reaches half maximum and, presumably, is related to the number of surface sites available to bind oxygen. In these studies, the value of k_a^{-1} remained relatively constant at 475 ± 25 torr.

The spectra obtained are completely reversible during observation times of under one hour. For longer time periods of oxygen exposure in the dark, the DC offset signal increases slowly, ~5-10% of the Langmuir isotherm value per hour. It increases more quickly if the porphyrin films are heated for several minutes with a heat gun, while under oxygen.

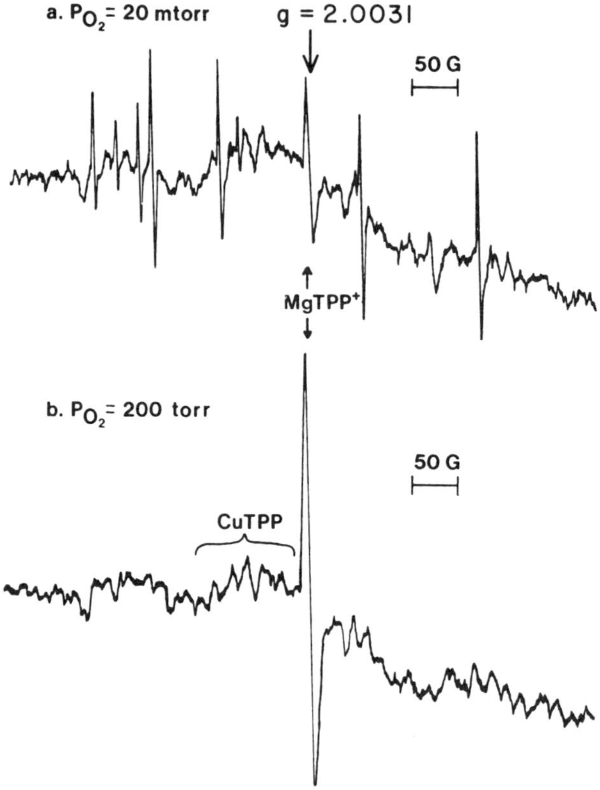

Figure 3. EPR spectra of a thin film MgTPP sample showing; (a) residual MgTPP cation radical and gas phase oxygen at an oxygen pressure of 20 mtorr and (b) additional MgTPP cation radical formed at an oxygen pressure of 200 torr. Also shown is the internal standard's, CuTPP, peaks.

<u>EPR changes with oxygen and light</u>. An increase in the radical cation signal occurs when the MgTPP films exposed to oxygen are additionally exposed to light. This light/oxygen exposure shows a reversible and an irreversible component, as shown in Figure 5. The initial oxygen exposure in the dark immediately raises the EPR signal strength to a constant level, A in Figure 5, as given by the isotherm equation above. With a 15 second light exposure, the signal increases to a new level, B_1 in Figure 5. When the light is turned off, the signal decays to level C_1. A second light pulse causes a rise to B_2 and decays to C_2. A third light pulse gives B_3, C_3 etc. Thus, each light pulse produces a reversible increase, $B_i - C_i$, and an irreversible increase, $C_{i+1} - C_i$. With increasing i, the $C_{i+1} - C_i \rightarrow 0$ while the $B_i - C_i$ rise to a maximum.

Figure 6 shows the results obtained when a 200 ms light pulse was applied to a sample of MgTPP after $B_i - C_i$ was maximized. The data for the rise and decay curves were fit to exponential decays using a program developed by Provencher (<u>31</u>). The program chose the number of components present in the curves utilizing a best fit criteria. In every case in which the computer calculated a signal to noise ratio greater than 10, the computer returned a best fit with two exponentials. The best signal to noise ratio, 27, was obtained using data summed from many experiments using one particular film. The computer returned, in this case, a double exponential decay of 273 and 33 ms. Also, the rise time fits, in each case, returned a double exponential best fit. [The rise time results are not presented because the lifetime determined by this method requires that the exact excitation rates be known.]

Using the lifetime data obtained from several different temperatures, an Arrhenius plot of $\ln\tau$ vs. $1/T$ was made. From the least squares fit, the energy of activation was calculated to be $\sim 580 cm^{-1}$.

<u>Infrared changes</u>. The films were observed by FTIR in the spectral region 600 -4000 cm^{-1}, using various pressures of oxygen and exposure times to white light. If the films were exposed to just oxygen, in the absence of light, no new strong absorption bands were observed. However, changes were observed in this spectral range when difference absorption spectra were used. There appears a small shift in many of the porphyrin vibration modes.

Figure 7a shows the absorption spectra obtained from an evacuated MgTPP film, and Figure 7b shows the same film after exposure to 750 torr of oxygen and 48 hours of white light. The new bands that appear are indicated by arrows. These new bands are better

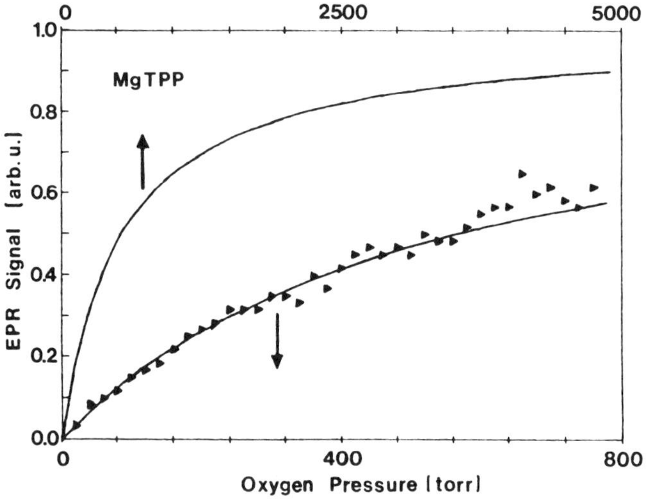

Figure 4. EPR signal (▲) and least squares fit of MgTPP cation radical peak height as a function of oxygen pressure. The upper curve shows an extension of the least squares fit to higher pressures. DC offset has been substracted from each point prior to least squares calculation.

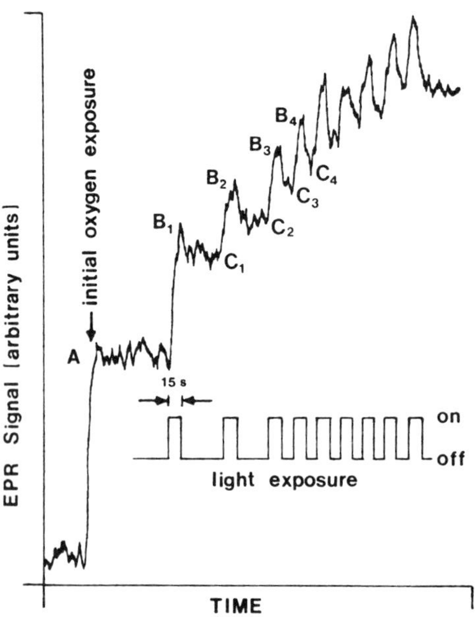

Figure 5. MgTPP cation radical EPR peak height measured while the thin film is exposed to oxygen and nine, 15 second, pulses of white light.

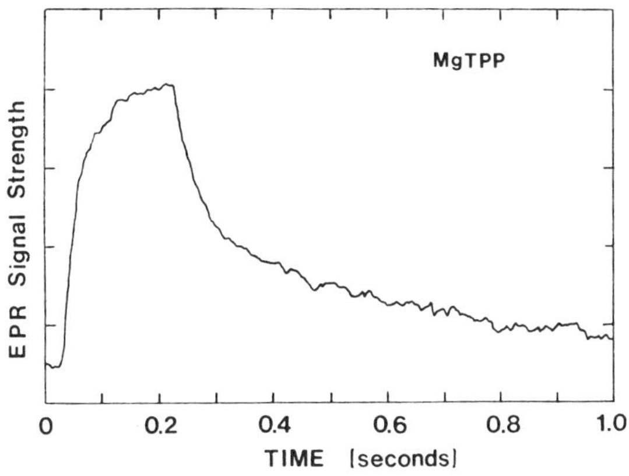

Figure 6. EPR signal strength vs. time for a thin film MgTPP sample showing the lifetime of the radical formed upon exposure to 200 ms of white light. The double exponental decay is fit with lifetimes of 273 and 33 ms.

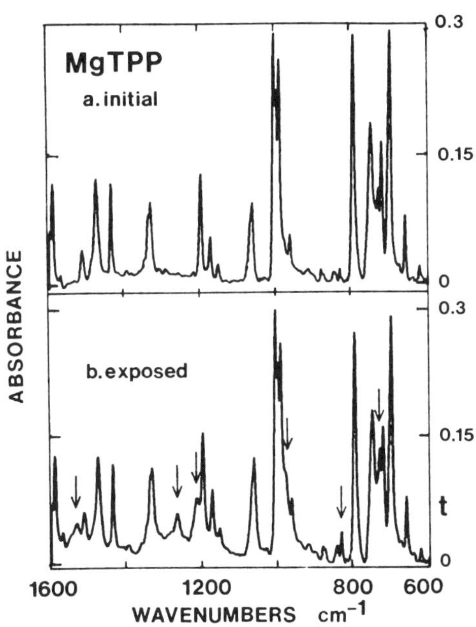

Figure 7. FTIR absorbance spectra, 600–1600 cm^{-1}, of MgTPP thin film (a) initial evacuation and (b) evacuated after exposure to 750 torr of oxygen and light for 48 hours.

seen in the absorption difference spectra shown for MgTPP in Figure 8a. New absorption peaks appear at 729, 833, 982, 1060, 1217, 1269, and 1537 cm^{-1}. The new band at 1269 cm^{-1} is assigned to the IR-active ring-mode for the cation species, as reported by Shimomura et al. (21) and is notably absent in Figure 8b. This spectrum shows several of the same bands appearing as the MgTPP sample, Figure 8a. The band expected for the radical cation of MgEtio should be similar to that reported for metallated OEP cation radicals, near 1530 cm^{-1} (21). There does appear to be a small peak at this energy, but adsorbed oxygen also absorbs in this energy range, such that a definite assignment cannot be made. However, due to the close proximity of the peaks for the MgTPP and MgEtio, at 1535 and 1560 cm^{-1}, respectively, we assign these peaks to absorbed molecular oxygen (1554.7 cm^{-1}) (32). The peaks observed at 1060 and 1058 cm-1 for the two complexes, we assign to superoxide, O_2^- (1053 cm-1 (22)). We, tentatively, assign the peaks at 833 and 827 cm-1 to a reduced ozonide species (802-831 cm^{-1} (33). The other FTIR absorption bands cannot be assigned at this time.

Another absorption that becomes apparent with exposure of the thin films of MgEtio and MgTPP to oxygen and light are presented in Figure 9a and 9b. At least three, very broad, peaks are observed to grow with increasing oxygen and light exposure in the IR absorbance range of 600-4000 cm^{-1}. The two spectra suggest that there are peaks just above and just below the region spanned by the FTIR instrument; a third broad absorption band is clearly observed at ~2500 cm^{-1} for the MgTPP thin film and at ~2300 cm^{-1} for MgEtio, with FWHM ~800 cm^{-1}.

Emission spectral changes. Figure 10 shows the emission spectrum from a thin film sample of MgTPP and from the same film on exposure to oxygen. The film emission consists of three bands at 618, 666, and 690 nm. The bands at 618 and 666 nm are the expected Q(0,0) and Q(0,1) emission bands and the band at 690 nm is an impurity. Porphyrin films quite generally show impurity emissions (34). It is presumed that on excitation of the films, the excited state can move through the solid and transfer its energy to an impurity guest, which then emits as a trap. Figure 10 shows that the emission drops on exposure to oxygen, with the impurity emission dropping more than the emission from the MgTPP. This is consistent with a quenching mechanism that we ascribe to,

$$MgTPP^* + MgTPP^+ \longrightarrow MgTPP^+ + MgTPP \qquad (2)$$

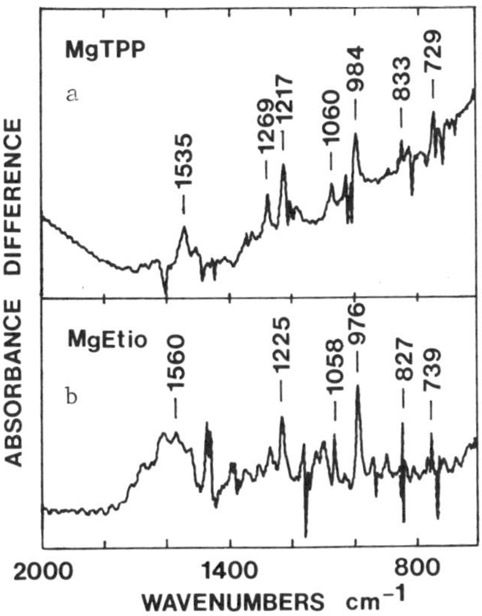

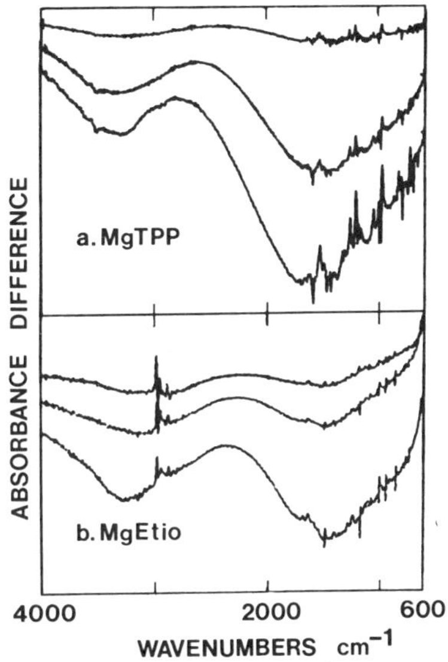

Figure 8 (left). FTIR absorbance difference spectra, 600-2000 cm^{-1}, of evacuated MgTPP and MgEtio thin films, previously exposed to oxygen and white light. See text for peak assignment.

Figure 9 (right). FTIR absorbance difference spectra, 600-4000 cm^{-1}, of evacuated MgTPP and MgEtio thin films, previously exposed to oxygen and white light.

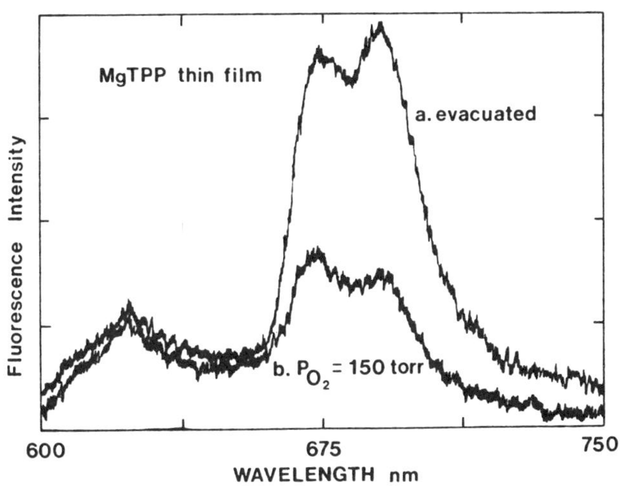

Figure 10. Emission spectra, 600-750 nm, obtained from MgTPP thin film (a) evacuated and (b) exposed to 150 torr of oxygen. Excitation wavelength was 400 nm.

Here $MgTPP^*$ is the excited singlet. It is known that the excited states of porphyrin are both good oxidizing and good reducing agents. The cation is, of course, a good oxidizing agent. So therefore, quenching of the excited state by transfer of the excited electron to the cation is quite reasonable. Quenching by Eq. (2) would presumably lower the guest emission more than the host emission because the probability of the excitation reaching a guest would be decreased with shorter lifetime of $MgTPP^*$.

Discussion

MgTPP. The most data were taken on MgTPP, and the results are quite unambiguous. On exposure to oxygen and light there are spectral changes in absorption, emission, FTIR and EPR that are consistent with the change MgTPP → $MgTPP^+$. Since, presumably, this reaction depends on the presence of oxygen, it is useful to consider the free energy changes for these two reactions:

$$\text{dark: } P + O_2 \rightarrow P^+ + O_2^-, \ \Delta F = 1.2 \text{ eV} \quad \text{(I)}$$

$$\text{light: } P^* + O_2 \rightarrow P^+ + O_2^-, \ \Delta F = -0.8 \text{ eV} \quad \text{(II)}$$

Here, P stands for the MgTPP ground state and P^* for its excited state. The ΔF values are based on a half cell potential vs. hydrogen of +0.75 V for MgTPP oxidation (15) and a half cell potential of −0.43 V for O_2 reduction (35). We assume that in its excited state, the half cell potential of $MgTPP^*$ should be ~2 eV more negative, due to the excitation energy. Thus, dark reaction (I) is not thermodynamically allowed while the light reaction (II) is. However, we would expect the light reaction products of (II) to revert to neutral species by reversing reaction (I).

Our observations on MgTPP with oxygen and light show the following reactions occur; (Ia), a fast reversible dark reaction that follows a Langmuir adsorption isotherm with O_2 pressure; (Ib), a slow irreversible dark reaction; (IIb), an irreversible light driven reaction that builds up $MgTPP^+$ to some saturation level; and (IIa), as the irreversible light reaction goes to saturation, there is a reversible production of $MgTPP^+$ that reverts to MgTPP with a lifetime of ≤300 ms.

At this point it should be noted that the reversible dark reaction was observed only by EPR and not by optical absorption. We attribute this to the fact that the EPR is very sensitive to a very small percent of radical. That the EPR dark signal follows an adsorption isotherm with respect to oxygen suggests that oxygen molecules on the surface are involved. This can be understood in terms of a model presented by Weisz (36), in which the chemisorption process "polarizes" charges at the surface of the film. Such a polarization must redistribute charges in such a way as to drastically reduce ΔF for reaction (I) for O_2 adsorbed on the surface. Thus we explain reactions (Ia) as a surface phenomenon.

The irreversible light reaction (IIb), shown in Figure 2 should be favorable for any oxygen, in particular, O_2 in the bulk. The problem in this case is that the reaction does not reverse in the dark. The O_2^- must reside in a special site or must create a special state that makes reversal by (I) kinetically unfavorable. Alternately, perhaps the O_2^- reacts further to create a more stable species, e.g. O_2^{2-}. While further reaction with traces of water could produce OH^- as a stable counterion, our attempts to support a role for water by adding or excluding it were inconclusive. But since rigorous exclusion of all water was not possible, we cannot rule out a stabilization of $MgTPP^+$ that involves both O_2 and H_2O.

Having "saturated" the amount of $MgTPP^+$ that can be produced by O_2 and light, further light produces $MgTPP^+$ reversibly, reaction (IIa). Since the Langmuir isotherms suggest that the reversible dark reaction produces O_2^- on the surface, in the presence of light this reaction may go further. But the same may hold for O_2^- in the bulk. The double exponential decay might reflect the fact that both surface and bulk O_2^- is involved in the reversible light reaction. In analogy, we note that in studies on the electronic surface states of metal-oxide-semiconductor (MOS) devices, capacitance vs. voltage measurements turn up what is historically known as slow and fast states (37).

Finally, we turn to reaction (Ib), the slow reversible formation of $MgTPP^+$ with O_2 in the dark. While we have no detailed explanation, it is not so surprising that some $MgTPP^+$ and O_2^+ formed reversibly in the dark crosses over to the irreversible form produced in light.

MgEtio. The FTIR spectra on exposure of MgEtio thin films to oxygen and light showed very similiar results as the MgTPP thin films, except for some shifts in frequency and intensities. Absorption and emission observations also reproduced the MgTPP results. The most curious observations came with the attempt to reproduce the EPR results. Only after prolonged exposure to both oxygen and light was there any noticeable EPR signal. Therefore, phototransient lifetime data were not taken. A possible explanation for the lack of EPR signal comes from the observation that a similiar compound, MgOEP, forms diamagnetic cation dimers in solution (19). In our case, the close proximity of the molecules in films might allow easy formation of the diamagnetic dimer. In a separate experiment, the thin films of MgEtio were exposed to iodine vapor and in this case the EPR signal grew extremely rapidly with exposure to aliquots of the vapor. Pumping on the iodine exposed films partially reverses the effect. Upon exposure to light the EPR signal would decrease and would rise back to the dark signal strength after the light was removed.

ZnTFPP. No substantial amount of cation could be observed with any of the spectroscopic techniques when ZnTFPP films were used.

Conclusion

Based on the data presented, the following four conclusions about the interactions of light, oxygen and porphyrin thin films can be made. (1) There exists a dark reaction of oxygen that has an oxygen pressure dependancy that follows a Langmuir adsorption isotherm. (2) Cations are formed irreversibly with exposure to both oxygen and light, although the anion site is still undetermined. (3) A transient cation radical is formed in the presence of oxygen and light with observed lifetimes of 33 and 273 ms saturate. (4) The irreversible counterions are associated with IR absorption bands at ~1225, ~1050 and ~830 cm^{-1}, indicative of superoxide, ozonide, and an adsorbed oxygenic species.

The reaction of solid porphyrin films with light in the presence of oxygen by producing $MgTPP^+$ must affect electrical properties, in particular semiconduction, photoconduction, and photovoltaic properties. We have provided evidence for "photo-doping" by light and oxygen, a phenomenon that must be clearly understood if these materials are to have device applications.

Acknowledgments

We thank Professor Bruce Robinson for valuable discussions on EPR measurements and Professor Alvin Kwiram for use of his EPR instrument. Michael Russell provided computer fitting routines and Eric Shankland provided purified samples. This research was supported in part by NSF grant DMR-7823958.

Literature Cited

1. Heilmeir, G. H.; Harrison, S. E. Phys. Rev. 1963, 132, 2010.
2. Assour, J. M.; Harrison, S. E. J. Phys. Chem. 1964, 68, 872.
3. Meier, K. "Organic Semiconductors"; Monographs in Modern Chemistry: Verlag Chem, 1974; Vol. II, p. 130, 150, 154, 230, 473.
4. Gutmann, F.; Keyzer, H.; and Lyons, L. E. "Organic Semiconductors" Part B, Robert Krieger Publishing Co.: Malabar, FL, 1983.
5. Day, P.; Price, M. G. J. Chem. Soc. A, 1969, 236.
6. Fan, F. R.; Faulkner, L.R. J. Chem. Phys. 1978, 69, 3334.
7. Meshkova, G. N.; Meshkov, A.M.; Vartanyan, S. T. Sov. Phys. Semicond. 1969, 3, 605.
8. Vartanyan, A. T. Optical Tech. 1970, 37, 279.
9. Murti, D. K.; Brillison, L. J.; Slowik, J. H. J. Vac. Sci. Technol. 1982, 20(2), 233.
10. Langford, C. H.; Hollebone, B. R.; Nadezhdin, D. Can. J. Chem. 1981, 59, 652.
11. Bree, A.; Lyons, L. E. J. Chem. Soc. 1960, 52, 5179.
12. Stanbery, B.J.; Gouterman, M.; Burgess, R. M. J. Phys. Chem. 1985, 89, 4590.
13. Clarkson, R. B.; McClellan, S. J. Catal. 1980, 61, 552.
14. Kazusaka, A.; Yong, L.K.; Hower, R.F. Chem. Phys. Lett. 1978, 57, 592.
15. Felton, R. H. In "The Porphyrins"; Dolphin, D., Ed.; Academic Press: New York, 1978; Vol. 5, Chapter 3.
16. Gouterman, M.; Kim, S. Unpublished work.
17. Lauher, J. W. J. Am. Chem. Soc. 1973, 95, 5148.
18. Silvers, S. J. Am. Chem. Soc. 1967, 89, 3331.
19. Fajer, J.; Davis, M. S. In "The Porphyrins"; Dolphin, D., Ed.; Academic Press: New York, 1978; Vol 4, Chapter 4.
20. Spellane, P. J.; Gouterman, M.; Antipas, A.; Kim, S.; Liu, Y. C. Inorg. Chem. 1980, 19, 386.
21. Shimomura, E. T.; Philippi, M. A.; Goff, H. M. J. Am. Chem. Soc. 1981, 103, 6778.
22. Petterkofer, C.; Pockrand, I.; Otto, A. Surf. Sci. 1983, 135, 52.
23. Burgess, R. M. Doctoral Dissertation, University of Washington, Seattle, WA, 1985.
24. van Zee, J. Doctoral Dissertation, Harvard University, Cambridge, MA, 1985.
25. Longo, F. R.; Finarelli, M. G.; Kim, J. B. J. Heterocyclic. Chem. 1969, 6, 927.

26. Buchler, J. W., : Puppe, J. Justus Liebigs. Ann. Chem. 1970, 740, 142.
27. Fuhrhop, J. H.; Mauzerall, D. J. Am. Chem. Soc. 1969, 91, 4174.
28. Fajer, J.; Borg, D.C.; Forman, A.; Dolphin, D.; Felton, R. H. J. Am. Chem. Soc. 1970, 92, 3451
29. Bowers, K. D.; Kampar, R.A.; Lustig, C. D. Proc. Roy. Soc. 1959, p. 565.
30. Bevington, P. R. In "Data Reduction and Error Analysis for the Physical Sciences"; McGraw-Hill: New York, 1969.
31. Provencher, S. W. J. Chem. Phys. 1976, 64, 2772.
32. Jones, R. D.; Summerville, D. A.; Basolo, F. Chem. Rev. 1979, 79, 139.
33. Prochaska, E. S.; Andrews, L. J. Chem. Phys. 1980, 72, 6782.
34. Bonham, J. S.; Gouterman, M.; Howell, D. B. J. Lumin. 1975, 10, 295.
35. Pack, J. C.; Phelps, A. V. J. Chem. Phys. 1966, 44, 1870.
36. Weisz, P. B. J. Chem. Phys. 1953, 21, 1531.
37. Sze, S. M. In "Physics of Semiconductor Devices"; Wiley: New York, 1969, p. 445.

RECEIVED June 19, 1986

23

Chemistry and Structure of the Principal Tumor-Localizing Porphyrin Photosensitizer in Hematoporphyrin Derivative

Chi K. Chang[1], Shinji Takamura[1], Brian D. Musselman[2], and David Kessel[3]

[1]Department of Chemistry, Michigan State University, East Lansing, MI 48824
[2]Department of Biochemistry, Michigan State University, East Lansing, MI 48824
[3]Departments of Pharmacology and Medicine, Wayne State University, Detroit, MI 48201

The principal tumor-localizing component of Hematoporphyrin Derivative (HPD) has been demonstrated to be dimeric and trimeric hematoporphyrins (HP) interconnected with ester groups. Synthetic analogs as well as model compounds are used in our study to conclude that the reaction conditions employed in the traditional HPD preparation promote a nucleophilic substitution of the acetate group of one HP-acetate molecule by an propionate anion of another HP molecule. The effect of solvent on the stability and structural conformation of the diporphyrin esters have also been examined by spectroscopic methods.

The selective eradication of neoplastic lesions, with minimal damage to normal host tissues, has been the goal of antitumor therapeutics. The combination of porphyrin and light has displayed this degree of selectivity and formed the basis of the emerging "photodynamic therapy (PDT)." The porphyrin most often used in PDT has been a crude product from reaction of hematoporphyrin with sulfuric acid and acetic acid, followed by alkaline treatment--appropriately termed hematoporphyrin derivative (HPD).(1) The basic scheme utilized in PDT is to activate the tumor by injection of HPD. Preferential accumulation of an HPD component at neoplastic loci occurs between 24-96 hours after this injection, at which time exposure of the activated tumor to red light results in cell death. Fluorescence by the localized HPD components also facilitates delineation of the tumor via fluorescence endoscopy. Clinical trials have indicated that the drug is effective, as well as highly selective against a variety of solid tumors.(2-4) The major side effect of the drug is manifest in patient hypersensitivity to sunlight, which may result in development of skin photodamage after treatment is completed.(3) The photodynamic effect of HPD is believed to involve the generation of singlet

0097-6156/86/0321-0347$06.00/0

oxygen via energy transfers from the triplet state of the photosensitizer to molecular oxygen.(5,6)

Extensive analyses of HPD have been carried out in attempts to identify the active agents in this drug.(7,8) Numerous porphyrin components isolated from HPD (Fig. 1) have been assayed for photosensitization, both individually and in mixtures, while some activity can be attributed to each individual component, none are as effective as HPD itself. More recently, investigators have begun to investigate the possibility that oligomers of hematoporphyrin origin are present in HPD.(9,10) While chromatographic procedures designed to separate these oligomers from the monomers of HPD have been developed, the chemical nature of the high molecular weight molecules has not been characterized.(11) In this paper, we present evidence demonstrating that the active components of HPD are dimeric and trimeric hematoporphyrins linked via ester bonds.

RESULTS

Analysis of HPD. HPD was prepared from hematoporphyrin according to the standard Lipson procedure.(1) The product was applied to a Sephadex LH-20 column in a tetrahydrofuran:methanol:5mM sodium phosphate solvent (2:1:1) solvent buffered to pH 7. The first fraction eluted was shown to contain a broad featureless band when analyzed by a reverse phase HPLC method (Fig. 2). This band did not contain the starting material hematoporphyrin, or the typical hematoporphyrin dehydration products, hydroxy-ethyl vinyl deuteroporphyrin and protoporphyrin. This fraction constituted approximately 35% of the HPD mixture applied to the column. Approximately 90% of the tumor-localizing capacity of HPD is obtained using this fraction.(12) Utilizing fast atom bombardment (FAB) mass spectrometry (MS) we recently developed a procedure suitable for the analysis of porphyrin oligomers.(13) Direct analysis of crude HPD product was completed using a FAB matrix of dithioerythritol:dithiothreitol:thioglycerol (1:1:2) which contained 0.01M trifluoroacetic acid. FAB analyses of porphyrins using this acidified matrix typically result in the generation of ions whose mass corresponds to that of the protonated porphyrin molecule. Analysis of several HPD samples by FAB resulted in the generation of relatively intense ion species at m/z 1179, 1161, and 1143 (Fig. 3). The mass of two hematoporphyrin molecules is 1196, covalent bonding of these molecules via an ether or ester bond would result in the loss of water to give a mass of 1196-18=1178, protonation of this porphyrin dimer yields an ion of mass 1179 which corresonds to the m/z 1179 component identified in the mass spectrum. Ions at m/z 1161 and 1143 may result from successive loss of water molecules from the molecular ion species. This loss of water molecules may arise from either the elimination of a water molecule after ionization by FAB or from the combination of HP with partially dehydrated hematoporphyrin molecules present in the HP acetate solution during the synthesis of the oligomers. A second, less intense series of ion clusters were present in the spectrum at m/z 1201, 1183, and 1165. These ions arise due to the presence of sodium salt in the sample carried over from the buffer solvent, and in this case the sodium adduct (Na^+) rather than protonated species

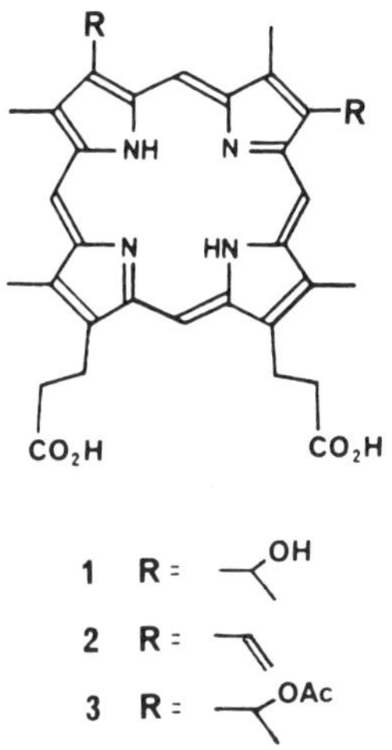

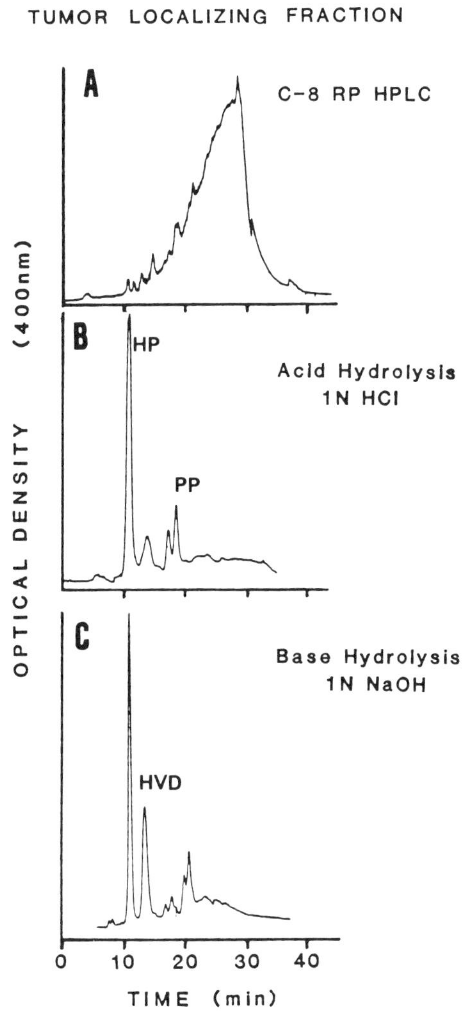

Figure 1 (left). Structure of **1**: hematoporphyrin; **2**: protoporphyrin; and **3**: hematoporphyrin diacetate. These are common constituents of HPD. In addition, the two R groups may have all permutations of these functional groups resulting in a large variety of porphyrins in HPD.

Figure 2 (right). Reverse-phase HPLC elution profile of the tumor-localizing fraction of HPD isolated by non-aqueous gel exclusion chromatography. (B) Hydrolysis of this material in 50% aqueous THF containing 1M HCl, at 37 °C for 24 hours. (C) After hydrolysis in 1M NaOH, 50% aqueous THF under the same condition. The major porphyrins resulted are: hematoporphyrin (HP), hydroxyvinyl deuteroporphyrin (HVD), and protoporphyrin (PP).

of each dimer species typically produce during FAB ionization conditions. The appearance of ion clusters at masses corresponding to protonated porphyrin trimers was also evident. These ions correspond to the molecular weight of protonated hematoporphyrin trimer (1775 daltons) minus 2, 3, and 4 water molecules respectively. Sodium cation-containing species were also present at m/z 1745 and 1763.

Despite our success in obtaining the precise molecular weight of the oligomers, FAB-MS could not distinguish between the two possible inter-porphyrin linkages, ether or ester, proposed for the dimer. The ether structure was favored by Dougherty, (10) based on the finding that the tumor-localizing component of HPD was hydrolyzed in aqueous acid but not in alkali. This behavior is consistent for a HP dimer covalent linked via an ether group, which would be stable in base. An ester linked dimer should not be stable in base or acid. However, possibilities exist for an ester group linking two porphyrins to be enshrouded between the large hydrophobic rings in aqueous base such that it may not be amendable to hydrolysis at an appreciable rate. Examination of the hydrolysis of the purified tumor-localizing component in a 50% tetrahydrofuran:water solvent was completed in order to lessen the contribution of such putative steric factors to the resistance of the HP dimer to base hydrolysis in the aqueous solvent. Indeed, in this partially non-aqueous solvent, hydrolysis of the porphyrin dimer in both acid and base was rapid. These results indicate that the covalent bond linking the HP-dimer is susceptible to cleavage by both acid and base, thereby invalidating the exclusive argument for an ether linkage.

Mixed Porphyrin Esters. If an ester bond is formed during the HPD-forming procedures, we theorized that analogs of HPD should be possible by cross coupling a molecule of hematoporphyrin acetate and another porphyrin that has a propionic acid but does not contain a secondary alcohol functional group. Such a mixed dimer, of course, will not be possible if coupling occurs only with the ether linkage. One such compound is deuteroporphyrin (DP): the periphery of the DP ring contains only 2 carboxyl and 4 alkyl groups. To test this hypothesis, we dissolved 10 mg of previously prepared HP acetate in 1ml of 0.1M NaOH containing 10 mg of dissolved deuteroporphyrin. After 60 minutes the reaction mixture was neutralized and analyzed by reverse-phase HPLC. The elution profile is similar to that of HPD with the addition of a peak corresponding in retention time to unreacted DP. Elution through a 0.8 x 40 cm LH-20 column resulted in the separation of porphyrin oligomers from DP and other porphyrin monomers (Fig. 4). The product of this reaction contains covalently-bound DP as determined in the following experiment. Alkaline hydrolysis in 50% tetrahydrofuran resulted in the formation of hematoporphyrin and deuteroporphyrin from the oligomer mixture. This hydrolysis pattern is only possible with a diporphyrin ester structure wherein a carboxyl group of deuteroporphyrin is esterified with a sec-OH group of hematoporphyrin. FAB-MS analysis of this product showed ion species previously identified for HP dimer, at m/z 1179, 1161, and 1143. These ions are present since HP can still react with itself to yield the HP-HP dimers. An additional series of ions

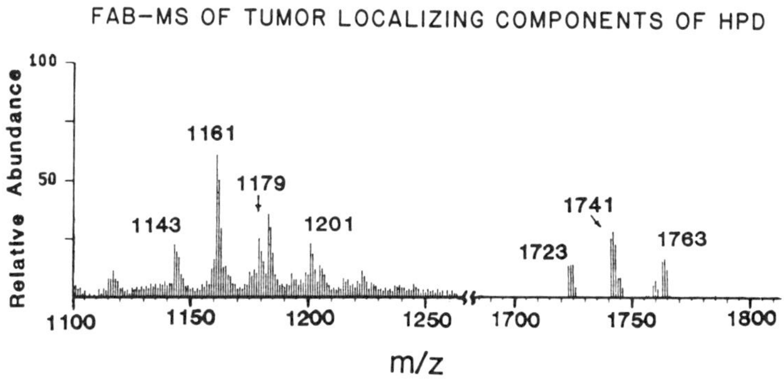

Fig. 3. High mass region of the FAB mass spectrum of the tumor-localizing component of HPD isolated from Sephadex LH-20 column.

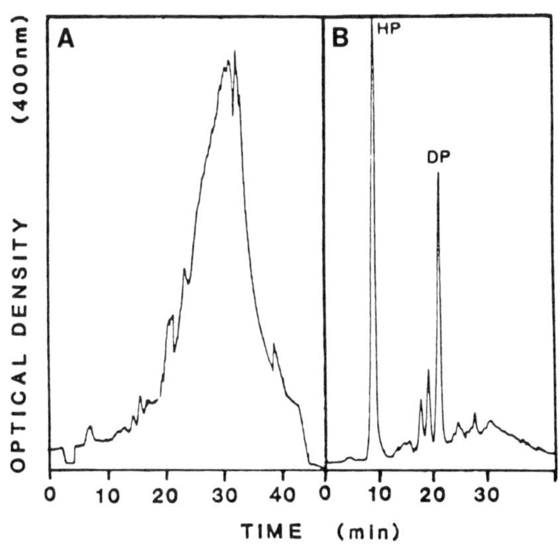

Fig. 4. HPLC analyses of a deuteroporphyrin-containing, mix-coupling product. (A) Monomer-free material separated using a Sephadex LH-20 column. (B) After hydrolysis in 1M NaOH, 50% aqueous THF, showing formation of hematoporphyrin (HP) and deuteroporphyrin (DP).

were identified which correspond to the mass of protonated DP-HP ester at m/z=1091 and 1073, with the latter mass resulting from dehydration of one *sec*-OH of the HP moiety present in the dimer.

Model Studies. The major difficulty associated with separation and analysis of HPD arises from the fact that there are two hydroxyl and two carboxyl groups positioned unsymmetrically on the porphyrin ring. Statistically, there are 3 ether structural isomers and 4 esters possible between two hematoporphyrin molecules; and each of these is a diastereomeric mixture. No doubt, this situation has made the isolation and characterization for a specific structure a nightmare.

In our goal to answer the question concerning the structure of the active component of HPD, we have decided to simplify the situation by examining a model compound. We have synthesized(14) a model porphyrin (Fig. 5) which contains the necessary functional groups for coupling but has no possibility of forming isomeric products. This model compound was treated using the standard Lipson procedure to yield a mixture which was separated as previously described with a Sephadex LH-20 column as well as by chromatography on silica gel using methylene chloride/methanol eluent. FAB-MS of the high molecular weight fraction indeed confirmed the presense of dimers having the -OH and the vinyl group in their structure (Fig. 6). The covalently linked hydroxy porphyrin dimer has a mass of 1058 and protonation of this dimer gives an ion corresponding to the m/z 1059 peak. Successive loss of two water molecules as well as cationization by sodium yields the other ions at m/z 1041, 1023, and 1081. However, neither the gel permeation nor silica gel chromatography could purify the dimer fraction sufficiently to allow an unambiguous ^{1}H NMR study of structure. Thus, the crude product was treated with diazomethane in methanol to methylate the carboxylic acid before it was subjected to TLC separation. In this manner we obtained two dimer methyl esters: one contains a vinyl group while the other has a methyl ether group on one of the porphyrin. The ^{1}H NMR spectrum of the methyl ether **7** is shown in Fig. 7. The formation of the methyl ether at the free hydroxy group was traced to a silica gel-catalyzed methanolysis of the acetate. Indeed, if the diporphyrin esters were left in contact with silica gel and methanol, the intramolecular ester linkage would slowly undergo methanolysis to yield monomeric porphyrin methyl ether, a situation which should be avoided.

DISCUSSION

The experiments presented above definitively established the fact the Lipson's HPD-forming procedure gives dimeric and trimeric porphyrin esters as the high-molecular weight component which is eminently better localized and retained in tumor cells and is responsible for most of the photodynamic activity associated with HPD. Mechanistically, the formation of ester linkage is not only understandable but expected as well. Under the alkaline condition employed in the synthesis of HPD, the nucleophilic substitution reaction illustrated in the following scheme is very similar to the

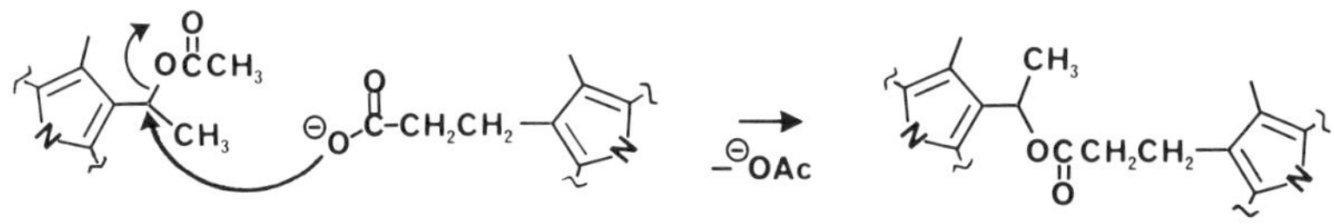

mechanism identified in classical experiments of Philips(**15**) for establishing the inversion of optical activity associated with S_N2 reactions. It becomes obvious with this reaction that optimization of the dimer yield may be achieved by increasing the concentration of the reactants and by the use of a near stoichiometric amount of alkali--sufficiently to convert all propionic acid to the anion but not in a large excess to hydrolyze off the acetate group. In reality, for solubility considerations, a mild base such as sodium bicarbonate present in slight excess can perhaps achieve the best result. Our observation with the model compounds also suggests that these porphyrin dimers are intrinsically unstable on silica gel; chromatography with that material, particularly in the presence of methanol should be avoided.

Proton NMR studies of the methylated model porphyrin compound yielded the following results. When dissolved in the non-polar solvent, $CDCl_3$, each of eight meso protons in the dimer appeared as a doublet so that a total of sixteen prominent resonance signals appeared in the spectrum with chemical shifts between 8.5 to 10.3 ppm, Fig. 8A. In contrast, the ^{1}H NMR spectrum of the monomer contains at maximum four signals in the range between 9.7 and 10.3 ppm. This difference may arise from interactions between the two ring systems in the diporphyrin molecule which create a unique chemical environment for each meso hydrogen. The good resolution of these signals and the large chemical shifts apparent in the spectrum are indicative of a well defined structure.

Analysis of the model compound in a more polar solvent was completed in order to examine the effects of different solvents on the conformation of the dimer. The ^{1}H NMR spectrum taken in acetonitrile is shown in Fig. 8B. The appearance of meso hydrogen signals from 7.0 to 10.2 ppm represents a doubling of the range of chemical shifts apparent in the non-polar solvent. This spectrum indicates that the meso hydrogens interact with each other to a greater extent in the polar solvent, and therefore, suggests that a more rigid, tightly bound dimer conformation exists in the polar environment.

Absorption studies of the model compound in polar and non-polar solvents support the finding that the dimer is present in a form which allows for close proximity between the two porphyrin rings. The absorption spectra of the dimer and the monomer are shown in Fig. 9. The spectra in methanol show a significant blue-shift of the Soret peak and a small red-shift in the visible bands for the diporphyrin, consistent with the spectral properties of previously synthesized cofacial diporphyrins(**16,17**) wherein exciton interactions can take place.(**18**) In methylene chloride, the Soret blue shift appears to be much less (<5 nm with reference to **4**).

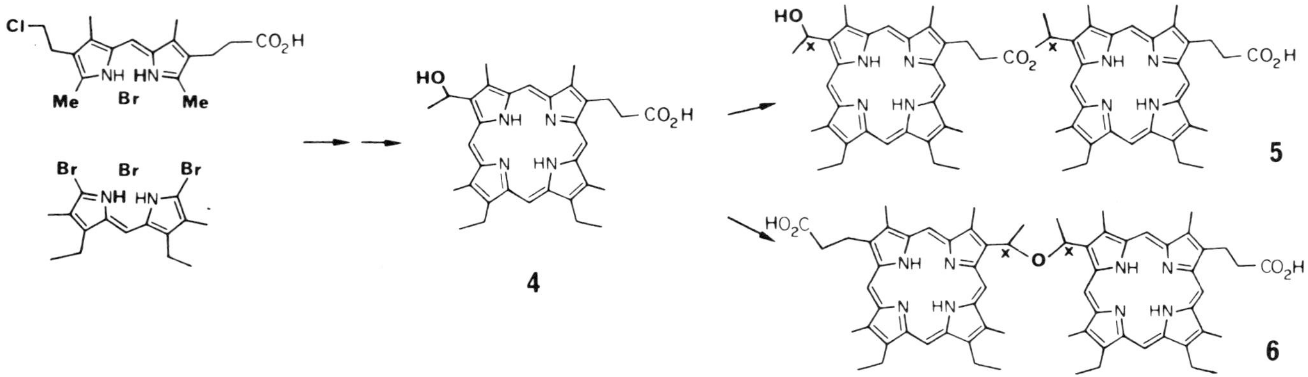

Fig. 5. Scheme of the synthesis of a "model hematoporphyrin" and the two possible dimers that can be derived from it. The ester dimer 5 has a very different chemical environment for the two α-H, marked x, while in the ether dimer 6, the two α-H would have the same chemical shift in ^{1}H NMR.

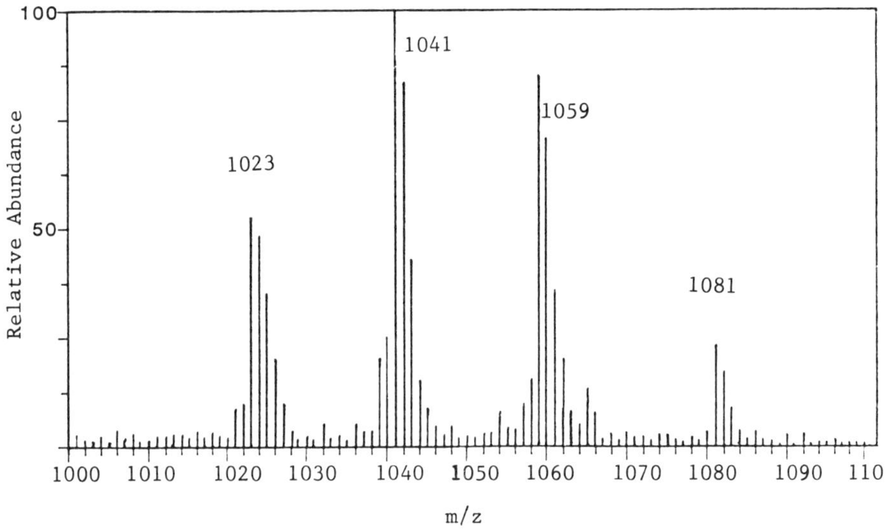

Fig. 6. Partial FAB-MS spectrum of the high molecular weight fraction of the synthetic "model HPD" eluted from a Sephadex LH-20 column.

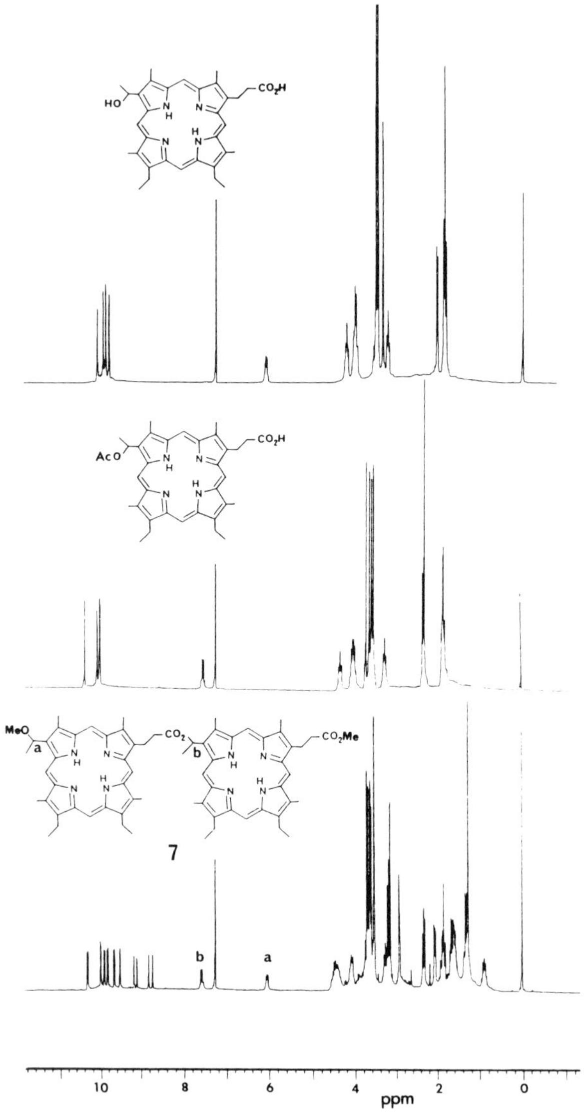

Fig. 7. ^{1}H NMR of the hematoporphyrin model (A), its acetate (B), and the purified but methylated diporphyrin 7 (C). The two α-H can be seen clearly at 6.1 ppm (adjacent to MeO) and at 7.6 ppm (adjacent to the ring-connecting ester). The α-H next to the methoxy group has a chemical shift very similar to that of the hydroxy porphyrin as verified by monomeric porphyrins.

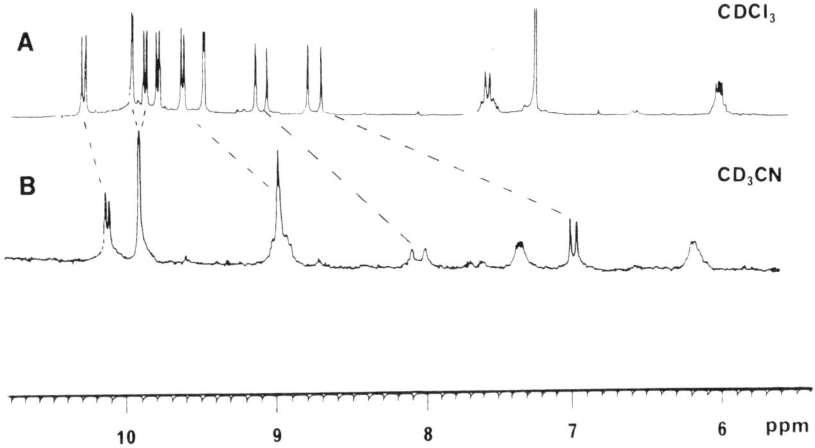

Fig. 8. NMR spectra at the meso proton region of the diporphyrin in $CDCl_3$ (A), and in CD_3CN (B). The greater scattering of the meso protons simply reflects a tighter interaction of the two porphyrin rings in the more polar CD_3CN solvent.

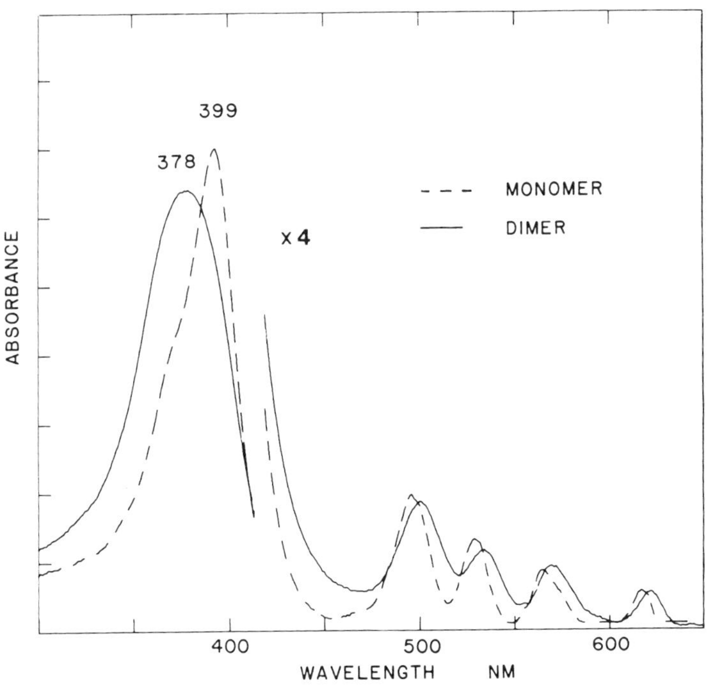

Fig. 9. Visible absorption spectra of the diporphyrin **7** (solid line) versus the starting material **4** (broken line), measured in methanol.

These observations suggest that while the two porphyrin rings of the dimer are closely associated in most solvents, their interaction in a polar solvent is tighter and more extensive than that observed in a non-polar solvent.

The spectroscopic properties of the dimeric porphyrins could provide a possible explanation for HPD localization in tumor tissue. A major contribution to the elucidation of porphyrin-tissue interactions in vivo was provided by two reports describing porphyrin-lipoprotein affinities.(19,20) These data indicate binding of HP and PP to low density lipoprotein (LDL) and high density lipoprotein (HDL) fractions of plasma. While the phenomena of binding and transport of HPD by lipoproteins have not been examined in detail, the pattern of HPD localization in the mouse is correlated with numbers of LDL receptors in different tissues.(21,22) We postulate that the dimer conformation present in LDL is similar to that in a non-polar solvent. After the LDL enters the cell, the dimer is released into the cytosol, where the polar environment directs a change in the dimer conformation to the more restricted structure observed in a polar solvent. The dimer structure is selectively retained in the cell while monomers readily diffuse out of the cell. It is not known at present zhether the dimer may become protonated or metallated, resulting in further conformational changes which could be important in controlling the rate of hydrolysis and the lifetime of the porphyrin in the cell. However, with the understanding of the chemical structure of the active component of the tumor localizing agent, we are now able to define approaches which may answer important questions regarding the mechanism of localization and photodynamic activity of HPD.

EXPERIMENTAL

General. NMR spectra were obtained on a Bruker WM-250 instrument with tetramethylsilane (TMS) as internal standard. UV-visible spectra were taken on a Varian Cary 219. Fast-atom bombardment (FAB) mass spectrometry (MS) analysis were completed on a Vacuum Generators ZAB-2F double-focusing or a Varian MAT CH5 mass spectrometer equipped with an Ion Tech FAB Gun. Solvents for FAB matrix were made up of thioglycerol, dithiothreitol, and dithioerythretol (2:1:1); addition of 0.1M trifluoroacetic acid to the matrix facilitated the ionization of the porphyrins during FAB analysis.(13)

Preparation and Characterization of HPD. HPD was prepared by dissolving HP (100mg) in a mixture of 1.9ml of glacial acetic acid and 0.1ml of H_2SO_4. After 60 minutes of stirring at 20°C the solution was poured slowly into 15ml of 3% sodium acetate solution. The precipitate of HP acetates was collected by centrifugation, and washed with water to pH 6, and lyophilized. Addition of 20mg of this material to 1ml of 0.1M NaOH was completed with stirring. After 60 minutes the solution was brought to pH 7.5 with 0.1M HCl and the volume adjusted to 2ml. Preparations of HPD containing a higher concentration of tumor-localizing material was completed by doubling the amount of HP acetates added to the 0.1M NaOH in the hydrolysis step. Synthesis of the mixed deuteroporphyrin-hematoporphyrin (DP-HP) dimer was completed by dissolving 10mg of the HP acetates as prepared above in 1ml of 0.1M NaOH containing 10mg of dissolved deuteroporphyrin. After 60 minutes the reaction mixture was brought to pH 7.5 with 0.1M HCl and the volume adjusted to 2ml.

Preparative fractionation of 100mg sample of HPD and the product of the DP-HP synthesis were carried out on a 5x30cm Sephadex LH-20 column (Pharmacia, Piscataway, NJ.). The porphyrin components were first dissolved in 40ml of tetrahydrofuran-methanol-5mM aqueous phosphate buffer, pH 7, (2:1:1), and subsequently eluted with the same solvent. The use of this methodology for the isolation of the tumor-localizing fraction of HPD from other porphyrin components present in the drug mixture has been discussed in detail in the literature.(**11**) The component of HPD which elutes very close to the void volume of the LH-20 column has been characterized as the tumor-localizing fraction of HPD.

Separate analyses of the stability of HPD synthesis products to acid were completed in 50% aqueous tetrahydrofuran containing 1M HCl. The same solvent system containing 1M NaOH was used to determine the stability of HPD products in base. Assays to determine the extent of dissociation of the oligomers in both acid and base were completed using an analytical HPLC system consisting of a reverse phase C_8 column and a CN guard column in a Waters Z-module radial compression housing. The column was eluted with a 30 minute gradient of 70-100% methanol (remainder aqueous 5mM t-butyl ammonium phosphate pH 3.5). The flow rate was 1ml/min. After 30 min the flow rate was increased to 1.5ml/min with 100% methanol. Absorbance at 400nm was monitored with a Spectraflow Model 773 HPLC detector.

Model HPD Preparation. **Method 1.** 3-(1-Acetoxyethyl)-17-(2-carboxyethyl)-8,13-diethyl-2,7,13,18-tetramethyl-21H,23H-porphine(**14**) (0.020g, 0.034 x 10^{-3} mol) was dissolved in tetrahydrofuran (0.5 ml) and 0.1 N NaOH solution (0.5 ml) was added. The mixture was stirred in subdued light at room temperature (ca. 20°C). The reaction was monitored by TLC (silica gel) using 5-10% methanol in methylene chloride as eluent and stopped after no more increase of the rapid-moving fraction on TLC was recognized. It generally took 2-6 hours. The solution was brought to pH 5 by dropwise addition of dilute hydrochloric acid and shaken with tetrahydrofuran (20 ml) and saturated sodium chloride solution (20 ml). Ethyl acetate (30 ml) was added, and the mixture was shaken. The organic layer was washed with water twice, separated and taken to dryness.

Method 2. Acetylated porphyrin (0.020g, 0.034 x 10^{-3} mol) was dissolved in DMSO (0.5 ml) and 0.1 N $NaHCO_3$ solution (0.5 ml) was added. The reaction was carried out as described above. After acidification, excess ice-cold water was added (ca. 5 ml) until the porphyrin was precipitated. (A small amount of sodium chloride was added when the precipitation did not occur.) The solid was collected by filtration, washed with water and dried *in vacuo*. The reaction mixture gave four different spots on TLC.

Fractionation of Porphyrin-free Acid. The product prepared in the preceding section was chromatographed on silica gel using 5% (v/v) methanol in methylene chloride as eluent. From this column, the slowest fraction was identified as hydroxyethyl porphyrin **4**. The early fractions were separated on a plate of silica gel and found to be a mixture of vinyl- and acetoxyethyl porphyrins.

Fractionation of Porphyrin Methyl Ester. The product mixture (50 mg) was dissolved in methylene chloride (2.0 ml) and treated with excess ethereal diazomethane freshly prepared from p-toluenesulfonylmethylnitrosamide (Diazald). The mixture was stirred in subdued light at room temperature. The solvents were evaporated gradually in a fume cupboard. The residue was dried *in vacuo* and separated on preparative TLC plates to give the methoxyethyl diporphyrin methyl ester whose NMR is shown in Fig. 7 and 8.

ACKNOWLEDGMENTS

This work was supported in part by a Michigan State University All-University Research Initiation Grant, and by Grant #CA23378 from the National Cancer Institute, NIH, DHHS. The mass spectrometry instrumentation was in the NIH/MSU Mass Spectrometry Facility supported by the Biotechnology Resources Program of the Division of Research Resources of the NIH, RR00480-16. Additional mass spectra were obtained by Lubo Baczynskj, Upjohn Co., Inc., Kalamazoo, MI. HPLC analyses of HPD were carried out by May-Ling Cheng.

LITERATURE CITED

1. Lipson, R.; Bâldes, E.; Olson, A. (1961) *Natl. Cancer Inst.*, **26**, 1.
2. Dougherty, T.; Boyle, G.; Weishaupt, K. in "Porphyrin Photosensitization" (Kessel, D.; Dougherty, T., eds.), Plenum Press, New York (1983).
3. Dougherty, T., Photoradiation Therapy of Malignant Tumors. CRC Critical Reviews in Oncology/Hematology. CRC, New York (1984).
4. "Porphyrin Localization and Treatment of Tumors" (Doiron, D. R.; Gomer, C. J., eds.), A. R. Liss, New York (1984).
5. Spikes, J. D. in reference 4, p. 19. 6. Foote, C. S. in "Free Radicals in Biology" (Pryor, W. A., ed.), Academic Press, New York (1976), p. 85.
7. Bonnett, R.; Ridge, R. J.; Scourides, P. A. (1981) *J. Chem. Soc. Perkin I*, 3135.
8. Moan, J.; Sanderg, S.; Christensen, T.; Elander, S. in reference 2, p. 165.
9. Berenbaum, M. C.; Bonnett, R.; Scourides, P. A. (1982) *Br. J. Cancer*, **45**, 571.
10. Dougherty, T. J.; Potter, W. R.; Weishaupt, in reference 4, p. 301.
11. Kessel, D.; Cheng, M. L. (1985) *Cancer Res.*, **45**, 3053.
12. Kessel, D.; Chou, T. (1983) *Cancer Res.*, **43**, 1994.
13. Musselman, B. D.; Watson, J. T.; Chang, C. K. *Org. Mass Spect.*, in press.

14. Chang, C. K.; Takamura, S., manuscript in preparation.
15. Philips, H. (1923) J. Chem. Soc., **123**, 44.
16. Chang, C. K. (1979) ACS Adv. Chem. Ser., **173**, 162.
17. Chang, C. K.; Abdalmuhdi, I. (1984) Angew. Chem. Int. Ed. Engl., **23**, 164.
18. Kasha, M.; Rawls, H. R.; El-Bayoumi, M. A. (1965) Pure Appl. Chem., **11**, 371.
19. Reyftmann, J.; Morliere, P.; Goldstein, S.; Santus, R.; Dubertret, L.; Lagrange, D. (1984) Photochem. Photobiol., **40**, 721.
20. Jori, G.; Beltramini, M.; Reddi, E.; Salvato, B.; Pagnan, A.; Ziron, L.; Tomio, L.; Tsanov, T. (1984) Cancer Lett. **24**, 291.
21. Gomer, C. J.; Dougherty, T. J. (1979) Cancer Res. **39**, 146.
22. Brown, M. S.; Kovanen, P. T.; Goldstein, L. J. (1980) Ann. N. T. Acad. Sci., **348**, 48.

RECEIVED April 5, 1986

INDEXES

Author Index

Subject Index

Production by Cara Aldridge Young
Jacket design by Pamela Lewis

Elements typeset by Hot Type Ltd., Washington, DC
Printed and bound by Maple Press Co., York, PA

RECENT ACS BOOKS

Phenomena in Mixed Surfactant Systems
Edited by John F. Scamehorn
ACS Symposium Series 311; 350 pp; ISBN 0-8412-0975-8

Chemistry and Function of Pectins
Edited by Marshall Fishman and Joseph Jen
ACS Symposium Series 310; 286 pp; ISBN 0-8412-0974-X

Fundamentals and Applications of Chemical Sensors
Edited by Dennis Schuetzle and Robert Hammerle
ACS Symposium Series 309; 398 pp; ISBN 0-8412-0973-1

Polymeric Reagents and Catalysts
Edited by Warren T. Ford
ACS Symposium Series 308; 296 pp; ISBN 0-8412-0972-3

Excited States and Reactive Intermediates: Photochemistry, Photophysics, and Electrochemistry
Edited by A. B. P. Lever
ACS Symposium Series 307; 288 pp; ISBN 0-8412-0971-5

Artificial Intelligence Applications in Chemistry
Edited by Bruce A. Hohne and Thomas Pierce
ACS Symposium Series 306; 408 pp; ISBN 0-8412-0966-9

Organic Marine Geochemistry
Edited by Mary L. Sohn
ACS Symposium Series 305; 440 pp; ISBN 0-8412-0965-0

Fungicide Chemistry: Advances and Practical Applications
Edited by Maurice B. Green and Douglas A. Spilker
ACS Symposium Series 304; 184 pp; ISBN 0-8412-0963-4

Petroleum-Derived Carbons
Edited by John D. Bacha, John W. Newman and J. L. White
ACS Symposium Series 303; 416 pp; ISBN 0-8412-0964-2

Historic Textile and Paper Materials: Conservation and Characterization
Edited by Howard L. Needles and S. Haig Zeronian
Advances in Chemistry Series 212; 464 pp; ISBN 0-8412-0900-6

Multicomponent Polymer Materials
Edited by D. R. Paul and L. H. Sperling
Advances in Chemistry Series 211; 354 pp; ISBN 0-8412-0899-9

For further information and a free catalog of ACS books, contact:
American Chemical Society, Sales Office
1155 16th Street N.W., Washington, DC 20036
Telephone 800-424-6747